NON-LINEAR DYNAMIC PROBLEMS FOR COMPOSITE CYLINDRICAL SHELLS

NON-LINEAR DYNAMIC PROBLEMS FOR COMPOSITE CYLINDRICAL SHELLS

Alexander Bogdanovich
Engineering and Technology Centre
Latvian Academy of Sciences
Riga, Latvia

Translation Editor
C. W. BERT
University of Oklahoma, USA

ELSEVIER APPLIED SCIENCE
LONDON and NEW YORK

ELSEVIER APPLIED SCIENCE PUBLISHERS LTD
Crown House, Linton Road, Barking, Essex IG11 8JU, England

Updated and expanded English language edition
of the Russian "Non-linear Dynamic Problems for Composite
Cylindrical Shells", Zinatne, Riga

WITH 78 ILLUSTRATIONS AND 20 TABLES

British Library Cataloguing in Publication Data

Bogdanovich, Alexander
Nonlinear dynamic problems for composite cylindrical shells.
1. Dynamics
I. Title
620.1054

ISBN 1-85166-653-2

Library of Congress Cataloging-in-Publication Data

Bogdanovich, Alexander.
Nonlinear dynamic problems for composite cylindrical shells / by Alexander Bogdanovich: translation editor C. W. Bert
p. cm.
Translated from Russian.
Includes bibliographical references.
ISBN 1-85166-653-2
1. Shells (Engineering) 2. Cylinders. 3. Structural analysis (Engineering) 4. Shells (Engineering) — Vibration. I. Bert, Charles Wesley, 1929– . II. Title
TA660.S5B55 1991
624.1′7762–dc20 91-13133
CIP

Photoset by Alden Multimedia.
Printed in Great Britain at the Alden Press, Oxford.

Preface

Thin-walled shells are used in a multitude of load-bearing structural parts for various modern vehicles (aerospace, aircraft, marine, automotive) and civil engineering structures. This wide range of practical applications demands a fundamental understanding of their deformation, vibration, static and dynamic stability, strength, and reliability. For traditional metal shells, these problems have been studied exhaustively. However for thin-walled composite shells, many of these problems are still unsolved. The following are among the most important unsolved problems for composite shells: nonlinear theory of composite shells, nonlinear vibrations, static and dynamic buckling, transient deformation under short-time impulsive and impact loads, strength and failure predictions. These problems are the focus of the monograph. The basic idea passing through all six chapters is to connect the problems of deformation, buckling, and failure, usually treated separately, and to demonstrate their mutual interaction on some examples of dynamically loaded laminated composite cylindrical shells.

In the general formulation, a dynamic analysis of heavily loaded thin-walled shells of revolution has to include the following topics: pre-buckling transient deformation, axisymmetric buckling in the zone of edge effect, nonaxisymmetric buckling in the sites of rapidly growing initial imperfections, and nonlinear transient post-buckling vibrations. This should be accompanied by failure analysis in order to track changes in material properties. Usually each part of the general problem is treated separately. This approach is permissible, but only under certain conditions. One of the goals of this monograph is to identify conditions for which this approach holds.

In the particular case of a cylindrical shell loaded by axial impact, the process of deformation starts with a stress wave propagation from the impacted edge surface into the shell body. In the case of a thin-walled shell it can be considered as a process of stress wave propagation along the medial surface. Under the boundary conditions restricting edge mobility in the radial direction, an intensive axisymmetric buckling starts just after the impact load attains some critical value. Furthermore, nonaxisymmetric buckling is initiated at other shell sites. This is regulated by a distribution of initial nonaxisymmetric imperfections. These two buckling processes are mutually interacting, and also interact with the continuing stress wave propagation. These interactions can be handled in an analytical model only if the geometric nonlinearity is taken into account. The general formulation and solution of this problem is presented in Chapter 5.

Shell failure processes can already start at the stage of stress wave propagation, or can be initiated by intensive axisymmetric or nonaxisymmetric buckling processes. It depends on the specific combination of mass and velocity of the impactor, or, correspondingly, the amplitude and duration of the impulse in the case of impulsive loading. The failure process also depends on the edge clamping conditions, shell thickness ratio, characteristics of the field of initial imperfections, and stiffness and strength of the material. For laminated composite shells, the reinforcement angles and ply sequence also play an important role. The effects of these factors are illustrated through a number of examples in Chapters 5 and 6.

The initial failure zones (first ply failure zones) are extending and multiplying during the continuing impact event. Finally, they can occupy the whole through-thickness section and lead to a total exhaustion of a shell's load bearing capacity. The corresponding last ply failure load value can be estimated by the use of some ply-by-ply failure model. One of the possible models is described in Chapter 6.

The computational procedure proposed and results presented give some ideas about the effect of partial or total ply failure on the lower-bound and upper-bound critical dynamic load values for imperfect composite cylindrical shells. It is worth pointing out that the failure process influences the deformation process. In the simplest analytical approach this is accounted for by reducing certain ply stiffnesses after each failure event is identified.

The effect of geometric nonlinearity is very important in the problems of deformation and failure of thin-walled shells subject to high load vibrations. For example, in the case of forced axial vibration of a shell of revolution, a specific type of transverse bending vibration (called parametric vibration) can be excited at certain combinations of load amplitude and frequency. Parametric vibrations are characterized typically with rather high deflection and corresponding stress amplitudes. The amplitudes can be calculated in principal only through a geometrically nonlinear formulation. Moreover, geometric nonlinearity leads to some non-trivial effects of vibration mode interaction, as revealed in Chapter 4. These effects obviously play an important role in accurately predicting the durability of a vibrating structural part.

As illustrated in Chapter 4, the distinction between "good" and "bad" vibration regimes can be distinguished through a simple analysis of dynamic instability region spectra. In its turn, the calculation of these spectra is provided in terms of the spectrum of natural frequencies and the spectrum of static buckling loads. Therefore, it is an important task to calculate both of them with sufficiently high precision. At this point, some effects specific to composites (e.g., low resistance to transverse shear deformation) have to be evaluated. This requires a special investigation of the basic linear problems of free vibrations and static stability, which are presented in Chapter 3.

The analysis of all of the main problems examined in this monograph is based on geometrically nonlinear equations of motion for orthotropic shells. Therefore it is necessary to obtain equations themselves. Up to this time, it seems to be a non-trivial task; usually a number of intermediate simplifications are introduced when proceeding from the equations of geometrically nonlinear 3-D theory of elasticity to the equations of geometrically nonlinear 2-D theory of thin elastic shells. Great care must be taken in evolving these equations. Inconsistent simplifications which neglect small order terms in one part of the set of equations and/or retain them in another part will yield results that are even qualitatively wrong when applied to the solution of some particular nonlinear problem.

In Chapters 1 and 2, a procedure for evolving several versions of geometrically nonlinear equations for orthotropic thin-walled shells is proposed. The final sets of equations are compared to the well known ones.

Each of the six chapters in the monograph is prefaced by an extensive survey of the most important works published on the corresponding problem. The author tried to balance a contribution of Soviet and foreign research. Most of the referred Soviet books and papers are available only in Russian, therefore the monograph can be a useful guideline for English speaking scientists to become acquainted with a history and state-of-the-art in Soviet research on nonlinear shell theory, free and parametric vibrations, dynamic stability, transient deformation, and failure of thin-walled shells, as well as related deformation and failure problems for composite materials.

ALEXANDER BOGDANOVICH

Contents

Introduction

Over several decades, the thin-walled structural elements in the shape of cylindrical shells have been the object of considerable research. The interest in the problems of deformation, strength, vibrations, static and dynamic stability of cylindrical shells can be attributed, firstly, to the fact that they are the primary load-bearing structural elements being widely used in aviation and aerospace engineering, submerged apparatuses, pipelines and today's energetic installations.

In the last two decades, the mechanics of composites has undergone intensive growth, i.e. the development of this area of mechanics was prompted by the needs for materials with predictable in advance behavior, materials which might meet specific extremal service conditions. The widespread use of filamentary and laminated composites in various items requires that not only the conventional methods of numerical analysis for thin-walled shells should be refined, but also puts forth new problems. These problems are revealed, firstly, by the necessity of principally considering new factors, determining the load-carrying capacity of structures. Such factors are the strong anisotropy of stiffness and strength characteristics, a considerable scatter of its elastic and strength properties, viscoelastic behavior, a range of different fracture modes and an increased danger of failure along interfaces in multilayered composites. The above makes the calculation of stress–strain state as well as strength, stability and reliability studies of composite structures more complicated. On the other hand, the use of composites makes it possible to vary purposefully the main characteristics of structural elements by invariably retaining its shape and size. This allows to considerably increase the load level at which fail-safe operation of structure is ensured without mass increase or, on the contrary, to achieve a reasonable reduction in mass at fixed restrictions imposed on the characteristics of stress–strain state or other physical restrictions.

Thus, two classical aspects of mechanics of deformable solids, i.e. mechanics of materials and mechanics of structures, form a unity in treatment of composite items. This unity is appreciably specified also by the condition that the composite material in its final form and the structural element itself are being developed simultaneously in a unified technological process.

The basic stages in the development of mechanics of composite materials and results obtained so far have been sufficiently described in monographs[42, 76, 90, 145, 147, 185, 225, 240] and reviews.[23, 69, 71, 73, 220, 226, 237, 255, 341] The design of thin-walled structural elements of anisotropic, reinforced and multilayered materials was dealt with in monographs[21, 24, 76, 91, 92, 143, 146, 147, 173, 180, 185, 213, 224, 239, 245] and reviews.[22, 93, 136, 154, 199, 215, 238, 435, 436] We can single out[92, 199, 224, 245, 435, 436] where the main attention was focussed on the problems of static stability for composite shells. Some results for problems of vibrations and non-stationary deformation have found their reflection in reviews.[289, 291] On the whole, it can be asserted that such static problems such as calculation of a stress–strain state, stability, strength and optimal design have been fully resolved for shells of composite materials.

From among a rather small number of studies in dynamics of cylindrical composite shells the majority deals with the linear problems of harmonic wave propagation in infinitely long shells[76, 206, 290, 381, 390, 433, 455] and natural vibrations of shells of finite length.[52, 176, 177, 307–310, 330–333, 347, 354, 405, 406, 420, 424, 425, 448, 453] The theoretical studies of the problems of forced,[31, 432, 451] parametric[32, 45, 46, 50, 103] and non-linear natural[167, 316] vibrations have been under consideration in a number of papers. We should also note the experiments on natural and parametric vibrations of glass fiber plastic shells.[77, 144, 169] The optimization problems with restrictions imposed on the frequencies of natural vibrations have been treated in various papers.[7,8,40,200,229,393,404]

Until recently less attention has been paid to the problems of deformation, strength, stability of composite cylindrical shells under short-term impulsive and impact loads. A series of solutions for linear problems of non-stationary deformation under axial loading, external and internal pressures[2, 12, 111, 112, 130, 236, 298, 342, 359, 416, 431, 457] can be mentioned. Interesting results have been experimentally obtained in studying the deformation and strength of laminated glass fiber plastic shells under internal detonation load[227, 250, 256] and in experiments devoted to the effect of impact waves in multilayered cylindrical panels.[29, 144] The simple solutions of geometrically non-linear problems of dynamic stability for orthotropic cylindrical shells have been obtained by others.[87, 104, 105, 198, 270]

The methodology having been laid in the basis of these solutions, coincides fully with that used previously[4,95,100] for isotropic shells. This methodology was the subject matter for a thorough discussion in a review paper.[47] It is important to underline that the problems of structural optimization under dynamic restrictions, other than the restrictions imposed on the natural frequencies, have been treated in only a few works.[33,51,264,271]

Over the last years the actuality of solving the deformation and strength problems for composite shells under dynamic compressive loadings has greatly increased. This can be ascribed, first, to the constantly increasing (both as to the range of products and volume of the material used) application of composites in load-bearing structural elements operating in intensive dynamic regimes. It should be borne in mind that the practical needs require that high reliability should be ensured in primary structures with several elements made of composite materials. The full-scale dynamic tests performed for this purpose are becoming more and more complex and expensive. This problem can be efficiently overcome only on the basis of comprehensive theoretical and experimental studies. Their main objective should be uncovering the physical essence of the processes occurring both in the composite material and in the structural element under specified loading conditions.

As to the theoretical aspect of the general problem, the foremost difficulty in the treatment of these complex structures lies in the evolvement of efficient mathematical models for the structures under investigation which would satisfy not only the requirements for informativeness and accuracy of investigations, but at the same time be economic. The mathematical models of the phenomena under consideration and mathematical methods used for numerical analysis, in their ideal case, should be exact, reliable and at the same time universal. However, in practice it is impossible to satisfy all these requirements when solving the problems of dynamics for thin-walled shells. This can be attributed to a very complicated physical essence of transient dynamics of thin-walled structures. For instance, the related, at first sight, problems about deformation under axial periodic and short-term impulsive loadings of a cylindrical shell, described by outwardly scarcely differing equations, require quite different approaches (see Chapters 4 and 5). The same can be said about the solutions of non-linear buckling problems for cylindrical shells under static and comparatively short-term impulsive axial loadings (see Chapter 5).

At the present time, the 'universal' solutions of very different non-linear dynamic problems obtained by Bubnov-Galerkin's method in terms of one and the same traditional approximations for deflection cannot, obviously, be considered as satisfactory any more. As to the solutions by the methods of finite differences and finite elements, then upon formal developing of numerical algorithms, without due regard for physical peculiarities of the particular deformation process, the calculation methodology either turns out to be extremely ineconomic or does not allow to go beyond the solution of elementary test problems. As stated by I. F. Obraztsov,[212] 'it would be naive to hope that satisfactory solutions might be obtained merely through parameters of high-efficiency computers'. This has been fully confirmed, for instance, by the character of results in a number of papers on dynamics of shells where the finite element technique was used.

In the design of thin-walled shells under dynamic compressive loads the following circumstances should be borne in mind. For actual shells having initial imperfections (even very small ones) it is often impossible to consider separately the classical problems of deformation, dynamic stability and strength. A comparatively slow growth of deformations (momentless or moment axisymmetric in the zone of edge effect) during the initial stage of loading, a subsequent abrupt transition of this process into intensive non-axisymmetric buckling and the following development of the local inelastic deformations or local damages in the material represent interacting aspects of a unified process. We can speak only about which of the aspects in certain situations (at a specified kind of loading, interval of loading rates, field of initial imperfections, geometric parameters of a shell, stiffnesses and strengths of the material) will be the dominating one.

In the present monograph, the attempt at complex analysis of the problem of dynamic deformation and strength in terms of cylindrical shells of fibrous composite laminates has been made. For this reason, a series of particular problems are being solved. Firstly, selection of the initial mathematical model and substantiation of its applicability to the shells and loads are under investigation. When discussing this problem it should be kept in mind that if the thin-walled structure is under dynamic compression, the geometrically non-linear statement should be employed at least. Thus, in dealing with the axial vibratory loading, resulting in parametric resonance, the characteristics of the stress–strain state of a shell principally cannot be calculated on the basis of linearized equations of motion. Under impulsive or impact loading, the use of linearized equations of motion yields acceptable accuracy in characteristics of a stress–strain state only within a limited range of loading rates (the problem is treated in detail in Chapter 5). As to the necessity of considering the physical non-linearity, this effect is not of such essential significance and can be treated for each particular dynamic problem separately.

By assuming the geometrically non-linear statement of the problem for some well-known reasons (including purely technical ones), it is natural to strive for maximum simplicity of the basic system of equations of motion meant for practical use. In this sense the following reasonable simplifications of the mathematical model are possible: (i) reduction of a multilayered shell to a homogeneous one is done by adopting kinematic hypotheses of Timoshenko or Kirchhoff-Love for the entire package of layers. This procedure is of great significance for subsequent numerical realization, since the possibilities of bringing the solutions of non-linear dynamic problems to quantitative results on the basis of equations of motion whose order depends on the number of layers still seem to be rather problematic. Of course, this simplification imposes restrictions on both the class of multilayered shells to be investigated and the loading rates; (ii) making a number of characteristics for non-linear theory simplifying assumptions, pertaining to comparative values of elongations, shears and angles of rotation; (iii) introduction of additional simplifications in the sense of the technical theory (the theory of shallow shells); (iv) disregard of the transverse shear strains; and (v) disregard for some inertia terms in the equations of motion. In the first three chapters of the monograph, the results of studying the conditions under which the latter four of the cited simplifications are admissible are presented.

In the first chapter, the basic equations of non-linear theory for orthotropic shells, with due regard for transverse shear strains, are deduced. All the groups of equations have been obtained through subsequent use of the equations and relations of the non-linear three-dimensional theory of elasticity. In the process of their being reduced to a two-dimensional case all the simplifications are strictly specified. This enables us not only to avoid controversy of separate fragments of the theory, from the point of view of taking small terms into account, but also to formulate quantitative criteria of applicability for the eventual simplest version of equation of motion. These criteria can be practically verified after the solution of each particular problem has been obtained. Along with the equations of non-linear theory, taking into account the transverse shear strains, the equations based on Kirchhoff-Love model have also been deduced. Special attention was focused on the non-linear equations related to the technical theory for orthotropic shells. The respective variants of linearized equations are obtained too.

In the second chapter the non-linear and linearized equations of motion for orthotropic cylindrical shells are obtained on the basis of general equations presented in Chapter 1. Apart from the equations corresponding to Kirchhoff-Love and Timoshenko's models combined with the assumptions of the technical theory, the linearized equations of a more general type have been treated. The well-known approach employed by W. Flügge was used in their deduction. The second chapter comprises all the basic equations of motion for a cylindrical shell used in the following chapters.

In Chapter 3 the problems of free vibrations of simply supported orthotropic cylindrical shells are treated. The expressions for the frequencies of natural vibrations as well as critical static loads according to several versions of the theory are obtained. The analytical study of the expressions in various limitary cases and their comparison with the formulas given by the technical theory have been accomplished. The results of numerical calculations are presented. They illustrate the dependence of the error introduced by the assumptions of the technical theory into the magnitude of natural frequencies at various ratios between geometric parameters of a shell, and various modes of vibration. The dependencies of relative correction given to natural frequencies by taking into account transverse shears are presented at different radius-to-thickness ratios, in-plane shear-transverse shear moduli ratios for various modes of vibrations. A thorough quantitative study of the effects, associated with the in-plane inertia terms and inertia of rotation, is presented.

As a result of the study presented in Chapter 3, comprehensive quantitative estimates of the areas of applicability for the technical theory equations for orthotropic cylindrical shells based on Kirchhoff-Love model have been obtained. These equations are used in the following chapters for solution of the non-linear problems in dynamics for cylindrical shells.

Chapter 4 is dedicated to the problems of parametric vibrations of elastic and viscoelastic orthotropic cylindrical shells. The equations for boundaries of the main regions of dynamic instability have been deduced on the basis of linearized equations of motion. Thereby, the transverse shear strains are considered and the assumption is made that the material reveals viscoelastic properties only under shear deformations. Calculation of the spectrum of dynamic instability regions is the necessary preliminary stage in the solution of the problem of non-linear parametric vibrations.

The essence of the approach to the calculation of non-linear parametric vibrations consists in preliminary separation of resonance regimes from resonance-free regimes and selection of the approximation for deflection by proceeding from the principal fact, which spatial forms of flexural vibrations are of reasonance type for specified parameters (amplitude and frequency) of vibratory load. The stress–strain state aroused in a shell in each regime of parametric resonance can be calculated by this approach. A study of specific features of deflection and stress development in time, also at load parameters corresponding to intersection of several regions of dynamic instability, has been accomplished. The possibility of governing the initial part (corresponding to the lowest frequencies) of dynamic instability regions' spectra for fibrous laminated cylindrical shells has been illustrated. Some examples of solving the problems of optimal design with physical restrictions, characteristic for operation of the shell in the regime of parametric resonance, are presented.

In the fifth chapter the problems connected with different kinds of impulsive loading of orthotropic cylindrical shells are studied. In geometrically non-linear statement, the problem of axisymmetric deformation of a shell subjected to impulsive longitudinal compression at various boundary conditions is being solved in terms of a finite-difference method. The effect of axisymmetric initial imperfections on the deformation process is treated. The theoretical and experimental results are compared. For the same loading pattern the solution of deformation problem for the shell with non-axisymmetric initial imperfections is treated. The Bubnov-Galerkin's method with regard to the circumferential coordinate and the method of finite difference with regard to the axial coordinate have been used. The effects of end face restraint conditions and initial deflection on the interaction between the processes of axisymmetric and non-axisymmetric dynamic buckling are investigated. The solution of the problem of non-axisymmetric dynamic buckling by neglecting the longitudinal inertia term is also obtained. Thereby, the Bubnov-Galerkin's method is used for both spatial variables. The effect of inter-dependence of several circumferential buckling modes is being studied. Comparison of the results of solving the problem for non-axisymmetric dynamic buckling of a shell under axial compression, obtained by the two approaches described above, enables us to formulate the conditions for applicability of superposition principle to axisymmetric and non-axisymmetric buckling modes in calculation of a stress–strain state of an imperfect cylindrical shell.

Chapter 6, based on the methods and results of Chapter 5, is dedicated to studying the initial fracture and processes of layer-by-layer failure in cylindrical shells of fibrous laminated composites under impulsive loads (axial compression, external pressure). The first problem is formulated as follows: to determine the load (impulse), at which for specified strength characteristics of the orthotropic monolayer at least in one point of the shell the strength condition is violated. As a result of its solution the first ultimate magnitude of the load (impulse) is found, the mode of initial failure in a layer identified and the location of initial failure zone assessed. Besides, the analysis of the magnitudes attained by transverse and interlaminar stresses in the shell at the moment of initial failure has been performed. The solution of the second problem, i.e. calculation of the process of layer-by-layer failure of the shell beyond the instant of initial failure, is based on the model of reduced stiffness. This solution permits to obtain the second ultimate magnitude of the load (impulse), corresponding, to a greater extent, to the total exhaustion of load-carrying capacity of the shell. As a result of a great amount of numerical data analyzed, the practical conclusions pertaining to selection of the structure of a multi-layered shell, ensuring the highest resistance to dynamic compressive loads, are formulated.

Thus, the general approach in this monograph, allows one to investigate jointly the processes of non-stationary deformation, dynamic buckling and layer-by-layer failure of cylindrical shells of fibrous laminated composites. For all the described algorithms the software for computer-assisted design is available.

1

Basic Equations of the Non-linear Theory for Thin Orthotropic Shells in Orthogonal Curvilinear Coordinates

The system of equations which describes in geometrically non-linear statement deformation of a thin-walled orthotropic shell, under the assumption of a linear-elastic material behavior, is deduced. In deduction of all the fragments of the theory, the equations and formulas of the non-linear theory of elasticity in orthogonal curvilinear coordinates have been successfully used. Such an approach permits to formulate a unified system of assumptions under which the order of one or another small term in strain-to-displacement relations, equilibrium equations and formulas of generalized Hooke's law are evaluated.

The equations of the non-linear technical theory (theory of shallow shells) are obtained from general equations. Along with the variant of equations, taking into account transverse shear strains, the variant corresponding to Kirchhoff-Love's model is considered. The linearized equations of stability are deduced from general non-linear equations. All the final results are written out in the orthogonal non-deformed system of coordinates.

Linear theory of thin-walled homogeneous isotropic shells, the basis of which was laid by H. Aron and A. Love, in the 1930–1940s was the subject of numerous and variegated studies of Soviet and foreign scientists. A thorough review and evaluation of the advances of Soviet mechanicians in developing classical theory of shells is given in V. V. Novozhilov's paper.[211] It is important to note that more earlier works of B. G. Galerkin, P. F. Papkovich, Yu. A. Shimanskii and S. P. Timoshenko were of fundamental significance for this research. In particular, B. G. Galerkin has evolved an original method for obtaining all the formulas in a shell theory from general equations of the theory of elasticity. From the viewpoint of evolving mathematically consistent linear theory for thin shells his works were of principal significance, they were widely used by A. I. Luriye, V. Z. Vlasov, A. L. Gol'denveiser, V. V. Novozhilov and H. M. Mushtari. It was A. I. Luriye who first deduced equations of the theory of thin shells[181] by this method. The fundamental principles laid by these scientists into the basis of the linear theory for homogeneous isotropic shells were of great methodical significance for evolving sophisticated theories as well as theories of anisotropic and multi-layered shells. They are of fundamental significance for evolving the non-linear theories of shells too.

For anisotropic shells the first study was, evidently, performed by I. Ya. Shtaerman.[269] In the 30s, W. Flügge, D. D. Dschow, J. L. Taylor and H. M. Mushtari performed stability analysis for structurally orthotropic shells. However, the theory of anisotropic shells was elaborated in a most complete form by S. A. Ambartsumian.[15–18] As is noted in Ref. 22, this research presents a symbiosis of the classical theory of isotropic shells (in particular, the theory of shallow shells by V. Z. Vlasov[94]) and the theory of anisotropic laminated plates by S. G. Lekhnitskii.[179,180] In Ref. 15, the equations have been obtained for 'rather shallow' (in V. Z. Vlasov's terminology) homogeneous through the thickness shell. In Refs 16 and 17 the theory of 'rather shallow' multilayered shells with symmetric layer layup through the thickness was suggested. In Ref. 18, the theory of multi-layered orthotropic shells with an arbitrary layer layup through the thickness has been elaborated. It should be added that in 1953 the paper by E. I. Grigolyuk[132] on the theory of bimetallic (isotropic, 2-ply) shells was published where the effects analogous to those treated in Ref. 18 were established. Further S. A. Ambartsumian's papers[19,20] were dedicated to several versions of sophisticated theories for anisotropic and laminated shells. Let us note specifically the paper by Ambartsumian[20] in which the refined theory of anisotropic shells was set forth. It was based on an *a priori* representation of parabolic distribution of transverse shear stresses through the thickness. This theory has been sufficiently widely used for numerous applications in bending, stability and vibration problems.

Papers by E. I. Grigolyuk and P. P. Chulkov[140–142] have been dedicated to the development of the theory for sandwich shells as well as for more general multi-layered shells, including viscoelastic ones. Thereby, kinematic hypotheses were assumed for each individual layer. One of the variants, being most widely used, was based on the model of a 'broken line' introduced by E. I. Grigolyuk for sandwich shells (1957–1958). A thorough analysis of the works in this branch of laminated shell theory was given in a review paper.[136]

To evolve a theory of shells with regular through the thickness structure V. V. Bolotin used the method of 'energetic smearing'.[76] The hypotheses of

Kirchhoff-Love were used for load-bearing layers, while binding layers were considered as soft and compressible.

During the last 20 years a great number of works has been published by Soviet and foreign authors on the theory of anisotropic and multi-layered shells. They employed the basic principles and methods evolved by S. A. Ambartsumian, E. I. Grigolyuk and V. V. Bolotin, as well as the main ideas of developing refined variants of a theory of homogeneous isotropic shells. These researches have been dealt with in monographs[24,143,173] and review papers.[115,136,138,154]

The fundamental principles and postulates elaborated in the linear mechanics of shells (i.e. selection of approximations for unknown functions through the thickness; checking of accuracy of all elements of the theory among themselves and with the accuracy of these approximations; methods of deduction of the equilibrium equations and elasticity relations; possibility of introducing stress functions and using equilibrium equations in the mixed form), served as a methodological basis in evolving the non-linear theory.

The first results of the theory of finite displacements for thin-walled elastic shells have been obtained in A. Love, S. P. Timoshenko and R. W. Southwell's works. Though a detailed theoretical development of geometrically non-linear theory was started by H. M. Mushtari,[193] having obtained the basic relations of the non-linear theory of thin orthotropic shells under the assumption that displacements of mid-surface points, comparable to the thickness, are small in comparison to other characteristic shell dimensions. In later works H. M. Mushtari evolved a more general non-linear theory as being valid for arbitrary bending of mid-surface, he also performed a qualitative study of stresses in a shell at small strains and arbitrary displacements and suggested a rigorous classification of the main problems in non-linear theory of shells. In Refs 194 and 195 he obtained (from a position of geometrically non-linear theory) the basic equations for shells, having arbitrary fields of initial stresses and geometric imperfections.

An outstanding contribution to the non-linear theory of shells was made by K. Z. Galimov. On the basis of general equations of non-linear theory of elasticity he obtained the tensor form of equilibrium equations, also formulated in a general form the static boundary conditions for the case of finite strains, and made use of variational methods to the problems of non-linear theory of shells.

Geometrically non-linear theory of shells, based on Kirchhoff-Love's hypotheses has been dealt with in H. M. Mushtari's and K. Z. Galimov's monograph[196] which has greatly influenced the following research in this field.

It is necessary to mention the results obtained by V. V. Novozhilov[208] which were of fundamental significance for evolving the geometrically non-linear theory of shells. He postulated the general approach to the problem of deformation of flexible bodies (thin rods, plates and shells). On the basis of the non-linear theory of elasticity, the equations for thin shells in orthogonal coordinates have been deduced by V. V. Novozhilov.[208] At the same time, the investigations in non-linear theory of shells have been undertaken by A. I. Luriye, V. Z. Vlasov, Yu. N. Rabotnov and N. A. Alumyae.

The foreign authors did not pay much attention to the non-linear theory of shells prior to the beginning of the 60s, although they have published some important results. Thus, L. H. Donnell[311,312] obtained non-linear medium bending equations for a cylindrical shell, subsequently generalized by K. Marguerre[371] for the case of shallow shells with arbitrary curvature. At the present time these equations are classical ones in the theory of shallow shells. The original research of W. Z. Chien[302] on the general non-linear theory of shells should be mentioned as well.

The classical trend in the non-linear theory of homogeneous isotropic shells, based on Kirchhoff-Love's model, has been further developed in works of W. Zerna,[465] P. M. Naghdi,[387,389] J. L. Sanders,[413] W. T. Koiter,[360] V. M. Darevskii,[148] L. I. Shkutin,[265] L. A. Shapovalov,[260–262] E. I. Grigolyuk and V. I. Mamai,[137] V. V. Kabanov[160,161] as well as in other papers.[362,363,391,397,398,445,447]

The non-linear theory of shells based on Timoshenko's kinematic model was first evolved in the tensor form by L. Ya. Ainola.[9] This variant of the improved non-linear theory of shells has been most thoroughly and fully developed by K. Z. Galimov. In monographs[113,114] its detailed interpretation in tensor form is given. A variant of non-linear theory of anisotropic shells with consideration of transverse shears was evolved by V. E. Spiro[232,233] In connection with evolvement of improved variants of the non-linear theory of shells let us note also L. I. Shkutin's paper[266] and N. P. Semenyuk's research.[92]

A comparatively small number of works is devoted to the non-linear theory of laminated shells. Among the first ones is the generalization of the linear theory,[19] accomplished in Ref. 25 for the case of taking into account the quadratic as to deflection terms as well as above mentioned investigations.[140–142] The variant of the non-linear theory of shallow laminated shells was worked out by A. P. Prusakov.[219] Yu. N. Novichkov's paper[207] deals with the tensor version of the non-linear theory of laminated shells with layers of alternating stiffness (for rigid layers Timoshenko's kinematic model was adopted, for soft layers only transverse deformations are taken into account). The non-linear equations for multi-layered shells of regular structure in tensor form have been deduced on the basis of mixed variational principle by M. S. Gershtein.[117,118] The non-linear theory for multi-layered shallow shells has been developed by G. M. Kulikov[175] (the earlier ideas of E. I. Grigolyuk and P. P. Chulkov's papers on sandwich shells have been further developed) and I. A. Buyakov,[88,89] where the model of a 'broken' line in combination with Reissner's variational principle was used to obtain the system of equations of motion for multi-layered anisotropic shells (the order of the system depends on the number of layers).

In summarizing the short bibliography above, we can say that at present the non-linear theory of homogeneous isotropic shells based on Kirchhoff-Love's model is most comprehensively developed. Complete systems of equations presented by various authors[134,137,148,160,232,260,360] permit, after uncomplicated additional calculations, to use them for solving practical problems. Besides, the equations of the Marguerre-Mushtari-Vlasov theory of shallow shells have found wide application in solving practical problems. For anisotropic shells, the practical solutions of non-linear problems have been realized mainly on the basis of the simplest variant, i.e. on the equations for 'rather shallow shells'.

The non-linear type theory of Timoshenko has been sufficiently well developed in general outlines. However, for orthotropic shells we failed to find in the literature correctly deduced equations of motion written in physical variables and related to the system of coordinates connected to the lines of curvature on the mid-surface.

There are several particular approaches in the non-linear theory of multi-layered shells, which differ essentially both as to the initial hypotheses and degree of their completion. The solutions of specific non-linear problems refer to either sandwich shells or are based on the models in which multi-layered shell is reduced to the homogeneous anisotropic one by using kinematic hypotheses for a package as a whole.

1.1 DEFORMATIONS

The relationships between deformations and displacements occupy the central place in the geometrically non-linear theory of thin-wall shells. Let us consider their deduction, proceeding from general relationships of the non-linear theory of elasticity in orthogonal curvilinear coordinates.[210]

Let α_1, α_2 and α_3 be the coordinates of a point in an arbitrary curvilinear orthogonal system of coordinates. Let us superpose the coordinate surface $\alpha_3 = 0$ on the mid-surface of a shell. Let the coordinate lines α_1 and α_2 lie along the lines of the main curvatures of this surface and denote the unit vectors corresponding to those coordinates by $\mathbf{e}_1$ and $\mathbf{e}_2$. Assume that the direction of a unit vector $\mathbf{e}_3$ coincides with the direction of the outer normal to the mid-surface of a shell.

Let us denote the projections of displacement vector of an arbitrary point (α_1, α_2) of the mid-surface on the directions $\mathbf{e}_1, \mathbf{e}_2, \mathbf{e}_3$ by $u_1^0(\alpha_1, \alpha_2)$, $u_2^0(\alpha_1, \alpha_2)$, $u_3^0(\alpha_1, \alpha_2)$ respectively, and assume that the projections $u_1(\alpha_1, \alpha_2, \alpha_3)$, $u_2(\alpha_1, \alpha_2, \alpha_3)$, $u_3(\alpha_1, \alpha_2, \alpha_3)$ of a displacement vector of an arbitrary point of a shell on the directions $\mathbf{e}_1, \mathbf{e}_2, \mathbf{e}_3$ are the linear functions of α_3 coordinate

$$u_i = u_i{}^0 + \alpha_3 v_i{}^0; \qquad u_3 = u_3{}^0 + \alpha_3 v_3{}^0, \tag{1.1}$$

where $v_1^0(\alpha_1, \alpha_2), v_2^0(\alpha_1, \alpha_2), v_3^0(\alpha_1, \alpha_2)$ are the unknown functions; the indices i, j everywhere later on pass the values 1 and 2. The linear approximation through the thickness, employed by V. V. Novozhilov,[208] was used in the majority of works on the non-linear theory of shells.

The components of deformation tensor are determined by the formulas (2.15) from Ref. 210†

$$\begin{aligned}
\varepsilon_{11} &= e_{11} + \frac{1}{2}\left[e_{11}{}^2 + \left(\frac{1}{2} e_{12} + \omega_3\right)^2 + \left(\frac{1}{2} e_{13} - \omega_2\right)^2 \right]; \\
\varepsilon_{12} &= e_{12} + e_{11}\left(\frac{1}{2} e_{12} - \omega_3\right) + e_{22}\left(\frac{1}{2} e_{12} + \omega_3\right) + \\
&\quad + \left(\frac{1}{2} e_{13} - \omega_2\right)\left(\frac{1}{2} e_{23} + \omega_1\right); \\
\varepsilon_{13} &= e_{13} + e_{11}\left(\frac{1}{2} e_{13} + \omega_2\right) + e_{33}\left(\frac{1}{2} e_{13} - \omega_2\right) + \\
&\quad + \left(\frac{1}{2} e_{12} + \omega_3\right)\left(\frac{1}{2} e_{23} - \omega_1\right).
\end{aligned} \tag{1.2}$$

The magnitudes $\varepsilon_{22}, \varepsilon_{33}, \varepsilon_{23}, \varepsilon_{31}, \varepsilon_{21}, \varepsilon_{32}$ are obtained from $\varepsilon_{11}, \varepsilon_{12}, \varepsilon_{13}$ through cyclic rearrangement of indices (1, 2, 3). In accordance with the formulas (2.6), (2.7) from Ref. 210:

†Here and later on the references to formulas of Chapter 4 of the indicated book are made if not mentioned especially.

$$
\begin{aligned}
e_{ii}&=\frac{1}{H_i}\frac{\partial u_i^0}{\partial \alpha_i}+\frac{1}{H_iH_j}\frac{\partial H_i}{\partial \alpha_j}u_j^0+\frac{1}{H_iH_3}\frac{\partial H_i}{\partial \alpha_3}u_3^0;\\
e_{33}&=\frac{1}{H_3}\frac{\partial u_3^0}{\partial \alpha_3}+\frac{1}{H_1H_3}\frac{\partial H_3}{\partial \alpha_1}u_1^0+\frac{1}{H_2H_3}\frac{\partial H_3}{\partial \alpha_2}u_2^0;\\
e_{12}=e_{21}&=\frac{H_2}{H_1}\frac{\partial}{\partial \alpha_1}\left(\frac{u_2^0}{H_2}\right)+\frac{H_1}{H_2}\frac{\partial}{\partial \alpha_2}\left(\frac{u_1^0}{H_1}\right);\\
e_{i3}=e_{3i}&=\frac{H_i}{H_3}\frac{\partial}{\partial \alpha_3}\left(\frac{u_i^0}{H_i}\right)+\frac{H_3}{H_i}\frac{\partial}{\partial \alpha_i}\left(\frac{u_3^0}{H_3}\right);\\
2\omega_i&=\frac{(-1)^j}{H_jH_3}\left[\frac{\partial}{\partial \alpha_j}(H_3u_3^0)-\frac{\partial}{\partial \alpha_3}(H_ju_j^0)\right];\\
2\omega_3&=\frac{1}{H_1H_2}\left[\frac{\partial}{\partial \alpha_1}(H_2u_2^0)-\frac{\partial}{\partial \alpha_2}(H_1u_1^0)\right],
\end{aligned}
\tag{1.3}
$$

where

$$
H_i=A_i(1+\alpha_3k_i);\quad H_3=1, \tag{1.4}
$$

and H_i are Lame's parameters; A_i^2 are coefficients of the first square form: k_i are normal curvatures of the undeformed mid-surface.

Let us note that the symmetricity of deformation tensor ε_{kl} follows from (1.2) and (1.3).

By substituting (1.1), (1.4) into (1.3), considering that $\alpha_3k_i \ll 1$, and confining ourselves to terms of the order α_3k_i, we obtain the following expressions:

$$
\begin{aligned}
e_{ii}&=(\varepsilon_i+\alpha_3\varkappa_i)(1-\alpha_3k_i);\\
e_{12}&=(\gamma_1+\alpha_3\tau_1)(1-\alpha_3k_1)+(\gamma_2+\alpha_3\tau_2)(1-\alpha_3k_2);\\
e_{i3}&=\theta_i(1+\alpha_3k_i)+v_i^0+\alpha_3\varkappa_{i3}-\alpha_3^2\frac{k_i}{A_i}\frac{\partial v_3^0}{\partial \alpha_i};\\
e_{33}&=v_3^0;\\
2\omega_i&=(-1)^j\left[\theta_j-v_j^0+\alpha_3(\varkappa_{j3}-k_j\theta_j)-\alpha_3^2\frac{k_j}{A_j}\frac{\partial v_3^0}{\partial \alpha_j}\right];\\
2\omega_3&=(\gamma_1+\alpha_3\tau_1)(1-\alpha_3k_1)-(\gamma_2+\alpha_3\tau_2)(1-\alpha_3k_2).
\end{aligned}
\tag{1.5}
$$

Here,

$$
\begin{aligned}
\varepsilon_i&=\frac{1}{A_i}\frac{\partial u_i^0}{\partial \alpha_i}+\frac{1}{A_iA_j}\frac{\partial A_i}{\partial \alpha_j}u_j^0+k_iu_3^0;\\
\gamma_i&=\frac{1}{A_i}\frac{\partial u_j^0}{\partial \alpha_i}-\frac{1}{A_iA_j}\frac{\partial A_i}{\partial \alpha_j}u_i^0;\\
\theta_i&=\frac{1}{A_i}\frac{\partial u_3^0}{\partial \alpha_i}-k_iu_i^0;
\end{aligned}
\tag{1.6}
$$

$$\varkappa_i = \frac{1}{A_i}\frac{\partial v_i{}^0}{\partial \alpha_i} + \frac{1}{A_iA_j}\frac{\partial A_i}{\partial \alpha_j} v_j{}^0 + k_i v_3{}^0;$$

$$\tau_i = \frac{1}{A_i}\frac{\partial v_j{}^0}{\partial \alpha_i} - \frac{1}{A_iA_j}\frac{\partial A_i}{\partial \alpha_j} v_i{}^0; \tag{1.7}$$

$$\varkappa_{i3} = \frac{1}{A_i}\frac{\partial v_3{}^0}{\partial \alpha_i} - k_i v_i{}^0.$$

In the formulas (1.3), (1.5)–(1.7) and the following the index j is equal to 2 at $i=1$ and to 1 at $i=2$.

As a result of substituting (1.5) into (1.2) we obtain the following expressions for deformations:

$$\varepsilon_{ii} = (\varepsilon_i + \alpha_3\varkappa_i)(1-\alpha_3k_i) + \frac{1}{2}[(\varepsilon_i+\alpha_3\varkappa_i)^2 +$$
$$+(\gamma_i+\alpha_3\tau_i)^2 + \theta_i{}^2](1-2\alpha_3k_i) + \frac{1}{2}\alpha_3{}^2\varkappa_{i3}{}^2 + \theta_i\varkappa_{i3}\alpha_3(1-\alpha_3k_i) -$$
$$-\theta_i\frac{k_i}{A_i}\frac{\partial v_3{}^0}{\partial\alpha_i}\alpha_3{}^2(1-\alpha_3k_i) - \varkappa_{i3}\frac{k_i}{A_i}\frac{\partial v_3{}^0}{\partial\alpha_i}\alpha_3{}^3; \tag{1.8}$$

$$\varepsilon_{12} = \varepsilon_{21} = (\gamma_1+\alpha_3\tau_1)(1-\alpha_3k_1) + (\gamma_2+\alpha_3\tau_2)(1-\alpha_3k_2) +$$
$$+[(\varepsilon_1+\alpha_3\varkappa_1)(\gamma_2+\alpha_3\tau_2) + (\varepsilon_2+\alpha_3\varkappa_2)(\gamma_1+\alpha_3\tau_1) +$$
$$+\theta_1\theta_2][1-\alpha_3(k_1+k_2)] + [\theta_1\varkappa_{23}(1-\alpha_3k_1) + \theta_2\varkappa_{13}(1-\alpha_3k_2)]\alpha_3 +$$
$$+\left[\varkappa_{13}\varkappa_{23} - \theta_1\frac{k_2}{A_2}\frac{\partial v_3{}^0}{\partial\alpha_2}(1-\alpha_3k_1) - \theta_2\frac{k_1}{A_1}\frac{\partial v_3{}^0}{\partial\alpha_1}(1-\alpha_3k_2)\right]\alpha_3{}^2 -$$
$$-\left(\varkappa_{13}\frac{k_2}{A_2}\frac{\partial v_3{}^0}{\partial\alpha_2} + \varkappa_{23}\frac{k_1}{A_1}\frac{\partial v_3{}^0}{\partial\alpha_1}\right)\alpha_3{}^3; \tag{1.9}$$

$$\varepsilon_{i3} = v_i{}^0 + [\theta_i(1+v_3{}^0) + v_i{}^0(\varepsilon_i+\alpha_3\varkappa_i) + v_j{}^0(\gamma_i+\alpha_3\tau_i)](1-\alpha_3k_i) +$$
$$+\alpha_3(1+v_3{}^0)\left(\varkappa_{i3} - \alpha_3\frac{k_i}{A_i}\frac{\partial v_3{}^0}{\partial\alpha_i}\right); \tag{1.10}$$

$$\varepsilon_{33} = v_3{}^0 + \frac{1}{2}(v_1{}^{0^2} + v_2{}^{0^2} + v_3{}^{0^2}). \tag{1.11}$$

In formula (1.8) the term

$$\frac{1}{2}\left(\frac{k_i}{A_i}\frac{\partial v_3{}^0}{\partial\alpha_i}\right)\alpha_3{}^4 \sim O(k_i{}^2\alpha_3{}^2)\cdot\frac{1}{2}\varkappa_{i3}{}^2\alpha_3{}^2,$$

and in (1.9) the term

$$\frac{k_1}{A_1}\frac{k_2}{A_2}\frac{\partial v_3{}^0}{\partial \alpha_1}\frac{\partial v_3{}^0}{\partial \alpha_2}\alpha_3{}^4 \sim O(k_1 k_2 \alpha_3{}^2)\varkappa_{13}\varkappa_{23}\alpha_3{}^2.$$

were dropped.

Let us note once more that in deducing the relationships (1.8)–(1.11) the following two assumptions have been made:

(1) the linear approximations of displacements through the thickness (1.1) were adopted; and
(2) the terms of a more higher order of smallness compared to a unity, than $k_i\alpha_3$ were omitted.

Further we shall make a series of consistent simplifications of the obtained expressions.

Firstly, considering the shell as being sufficiently thin, let us neglect the values $k_i\alpha_3$ compared to a unity in (1.8)–(1.10). Accordingly, we can omit those terms, which are of the order $O(k_i\alpha_3)$ compared to others. In formula (1.8) such terms are

$$\theta_i \frac{k_i}{A_i}\frac{\partial v_3{}^0}{\partial \alpha_i}\alpha_3{}^2 \sim O(k_i\alpha_3)\theta_i\varkappa_{i3}\alpha_3; \quad \varkappa_{i3}\frac{k_i}{A_i}\frac{\partial v_3{}^0}{\partial \alpha_i}\alpha_3{}^3 \sim O(k_i\alpha_3)\varkappa_{i3}{}^2\alpha_3{}^2;$$

in formula (1.9)

$$\theta_1 \frac{k_2}{A_2}\frac{\partial v_3{}^0}{\partial \alpha_2}\alpha_3{}^2 \sim O(k_2\alpha_3)\theta_1\varkappa_{23}\alpha_3;$$

$$\theta_2 \frac{k_1}{A_1}\frac{\partial v_3{}^0}{\partial \alpha_1}\alpha_3{}^2 \sim O(k_1\alpha_3)\theta_2\varkappa_{13}\alpha_3;$$

$$\varkappa_{13} \frac{k_2}{A_2}\frac{\partial v_3{}^0}{\partial \alpha_2}\alpha_3{}^3 \sim O(k_2\alpha_3)\varkappa_{13}\varkappa_{23}\alpha_3{}^2;$$

$$\varkappa_{23} \frac{k_1}{A_1}\frac{\partial v_3{}^0}{\partial \alpha_1}\alpha_3{}^3 \sim O(k_1\alpha_3)\varkappa_{13}\varkappa_{23}\alpha_3{}^2;$$

in formula (1.10)

$$\frac{k_i}{A_i}\frac{\partial v_3{}^0}{\partial \alpha_i}\alpha_3 \sim O(k_i\alpha_3)\varkappa_{i3}.$$

After simplification the expressions for deformations take the form:

$$\begin{aligned}
\varepsilon_{ii} &= (\varepsilon_i+\alpha_3\varkappa_i) + \frac{1}{2}[(\varepsilon_i+\alpha_3\varkappa_i)^2 + (\gamma_i+\alpha_3\tau_i)^2 + (\theta_i+\alpha_3\varkappa_{i3})^2];\\
\varepsilon_{12} &= (\gamma_1+\alpha_3\tau_1) + (\gamma_2+\alpha_3\tau_2) + (\varepsilon_1+\alpha_3\varkappa_1)(\gamma_2+\alpha_3\tau_2) +\\
&\quad + (\varepsilon_2+\alpha_3\varkappa_2)(\gamma_1+\alpha_3\tau_1) + \theta_1\theta_2 + \alpha_3(\theta_1\varkappa_{23}+\theta_2\varkappa_{13}) + \alpha_3{}^2\varkappa_{13}\varkappa_{23};\\
\varepsilon_{i3} &= v_i{}^0 + v_i{}^0(\varepsilon_i+\alpha_3\varkappa_i) + v_j{}^0(\gamma_i+\alpha_3\tau_i) + (1+v_3{}^0)(\theta_i+\alpha_3\varkappa_{i3}).
\end{aligned} \tag{1.12}$$

Secondly, let us introduce the indicators of variability along coordinate lines for the characteristics of the deformed state and for the coefficients of the first square form by means of the expressions:

$$\frac{\partial v_i^0}{\partial \alpha_i}=\lambda_{ii}v_i^0; \quad \frac{\partial v_i^0}{\partial \alpha_j}=\lambda_{ij}v_i^0; \quad \frac{\partial A_i}{\partial \alpha_i}=\mu_{ii}A_i; \quad \frac{\partial A_i}{\partial \alpha_j}=\mu_{ij}A_i.$$

Let us assume that the maximum values $\lambda_{ii}, \lambda_{ij}, \mu_{ii}, \mu_{ij}$ satisfy the conditions:

$$\frac{\lambda_{ii}}{A_i}v_i^0 \leqslant k_i; \quad \frac{\lambda_{ij}}{A_j}v_i^0 \leqslant k_i;$$

$$\frac{\mu_{ii}}{A_i}v_i^0 \leqslant k_i; \quad \frac{\mu_{ij}}{A_j}v_i^0 \leqslant k_i. \tag{1.13}$$

Then it is easy to make certain that in (1.12) many terms are of the order $O(k_i\alpha_3)$ compared to others. By omitting them in the indicated formulas, we find:

$$\varepsilon_{ii}=\varepsilon_i+\frac{1}{2}(\varepsilon_i^2+\gamma_i^2+\theta_i^2)+\alpha_3(\varkappa_i+\gamma_i\tau_i+\theta_i\varkappa_{i3});$$

$$\varepsilon_{12}=\gamma_1+\gamma_2+\varepsilon_1\gamma_2+\varepsilon_2\gamma_1+\theta_1\theta_2+\alpha_3(\tau_1+\tau_2+\varepsilon_1\tau_2+\varepsilon_2\tau_1+\theta_1\varkappa_{23}+\theta_2\varkappa_{13}); \tag{1.14}$$

$$\varepsilon_{i3}=v_i^0+(1+v_3^0)\theta_i+v_i^0\varepsilon_i+v_j^0\gamma_i.$$

The expressions based on the simplified variant of the non-linear theory of elasticity[208] have been widely used in the practical calculations of shells. This variant presupposes that the elongations and shear deformations are the values of higher order of smallness than the rotation angles and permits to omit the squares of ε_i, γ_i as well as their mutual products and the products with other unknown functions in formulas (1.14). As a result we obtain the following expressions for deformations:

$$\varepsilon_{ii}=\varepsilon_i+\frac{1}{2}\theta_i^2+\alpha_3(\varkappa_i+\theta_i\varkappa_{i3});$$

$$\varepsilon_{12}=\gamma_1+\gamma_2+\theta_1\theta_2+\alpha_3(\tau_1+\tau_2+\theta_1\varkappa_{23}+\theta_2\varkappa_{13}); \tag{1.15}$$

$$\varepsilon_{i3}=v_i^0+\theta_i(1+v_3^0).$$

Let us further consider the variant of the theory based on the kinematic hypotheses of Kirchhoff-Love. In V. V. Novozhilov's monograph[208] these hypotheses were interpreted in the sense of neglecting the transverse shear deformations in comparison to the angles of rotation of transverse sections, comprising unit vectors $\mathbf{e}_1, \mathbf{e}_3$ and $\mathbf{e}_2, \mathbf{e}_3$. It was emphasized that the conditions for the equality of deformations ε_{i3} to zero, being conceived as mathematically exact and, respectively, postulated as the basis of the theory, are absurd since they result in obvious contradictions (in particular, in Hooke's law). But if we interpret them as the possibility for disregarding the transverse shear deformations when determining the direction of the unit vector $\mathbf{e}_3$, then 'they take their stand and the conditions under which these equalities can be employed are cleared up'.[208] As regards these conditions, in the above interpretation the Kirchhoff–Love's model has outwardly a purely geometric character. However, it also undoubtedly bears certain mechanical contents in an unconspicuous form. S. A. Ambartsumian[22] has indicated this fact. The thing is that the possibility of disregarding the deformations ε_{i3} in finding out the unit vector direction $\mathbf{e}_3$ for an anisotropic material is determined to a considerable extent by the magnitudes of transverse shear moduli G_{i3}. In consequence, correctness of employing the Kirchhoff–Love's model is conditioned both by the character of the stressed state in the shell and the relative wall thickness, as well as relations between the elastic characteristics of a specific anisotropic material.

In such a way, following the indicated interpretation of the Kirchhoff–Love's hypothesis and considering (1.15), we obtain the condition

$$v_i{}^0 + \theta_i (1 + v_3{}^0) \ll \theta_i, \tag{1.16}$$

which can be satisfied by taking into account $v_3^0 \ll 1$ and putting

$$v_i{}^0 \approx -\theta_i. \tag{1.17}$$

The question of evaluating the function v_3^0 was discussed, for example, in Refs 137, 196, 208, 260, 262, 444. Two approximate formulas were used most often

$$v_3{}^0 \approx 0; \tag{1.18}$$

$$v_3{}^0 \approx -\frac{1}{2}(v_1{}^{0^2} + v_2{}^{0^2}). \tag{1.19}$$

If we assume (1.18), then according to (1.11)

$$\varepsilon_{33} \approx \frac{1}{2}(v_1{}^{0^2} + v_2{}^{0^2}) ; \tag{1.20}$$

but if (1.19) is assumed, then

$$\varepsilon \approx \frac{1}{8}(v_1{}^{0^2} + v_2{}^{0^2})^2 \tag{1.21}$$

As is seen, (1.20) and (1.21) have an evident disadvantage: the transverse normal deformation is positive beyond the dependency of the sign of deformation in the mid-surface of the shell. Therefore, a purely kinematic determination of value v_3^0, without employing the equations connecting stresses and strains (these equations will be treated in 1.2) is insufficiently correct. For completeness, let us consider here the problem of determining v_3^0 with allowance for the formula of the generalized Hooke's law

$$\varepsilon_{33}=-\frac{\nu_{13}}{E_1}\sigma_{11}-\frac{\nu_{23}}{E_2}\sigma_{22}+\frac{1}{E_3}\sigma_{33}. \tag{1.22}$$

Taking into consideration that the last term in (1.22) is of higher order of smallness than the first two terms,[197] taking into account the generalized Hooke's law and (1.15), by substituting σ_{11} and σ_{22} into (1.22), we obtain

$$\begin{aligned}\varepsilon_{33}=&-a_1\left[\varepsilon_1+\frac{1}{2}\theta_1{}^2+\alpha_3(\varkappa_1+\theta_1\varkappa_{13})\right]-\\&-a_2\left[\varepsilon_2+\frac{1}{2}\theta_2{}^2+\alpha_3(\varkappa_2+\theta_2\varkappa_{23})\right],\end{aligned} \tag{1.23}$$

where

$$a_1=\frac{\nu_{13}+\nu_{23}\nu_{21}}{1-\nu_{12}\nu_{21}};\quad a_2=\frac{\nu_{23}+\nu_{13}\nu_{12}}{1-\nu_{21}\nu_{12}}, \tag{1.24}$$

but $\nu_{12},\nu_{21},\nu_{13},\nu_{23}$ are Poisson ratios of the orthotropic material. Further by substituting (1.11) into (1.23) we obtain the equation

$$\begin{aligned}v_3{}^0+\frac{1}{2}(v_1{}^{0^2}+v_2{}^{0^2}+v_3{}^{0^2})=&-a_1\left[\varepsilon_1+\frac{1}{2}\theta_1{}^2+\alpha_3(\varkappa_1+\theta_1\varkappa_{13})\right]-\\&-a_2\left[\varepsilon_2+\frac{1}{2}\theta_2{}^2+\alpha_3(\varkappa_2+\theta_2\varkappa_{23})\right].\end{aligned} \tag{1.25}$$

Within the framework of the linear approximation of displacement u_3 (1.1) through the thickness it is impossible to fully satisfy eqn (1.25). Therefore, let us restrict ourselves to that part which is independent of α_3. Taking into account (1.17) we obtain

$$v_3{}^0+\frac{1}{2}v_3{}^{0^2}=-(a_1\varepsilon_1+a_2\varepsilon_2)-\frac{1}{2}\theta_1{}^2(1+a_1)-\frac{1}{2}\theta_2{}^2(1+a_2). \tag{1.26}$$

Taking into account that $v_3^0 \ll 1$, we obtain the following approximate solution (1.26):

$$v_3{}^0\approx-(a_1\varepsilon_1+a_2\varepsilon_2)-\frac{1}{2}\theta_1{}^2(1+a_1)-\frac{1}{2}\theta_2{}^2(1+a_2). \tag{1.27}$$

Let us note that at $a_1 = a_2 = 0$ (1.27) goes over into the formula

$$v_3{}^0 \approx -\frac{1}{2}(\theta_1{}^2+\theta_2{}^2), \tag{1.28}$$

coinciding with (1.19) under the condition (1.17). For highly anisotropic materials (for instance, unidirectionally reinforced fibrous composites) the Poisson ratios ν_{13} and ν_{23} (the axis 1 coincides with the fiber direction) are fairly low, and $a_1 \ll 1, a_2 \ll 1$. Taking into consideration the above assumption that elongations and squares of rotation angles for transverse sections are the magnitudes of one and the same order of smallness, we conclude that formulas (1.27) and (1.28) for such materials differ slightly. The deformation ε_{33} for such materials is considerably lower than ε_{11} and ε_{22}.

The procedure of obtaining the expressions for deformations (1.15), used in this paragraph permits its applicability to be checked in practice when solving concrete non-linear problems. Besides the evident conditions $k_i\alpha_3 \ll 1, \varepsilon_i \ll \theta_i, \gamma_i \ll \theta_i$ it is necessary to verify the fulfilment of relations (1.13) connecting the indicators of variability of stressed state and geometric parameters of mid-surface with the values of angles of rotation, but in the case of using the Kirchhoff-Love's hypotheses also the condition (1.16).

1.2 STRESSES

The equilibrium equations of the elementary oblique-angled parallelepiped, isolated from the deformed body, according to (3.14) from Ref. 210, take the form

$$\begin{gathered}
\frac{\partial}{\partial\alpha_i}(H_jH_3S_{ii})+\frac{\partial}{\partial\alpha_j}(H_iH_3S_{ji})+\frac{\partial}{\partial\alpha_3}(H_jH_iS_{3i})+H_3\frac{\partial H_i}{\partial\alpha_j}S_{ij}+\\
+H_j\frac{\partial H_i}{\partial\alpha_3}S_{i3}-H_3\frac{\partial H_j}{\partial\alpha_i}S_{jj}-H_j\frac{\partial H_3}{\partial\alpha_i}S_{33}+H_iH_jH_3F_i=0;\\
\frac{\partial}{\partial\alpha_1}(H_2H_3S_{13})+\frac{\partial}{\partial\alpha_2}(H_3H_1S_{23})+\frac{\partial}{\partial\alpha_3}(H_1H_2S_{33})+\\
+H_2\frac{\partial H_3}{\partial\alpha_1}S_{31}+H_1\frac{\partial H_3}{\partial\alpha_2}S_{32}-H_2\frac{\partial H_1}{\partial\alpha_3}S_{11}-H_1\frac{\partial H_2}{\partial\alpha_3}S_{22}+H_1H_2H_3F_3=0,
\end{gathered} \tag{1.29}$$

where F_i, F_3 are the projections of volumetric force on the unit vectors $\mathbf{e}_i, \mathbf{e}_3$ of undeformed system of coordinates. As before, $j = 2$ at $i = 1$ and $j = 1$ at $i = 2$.

The unknown functions entering (1.29) are expressed in terms of generalized stresses $\sigma^*_{ij} = \sigma^*_{ji}$ in accordance with formulas (3.8) from Ref. 210:

$$S_{11}=\sigma^*_{11}(1+e_{11})+\sigma^*_{12}\left(\frac{1}{2}e_{12}-\omega_3\right)+\sigma^*_{13}\left(\frac{1}{2}e_{13}+\omega_2\right);$$
$$S_{12}=\sigma^*_{11}\left(\frac{1}{2}e_{12}+\omega_3\right)+\sigma^*_{12}(1+e_{22})+\sigma^*_{13}\left(\frac{1}{2}e_{23}-\omega_1\right);$$
$$S_{13}=\sigma^*_{11}\left(\frac{1}{2}e_{12}-\omega_2\right)+\sigma^*_{12}\left(\frac{1}{2}e_{23}+\omega_1\right)+\sigma^*_{13}(1+e_{33}); \quad (1.30)$$

S_{22}, S_{33}, S_{23}, S_{31}, S_{21}, S_{32}, are obtained from S_{11}, S_{12}, S_{13} by cyclic rearrangement of indices (1, 2, 3).

The components of symmetric tensor of generalized stresses σ^*_{kl} are determined by formulas (3.9) from Ref. 210:

$$\sigma^*_{11}=\frac{\Sigma^*_1}{\Sigma_1}\frac{\sigma_{11}}{1+E_{\alpha_1}}; \quad \sigma^*_{12}=\frac{\Sigma^*_1}{\Sigma_1}\frac{\sigma_{12}}{1+E_{\alpha_2}}; \quad \sigma^*_{13}=\frac{\Sigma^*_1}{\Sigma_1}\frac{\sigma_{13}}{1+E_{\alpha_3}}, \quad (1.31)$$

The components σ^*_{22}, σ^*_{33}, σ^*_{23}, σ^*_{31}, σ^*_{21}, σ^*_{32} are obtained by cyclic rearrangement of indices (1, 2, 3). The values Σ_1, Σ_2, Σ_3 and Σ^*_1, Σ^*_2, Σ^*_3 are the areas of faces of a parallelepiped prior to and after the deformation. In accordance with formula (3.16) from Ref. 210

$$\frac{\Sigma^*_1}{\Sigma_1}=[(1+2\varepsilon_{22})(1+2\varepsilon_{33})-\varepsilon_{23}^2]^{1/2};$$
$$\frac{\Sigma^*_2}{\Sigma_2}=[(1+2\varepsilon_{33})(1+2\varepsilon_{11})-\varepsilon_{31}^2]^{1/2};$$
$$\frac{\Sigma^*_3}{\Sigma_3}=[(1+2\varepsilon_{11})(1+2\varepsilon_{22})-\varepsilon_{12}^2]^{1/2},$$

but for relative elongations there are expressions (2.18) from Ref. 210:

$$E_{\alpha_1}=(1+2\varepsilon_{11})^{1/2}-1; \quad E_{\alpha_2}=(1+2\varepsilon_{22})^{1/2}-1; \quad E_{\alpha_3}=(1+2\varepsilon_{33})^{1/2}-1.$$

In subsequent calculations, we shall assume that the deformation ε_{33} is negligible compared to a unity and use formulas (1.15). It is easy to check that in compliance with the assumptions lying in the basis of these formulas, the values ε_{ii}^2, $\varepsilon_{ii}\varepsilon_{jj}$ and ε_{12}^2 are negligible compared to a unity. Further, we also disregard the squares of transverse shear strains ε_{i3}^2. Then we have the following formulas for relative changes of areas and elongations

$$\frac{\Sigma_i^*}{\Sigma_i}=1+\varepsilon_{jj}; \quad \frac{\Sigma_3^*}{\Sigma_3}=1+\varepsilon_{11}+\varepsilon_{22}; \tag{1.32}$$

$$E_{\alpha_i}=\varepsilon_{ii}; \quad E_{\alpha_3}=0. \tag{1.33}$$

Substitution of (1.32) and (1.33) into (1.31) yields the following expressions:

$$\sigma^*_{ii}=\sigma_{ii}(1-\varepsilon_{ii}+\varepsilon_{jj}); \quad \sigma^*_{33}=\sigma_{33}(1+\varepsilon_{11}+\varepsilon_{22});$$
$$\sigma^*_{ij}=\sigma_{ij}; \quad \sigma^*_{i3}=\sigma_{i3}(1+\varepsilon_{jj}); \quad \sigma^*_{3i}=\sigma_{3i}(1+\varepsilon_{jj}). \tag{1.34}$$

Let us note that within the framework of the accepted accuracy it is inadmissible to disregard the changes in the areas Σ_1, Σ_2 under deformation, because the non-linear terms of the same order of smallness as the ones in the final formulas for generalized stresses, are lost.

To determine the functions S_{kl} the values e_{ii}, e_{12}, e_{i3}, e_{33}, ω_i, ω_3 should also be known. According to (1.5) and assuming the condition $k_i\alpha_3 \ll 1$, we have

$$e_{ii}=\varepsilon_i+\alpha_3\varkappa_i; \quad e_{12}=\gamma_1+\gamma_2+\alpha_3(\tau_1+\tau_2);$$
$$e_{i3}=\theta_i+v_i^0+\alpha_3\frac{1}{A_i}\frac{\partial v_3^0}{\partial \alpha_i}; \quad e_{33}=v_3^0;$$
$$2\omega_i=(-1)^j\left(\theta_j-v_j^0+\alpha_3\frac{1}{A_j}\frac{\partial v_3^0}{\partial \alpha_j}\right); \tag{1.35}$$
$$2\omega_3=\gamma_1-\gamma_2+\alpha_3(\tau_1-\tau_2).$$

As a result of substituting (1.32) and (1.35) into (1.30) we obtain

$$S_{ii}=\sigma_{ii}\left(1+\varepsilon_j+\frac{1}{2}\theta_j^2-\frac{1}{2}\theta_i^2+\alpha_3\varkappa_j\right)+\sigma_{ij}(\gamma_j+\alpha_3\tau_j)+\sigma_{i3}v_i^0;$$
$$S_{ij}=\sigma_{ij}(1+\varepsilon_j+\alpha_3\varkappa_j)+\sigma_{ii}(\gamma_i+\alpha_3\tau_i)+\sigma_{i3}v_j^0;$$
$$S_{i3}=\sigma_{i3}\left(1+\varepsilon_j+\frac{1}{2}\theta_j^2+v_3^0+\alpha_3\varkappa_j\right)+\sigma_{ii}\left(\theta_{ii}+\frac{\alpha_3}{A_i}\frac{\partial v_3^0}{\partial \alpha_i}\right)+$$
$$+\sigma_{ij}\left(\theta_j+\frac{\alpha_3}{A_j}\frac{\partial v_3^0}{\partial \alpha_j}\right); \tag{1.36}$$
$$S_{3i}=\sigma_{3i}\left[1+\varepsilon_i+\varepsilon_j+\frac{1}{2}\theta_j^2+\alpha_3(\varkappa_i+\varkappa_j)\right]+\sigma_{3j}(\gamma_j+\alpha_3\tau_j)+\sigma_{33}v_i^0;$$

$$S_{33}=\sigma_{33}\left[1+\varepsilon_1+\varepsilon_2+\frac{1}{2}\theta_1{}^2+\frac{1}{2}\theta_2{}^2+v_3{}^0+\alpha_3(\varkappa_1+\varkappa_2)\right]+$$
$$+\sigma_{31}\left(\theta_1+\frac{\alpha_3}{A_1}\frac{\partial v_3{}^0}{\partial\alpha_1}\right)+\sigma_{32}\left(\theta_2+\frac{\alpha_3}{A_2}\frac{\partial v_3{}^0}{\partial\alpha_2}\right).$$

The formulas (1.36) permit to start directly deducing the equilibrium equations for a shell form (1.29).

In the conclusion of this paragraph we shall dwell upon employing the relations of the generalized Hooke's law in geometrically non-linear theory of shells. According to Ref. 210, the constitutive equation in the non-linear theory of elasticity is written down between the symmetric tensor of generalized stresses σ^*_{kl} and the symmetric tensor of deformation ε_{kl}. Bearing in mind the condition $\varepsilon_{33} \ll 1$, in the case of a linearly elastic orthotropic material, we have

$$\sigma^*{}_{ii}=B_{ii}\varepsilon_{ii}+B_{ij}\varepsilon_{jj};\quad \sigma^*{}_{ij}=B_{66}\varepsilon_{ij};\quad \sigma^*{}_{i3}=B_{i3}\varepsilon_{i3}. \tag{1.37}$$

In (1.37) the following notations are used:

$$B_{ii}=\frac{E_i}{1-\nu_{ij}\nu_{ji}};\quad B_{ij}=B_{ji}=\frac{\nu_{ij}E_i}{1-\nu_{ij}\nu_{ji}}=\frac{\nu_{ji}E_j}{1-\nu_{ij}\nu_{ji}};$$
$$B_{66}=G_{12};\quad B_{13}=k'G_{13};\quad B_{23}=k''G_{23},$$

where k' and k'' are some constant coefficients.

Further we shall express from (1.34) the values σ_{kl} in terms of σ^*_{kl}, then employ the relations (1.37) and under the assumptions of this paragraph we shall restrict ourselves to the linear deformations. As a result we obtain

$$\sigma_{ii}=B_{ii}\varepsilon_{ii}+B_{ij}\varepsilon_{jj};\quad \sigma_{ij}=B_{66}\varepsilon_{ij};\quad \sigma_{i3}=B_{i3}\varepsilon_{i3}. \tag{1.38}$$

Let us summarize the contents of this paragraph. The final result is the formulas (1.36) based on expressions for deformations (1.15), (1.35) and neglecting of square components of deformation tensor composed with a unity. The latter condition is also expedient for writing down the generalized Hooke's law in the form (1.38).

1.3 TRACTIONS

The first three conditions of equilibrium for a shell can be arrived at by substitution of (1.4) and (1.36) into (1.29) and integration of the latter through the thickness in the interval from $-h/2$ to $h/2$ (h is the thickness of a shell). The generalized tractions determined by analogy with the linear theory of thin shells are introduced:[209]

$$T^*_{ii}=\int_{-h/2}^{h/2}(1+\alpha_3 k_j)S_{ii}d\alpha_3;\quad T^*_{ij}=\int_{-h/2}^{h/2}(1+\alpha_3 k_j)S_{ij}d\alpha_3;$$

$$T^*_{i3}=\int_{-h/2}^{h/2}(1+\alpha_3 k_j)S_{i3}d\alpha_3. \tag{1.39}$$

Their expressions in terms of tractions, corresponding to the linear theory

$$T_{ii}=\int_{-h/2}^{h/2}(1+\alpha_3 k_j)\sigma_{ii}d\alpha_3;\quad T_{ij}=\int_{-h/2}^{h/2}(1+\alpha_3 k_j)\sigma_{ij}d\alpha_3;$$

$$T_{i3}=\int_{-h/2}^{h/2}(1+\alpha_3 k_j)\sigma_{i3}d\alpha_3, \tag{1.40}$$

have the form

$$T^*_{ii}=T_{ii}\left(1+\varepsilon_j+\frac{1}{2}\theta_j^2-\frac{1}{2}\theta_i^2\right)+T_{ij}\gamma_j+T_{i3}v_i^0;$$
$$T^*_{ij}=T_{ij}(1+\varepsilon_j)+T_{ii}\gamma_i+T_{i3}v_j^0; \tag{1.41}$$
$$T^*_{i3}=T_{i3}\left(1+\varepsilon_j+\frac{1}{2}\theta_j^2+v_3^0\right)+T_{ii}\theta_i+T_{ij}\theta_j.$$

As a result we obtain the following three conditions of equilibrium for a shell:

$$\frac{\partial}{\partial\alpha_i}(A_jT^*_{ii})+\frac{\partial}{\partial\alpha_j}(A_iT^*_{ji})+\frac{\partial A_i}{\partial\alpha_j}T^*_{ij}-\frac{\partial A_j}{\partial\alpha_i}T^*_{jj}+$$
$$+A_iA_jk_iT^*_{i3}+A_iA_j(X^*_i+P^*_i)=0; \tag{1.42}$$
$$\frac{\partial}{\partial\alpha_1}(A_2T^*_{13})+\frac{\partial}{\partial\alpha_2}(A_1T^*_{23})-A_1A_2(k_1T^*_{11}+k_2T^*_{22})+$$
$$+A_1A_2(X^*_3+P^*_3)=0,$$

which coincide formally with the respective equations of the linear theory.[209] In (1.42) the notations are used:

$$P^*_i=\int_{-h/2}^{h/2}[1+\alpha_3(k_i+k_j)]F_id\alpha_3;\quad P^*_3=\int_{-h/2}^{h/2}[1+\alpha_3(k_1+k_2)]F_3d\alpha_3;$$
$$X^*_i=(X_i^+-X_i^-)\left(1+\varepsilon_i+\varepsilon_j+\frac{1}{2}\theta_j^2\right)+ \tag{1.43}$$

$$+(X_j^+-X_j^-)\gamma_j+(X_3^+-X_3^-)v_i^0+$$

$$+\frac{h}{2}[(X_i^++X_i^-)(\varkappa_i+\varkappa_j)+(X_j^++X_j^-)\tau_j]; \qquad (1.44)$$

$$X^*_3=(X_3^+-X_3^-)\left(1+\varepsilon_1+\varepsilon_2+\frac{1}{2}\theta_1^2+\frac{1}{2}\theta_2^2+v_3^0\right)+$$

$$+(X_1^+-X_1^-)\theta_1+(X_2^+-X_2^-)\theta_2+\frac{h}{2}\Big[(X_3^++X_3^-)(\varkappa_1+\varkappa_2)+$$

$$+(X_1^++X_1^-)\frac{1}{A_1}\frac{\partial v_3^0}{\partial\alpha_1}+(X_2^++X_2^-)\frac{1}{A_2}\frac{\partial v_3^0}{\partial\alpha_2}\Big],$$

where

$$X_i^+=\sigma_{3i}\Big|_{\alpha_3=\frac{h}{2}};\quad X_i^-=\sigma_{3i}\Big|_{\alpha_3=-\frac{h}{2}};$$

$$X_3^+=\sigma_{33}\Big|_{\alpha_3=\frac{h}{2}};\quad X_3^-=\sigma_{33}\Big|_{\alpha_3=-\frac{h}{2}}. \qquad (1.45)$$

In fulfilling the conditions (1.13), the terms in (1.44) proportional to $h/2$, can be omitted as values of the order $O(k_i h)$ compared with other terms. In such a way, within the accepted accuracy we have the following expressions

$$X^*_i=(X_i^+-X_i^-)\left(1+\varepsilon_i+\varepsilon_j+\frac{1}{2}\theta_j^2\right)+$$

$$+(X_j^+-X_j^-)\gamma_j+(X_3^+-X_3^-)v_i^0; \qquad (1.46)$$

$$X^*_3=(X_3^+-X_3^-)\left(1+\varepsilon_1+\varepsilon_2+\frac{1}{2}\theta_1^2+\frac{1}{2}\theta_2^2+v_3^0\right)+$$

$$+(X_1^+-X_1^-)\theta_1+(X_2^+-X_2^-)\theta_2.$$

By using the Kirchhoff-Love's model and as a result of substituting (1.17) into (1.41), we obtain the following expressions for generalized tractions:

$$T^*_{ii}=T_{ii}\left(1+\varepsilon_j+\frac{1}{2}\theta_j^2-\frac{1}{2}\theta_i^2\right)+T_{ij}\gamma_j-T_{i3}\theta_i;$$

$$T^*_{ij}=T_{ij}(1+\varepsilon_j)+T_{ii}\gamma_i-T_{i3}\theta_j; \qquad (1.47)$$

$$T^*_{i3}=T_{i3}\left(1+\varepsilon_j+\frac{1}{2}\theta_j^2+v_3^0\right)+T_{ii}\theta_i+T_{ij}\theta_j.$$

Substituting further (1.17) into (1.46) we obtain:

$$X^*_i = (X_i^+ - X_i^-)\left(1+\varepsilon_i+\varepsilon_j+\frac{1}{2}\theta_j^2\right) + (X_j^+ - X_j^-)\gamma_j - (X_3^+ - X_3^-)\theta_i;$$

$$X^*_3 = (X_3^+ - X_3^-)\left(1+\varepsilon_1+\varepsilon_2+\frac{1}{2}\theta_1^2+\frac{1}{2}\theta_2^2+v_3^0\right) + $$
$$+ (X_1^+ - X_1^-)\theta_1 + (X_2^+ - X_2^-)\theta_2. \qquad (1.48)$$

The function v_3^0 entering T^*_{i3} and X^*_3 is determined according to (1.27).

Reverting to relations of elasticity in (1.41) and (1.47), it should be noted that the terms, connected with subintegral functions proportional to α_3^2 have been omitted. This simplification meets the accepted accuracy of this theoretical variant provided the conditions (1.13) are observed. If it is necessary to satisfy certain formal requirements (in particular, the theorem of duality of stresses and the sixth equatiom of equilibrium), they can be modified. The simplest variant of such a modification consists, evidently, in obtaining the non-linear analogue of Novozhilov-Balabukh's formulas[209] treated in Ref. 260 and is widely used in the linear theory of shells.

Besides, there appear no principal difficulties (in spite of its obvious awkwardness) in generalization to the non-linear case of the well-known relations of elasticity obtained by A. I. Luriye[181] and A. L. Gol'denweiser.[123] Thereby, it should be borne in mind that the formulas for deformations are also in need of respective refinement. Incoordinated refinement of some elements of the theory within the framework of original hypotheses about the linear distribution of displacements through the thickness (1.1) does not ensure an improvement of the final result.

The problem of appropriate relations of elasticity in the linear theory of shells was discussed in detail in monographs[94, 123, 209] and in a number of other works. Obviously, the requirements formulated by V. V. Novozhilov[209] and A. L. Gol'denweiser[123] that the relations of elasticity are to be devoid of formal contradictions and should give the simplest possible final result (within the framework of the accepted accuracy), are valid also in the non-linear theory of shells, for which the second of these requirements is of special significance.

1.4 MOMENTS

Let us pass over to deduction of equations of equilibrium pertaining to moments. In the non-linear theory of shells it is the most complicated problem. There are several familiar ways of obtaining these equations.[135, 148, 196] Here, for consistency we shall proceed from the respective equations of the non-linear theory of elasticity by employing some ideas of deduction used in Ref. 196.

We shall obtain the vector equation revealing the condition of equality to zero of the principal moment of all forces, acting on the elementary oblique parallelepiped, by generalizing the deduction of eqn (8.1) of Chapter 2[210] for the case of a curvilinear system of coordinates:

$$H_1H_2H_3\left\{\left[(1+E_{\alpha_1})\mathbf{e}^*{}_1\times\mathbf{S}_{n_1}\frac{\Sigma^*{}_1}{\Sigma_1}\right]+\left[(1+E_{\alpha_2})\mathbf{e}^*{}_2\times\mathbf{S}_{n_2}\frac{\Sigma^*{}_2}{\Sigma_2}\right]+\right.$$
$$\left.+\left[(1+E_{\alpha_3})\mathbf{e}^*{}_3\times\mathbf{S}_{n_3}\frac{\Sigma^*{}_3}{\Sigma_3}\right]\right\}d\alpha_1 d\alpha_2 d\alpha_3=0, \qquad (1.49)$$

where $\mathbf{S}_{n_i}$, $\mathbf{S}_{n_3}$ are components of a stress vector acting on the faces perpendicular prior to deformation to the cordinate lines α_i, α_3; $\mathbf{e}_i^*$, $\mathbf{e}_3^*$ are the unit vectors of the deformed (oblique) system of coordinates. Equation (1.49) presupposes, in accordance with the statement of the classical theory of elasticity, the absence of the internal moments as well as the mass and distributed surface pairs (using (1.49) and the formulas of Chapter 4 from Ref. 210 it is easy to prove symmetricity of the tensor of generalized stresses σ_{kl}^*).

On substituting formulas (1.34) and (1.35) into eqn (1.49) and considering that $H_3 = 1$, $E_{\alpha_3} \approx 0$, we transform it as follows:

$$H^*{}_1H^*{}_2\{[\mathbf{e}^*{}_1\times\mathbf{S}_{n_1}]+[\mathbf{e}^*{}_2\times\mathbf{S}_{n_2}]+[\mathbf{e}^*{}_3\times\mathbf{S}_{n_3}]\}=0. \qquad (1.50)$$

Here,

$$H^*{}_i=H_i(1+\varepsilon_{ii}). \qquad (1.51)$$

Let us designate the radius-vectors of an arbitrary point A of a shell prior to and after deformation by $\mathbf{R}(A)$ and $\mathbf{R}^*(A)$. Let us also introduce the respective radius-vectors $\mathbf{r}(B)$ and $\mathbf{r}^*(B)$ of the point B on the mid-surface, lying on the same coordinate line α_3 with the point A. Due to the assumption of the deformation component ε_{33} smallness we shall disregard the change in the length of the segment AB during deformation. Taking into consideration also that according to the original hypothesis (1.1), the normal does not undergo curving, we can write

$$\mathbf{R}=\mathbf{r}+\alpha_3\mathbf{e}_3; \quad \mathbf{R}^*=\mathbf{r}^*+\alpha_3\mathbf{e}^*{}_3, \qquad (1.52)$$

By employing the formulas

$$\mathbf{e}^*_i = \frac{\dfrac{\partial \mathbf{R}^*}{\partial \alpha_i}}{\left|\dfrac{\partial \mathbf{R}^*}{\partial \alpha_i}\right|} = \frac{1}{H^*_i}\frac{\partial \mathbf{R}^*}{\partial \alpha_i}, \tag{1.53}$$

we can rewrite (1.50) as

$$H^*_2\left[\frac{\partial \mathbf{R}^*}{\partial \alpha_1}\times \mathbf{S}_{n_1}\right] + H^*_1\left[\frac{\partial \mathbf{R}^*}{\partial \alpha_2}\times \mathbf{S}_{n_2}\right] + H^*_1 H^*_2[\mathbf{e}^*_3\times \mathbf{S}_{n_3}] = 0. \tag{1.54}$$

If we assume $H_i^* \approx A_i^*$ in (1.54), then it coincides with eqn (5.3) from Ref. 196.

The vector equation characterizing the condition of equality to zero of the principal vector of forces acting on the elementary parallelepiped, we obtain from (3.4)[210] its form

$$\frac{\partial}{\partial \alpha_1}(H^*_2\mathbf{S}_{n_1}) + \frac{\partial}{\partial \alpha_2}(H^*_1\mathbf{S}_{n_2}) + \frac{\partial}{\partial \alpha_3}(H^*_1 H^*_2\mathbf{S}_{n_3}) + H^*_1 H^*_2\mathbf{F}^* = 0. \tag{1.55}$$

At the assumption $H_i^* \approx A_i^*$ (1.55) coincides with (5.2) from Ref. 196.

Further manipulations are analogous to the ones performed in Ref. 196. Equation (1.55) is multiplied vectorially by $\alpha_3\mathbf{e}_3^*$ and integrated with respect to α_3 from $-h/2$ to $h/2$. Then the condition (1.54) is being employed. Ultimately the following vector equation is obtained:

$$\int_{-h/2}^{h/2}\left\{\frac{\partial}{\partial \alpha_1}[\mathbf{e}^*_3\times \mathbf{S}_{n_1}H^*_2] + \frac{\partial}{\partial \alpha_2}[\mathbf{e}^*_3\times \mathbf{S}_{n_2}H^*_1] + H^*_1 H^*_2[\mathbf{e}^*_3\times \mathbf{F}]\right\}\alpha_3 d\alpha_3 + \int_{-h/2}^{h/2}\left\{\left[\frac{\partial \mathbf{r}^*}{\partial \alpha_1}\times \mathbf{S}_{n_1}H^*_2\right] + \left[\frac{\partial \mathbf{r}^*}{\partial \alpha_2}\times \mathbf{S}_{n_2}H^*_1\right]\right\}d\alpha_3 + [H^*_1 H^*_2\mathbf{e}^*_3\times \alpha_3\mathbf{S}_{n_3}]_{-h/2}^{h/2} = 0. \tag{1.56}$$

On putting down three components of (1.56) we can obtain three equations of equilibrium for a shell relative to bending and torque moments.

In further manipulations the formulas (1.51) and (1.15), the expressions (2.9) from Ref. 210 for unit vectors $\mathbf{e}_i^*$, $\mathbf{e}_3^*$ in the deformed oblique system of coordinates and the formulas (1.23) from Ref. 210 for differentiating unit vectors $\mathbf{e}_i$, $\mathbf{e}_3$ in the non-deformed orthogonal system of coordinates are used. Let us also note the useful equality

$$\mathbf{S}_{n_i} H^*{}_j = \mathbf{S}^*{}_{n_i} H_j,$$

satisfied upon the condition of neglecting ε_{kl}^2 in comparison to a unity introduced previously. Only the square terms as regards to unknown functions are preserved in all manipulations and it is remembered that ε_i, γ_i are of higher order of smallness than θ_i.

Let us present the final results. In projections on the axes of the non-deformed system of coordinates the fourth and fifth equations of equilibrium are written down as

$$\frac{\partial}{\partial\alpha_i}(A_j M^*{}_{ii}) + \frac{\partial}{\partial\alpha_j}(A_i M^*{}_{ji}) + \frac{\partial A_i}{\partial\alpha_j} M^*{}_{ij} - \frac{\partial A_j}{\partial\alpha_i} M^*{}_{jj} - \\ - A_i A_j T^{**}{}_{i3} + A_i A_j (F^*{}_j + \Phi^*{}_j) = 0. \tag{1.57}$$

They do not differ in their form from the respective equations of the linear theory of shells.[209] In (1.57) the following notations have been used

$$M^*{}_{ii} = M_{ii}\left(1 + \varepsilon_j + v_3{}^0 - \frac{1}{2}\theta_i{}^2 + \frac{1}{2}\theta_j{}^2 - v_i{}^0\theta_i\right) + \\ + M_{ij}(\gamma_j - v_i{}^0\theta_j) + N_{ii}\varkappa_j + N_{ij}\tau_j; \tag{1.58}$$

$$M^*{}_{ij} = M_{ij}(1 + \varepsilon_j + v_3{}^0 - v_j{}^0\theta_j) + M_{ii}(\gamma_i - v_j{}^0\theta_i) + N_{ii}\tau_i + N_{ij}\varkappa_j;$$

$$T^{**}{}_{i3} = (T_{i3} + k_i M_{i3})\left(1 + \varepsilon_i + \varepsilon_j + v_3{}^0 + \frac{1}{2}\theta_j{}^2 - v_i{}^0\theta_i\right) + \\ + (T_{j3} + k_j M_{j3})(\gamma_j - v_i{}^0\theta_j) + (T_{ij} - T_{ji} + k_i M_{ij} - k_j M_{ji})\theta_j + \\ + (M_{i3} + k_i N_{i3})(\varkappa_i + \varkappa_j) + (M_{j3} + k_j N_{j3})\tau_j - \\ - (M_{ij} - M_{ji} + k_i N_{ij} - k_j N_{ji})\left(\theta_i\tau_j - \frac{1}{A_j}\frac{\partial v_3{}^0}{\partial\alpha_j}\right) - \\ - k_j(M_{jj}v_i{}^0 - M_{ji}v_j{}^0 + N_{ji}\tau_i v_i{}^0 - N_{jj}\tau_j v_j{}^0); \tag{1.59}$$

$$F^*{}_i = \int_{-h/2}^{h/2} \left\{\left[F_j\left(1 + \varepsilon_i + \varepsilon_j + v_3{}^0 + \frac{1}{2}\theta_i{}^2 + \frac{1}{2}\theta_j{}^2\right) - F_3 v_j{}^0\right] \times \right. \\ \left. \times [1 + \alpha_3(k_i + k_j)] + \alpha_3 F_j(\varkappa_i + \varkappa_j)\right\} \alpha_3 d\alpha_3; \tag{1.60}$$

$$\Phi^*{}_i = \frac{h}{2}\left\{(X_j{}^+ + X_j{}^-)\left(1 + \varepsilon_i + \varepsilon_j + v_3{}^0 + \frac{1}{2}\theta_i{}^2 - \theta_j v_j{}^0\right) + \right. \\ + (X_i{}^+ - X_i{}^-)(\gamma_i - \theta_i v_j{}^0) + \frac{h}{2}[(X_j{}^+ - X_j{}^-) \times \\ \left. \times (\varkappa_i + \varkappa_j) + (X_i{}^+ - X_i{}^-)\tau_i]\right\}. \tag{1.61}$$

The following notations are also introduced:

$$M_{ii}=\int_{-h/2}^{h/2}\alpha_3(1+\alpha_3k_j)\sigma_{ii}d\alpha_3;\quad M_{ij}=\int_{-h/2}^{h/2}\alpha_3(1+\alpha_3k_j)\sigma_{ij}d\alpha_3;$$

$$N_{ii}=\int_{-h/2}^{h/2}\alpha_3^2(1+\alpha_3k_j)\sigma_{ii}d\alpha_3;\quad N_{ij}=\int_{-h/2}^{h/2}\alpha_3^2(1+\alpha_3k_j)\sigma_{ij}d\alpha_3; \tag{1.62}$$

where X_i^+ and X_i^- are determined according to (1.45).

On employing the Kirchhoff's Love's model and substituting (1.17) into (1.58)–(1.61) we find

$$M^*_{ii}=M_{ii}\left(1+\varepsilon_j+v_3^0+\frac{1}{2}\theta_i^2+\frac{1}{2}\theta_j^2\right)+M_{ij}(\gamma_j+\theta_i\theta_j)+N_{ii}\varkappa_j+N_{ij}\tau_i; \tag{1.63}$$

$$M^*_{ij}=M_{ij}(1+\varepsilon_j+v_3^0+\theta_j^2)+M_{ii}(\gamma_i+\theta_i\theta_j)+N_{ii}\tau_i+N_{ij}\varkappa_j;$$

$$\begin{aligned}T^{**}_{i3}&=(T_{i3}+k_iM_{i3})\left(1+\varepsilon_i+\varepsilon_j+v_3^0+\frac{1}{2}\theta_j^2+\theta_i^2\right)+\\&+(T_{j3}+k_jM_{j3})(\gamma_j+\theta_i\theta_j)+(T_{ij}-T_{ji}+k_iM_{ij}-k_jM_{ji})\theta_j+\\&+(M_{i3}+k_iN_{i3})(\varkappa_i+\varkappa_j)+(M_{j3}+k_jN_{j3})\tau_j-(M_{ij}-M_{ji}+k_iN_{ij}-k_jN_{ji})\times\\&\times\left(\theta_i\tau_j-\frac{1}{A_j}\frac{\partial v_3^0}{\partial\alpha_j}\right)+k_j(M_{jj}\theta_i-M_{ji}\theta_j+N_{ji}\tau_i\theta_i-N_{jj}\tau_j\theta_j);\end{aligned} \tag{1.64}$$

$$\begin{aligned}F^*_i=\int_{-h/2}^{h/2}&\left\{\left[F_j\left(1+\varepsilon_i+\varepsilon_j+v_3^0+\frac{1}{2}\theta_i^2+\frac{1}{2}\theta_j^2\right)+F_3\theta_j\right]\times\right.\\&\left.\times[1+\alpha_3(k_i+k_j)]+\alpha_3F_j(\varkappa_i+\varkappa_j)\right\}\alpha_3d\alpha_3;\end{aligned} \tag{1.65}$$

$$\begin{aligned}\Phi^*_i=\frac{h}{2}&\left\{(X_j^++X_j^-)\left(1+\varepsilon_i+\varepsilon_j+v_3^0+\frac{1}{2}\theta_i^2+\theta_j^2\right)+\right.\\&\left.+(X_i^+-X_i^-)(\gamma_i+\theta_i\theta_j)+\frac{h}{2}[(X_j^+-X_j^-)(\varkappa_i+\varkappa_j)+(X_i^+-X_i^-)\tau_i]\right\}.\end{aligned} \tag{1.66}$$

The sixth equation of equilibrium in projections on the axis of non-deformed systems of coordinates has the form:

$$\begin{aligned}&\frac{\partial}{\partial\alpha_j}[A_i(M_{jj}v_i^0-M_{ji}v_j^0+N_{ji}\tau_iv_i^0-N_{jj}\tau_jv_j^0)]-\\&-\frac{\partial}{\partial\alpha_i}[A_j(M_{ii}v_j^0-M_{ij}v_i^0+N_{ij}\tau_jv_j^0-N_{ii}\tau_iv_i^0)]+\\&+A_iA_j[k_iM^*_{ij}-k_jM^*_{ji}+(T_{ij}-T_{ji}+k_iM_{ij}-k_jM_{ji})(1+\varepsilon_i+\varepsilon_j)+\end{aligned} \tag{1.67}$$

$$+(T_{i3}+k_iM_{i3})v_j^0-(T_{j3}+k_jM_{j3})v_i^0+(M_{ij}-M_{ji}+k_iN_{ij}-k_jN_{ji})\times$$
$$\times(\varkappa_i+\varkappa_j)-(M_{i3}+k_iN_{i3})v_i^0\tau_i+(M_{j3}+k_jN_{j3})v_j^0\tau_j]+$$
$$+A_iA_j(F^*{}_3+\Phi^*{}_3)=0.$$

Here,

$$F^*{}_3=\int_{-h/2}^{h/2}\alpha_3[1+\alpha_3(k_1+k_2)](F_2v_1{}^0-F_1v_2{}^0)\,d\alpha_3;$$

$$\Phi^*{}_3=\frac{h}{2}\left\{(X_2^++X_2^-)v_1{}^0-(X_1^++X_1^-)v_2{}^0+\frac{h}{2}[(X_1^+-X_1^-)\tau_1v_1{}^0-\right.$$
$$\left.-(X_2^+-X_2^-)\tau_2v_2{}^0]\right\}. \tag{1.68}$$

It is obvious that eqn (1.67) is rather complex, and the question — to what degree of accuracy do the above presented expressions for deformations, tractions and moments satisfy this equation — is in need of special study.

Thus, the formulas connecting deformations and displacements, stresses and deformations, as well as elasticity relations and equations of equilibrium, related to the non-deformed orthogonal system of coordinates have been obtained. The equations of strain continuity in the non-linear theory of thin shells and boundary conditions have been treated in detail in monographs[114, 135, 196] and many other works, therefore, we shall not go into detail on these problems.

Henceforth, our aim is to pass over to the equations of equilibrium written down in respect of the unknown functions u_i^0, u_3^0, v_i^0.

1.5 EXPRESSION OF GENERALIZED TRACTIONS AND MOMENTS IN TERMS OF DEFORMATIONS

First of all, let us substitute stresses (1.38) and deformations (1.15) into the formulas for tractions (1.40). Upon integrating, just as earlier, in the deduction of the generalized tractions (1.41), we disregard the terms associated with subintegral functions, proportional to α_3^2.

As a result we obtain

$$T_{ii}=C_{ii}\left(\varepsilon_i+\frac{1}{2}\theta_i{}^2\right)+C_{ij}\left(\varepsilon_j+\frac{1}{2}\theta_j{}^2\right);$$
$$T_{ij}=C_{66}(\gamma_i+\gamma_j+\theta_i\theta_j);\quad T_{i3}=T_{3i}=C_{i3}(\theta_i+v_i{}^0). \tag{1.69}$$

Substitution of (1.38) and (1.15) into the formulas for moments leads to the following expressions:

$$M_{ii}=D_{ii}\left[\varkappa^*{}_i+k_j\left(\varepsilon_i+\frac{1}{2}\theta_i{}^2\right)\right]+D_{ij}\left[\varkappa^*{}_j+k_j\left(\varepsilon_j+\frac{1}{2}\theta_j{}^2\right)\right];$$
$$M_{ij}=D_{66}[\tau^*{}_i+\tau^*{}_j+k_j(\gamma_i+\gamma_j+\theta_i\theta_j)];$$
$$M_{i3}=D_{i3}k_j(\theta_i+v_i{}^0); \qquad (1.70)$$
$$N_{ii}=D_{ii}\left(\varepsilon_i+\frac{1}{2}\theta_i{}^2\right)+D_{ij}\left(\varepsilon_j+\frac{1}{2}\theta_j{}^2\right);$$
$$N_{ij}=N_{ji}=D_{66}(\gamma_i+\gamma_j+\theta_i\theta_j); \quad N_{i3}=D_{i3}(\theta_i+v_i{}^0).$$

Here $\varkappa_i^* = \varkappa_i + \theta_i\varkappa_{i3}$; $\tau_i^* = \tau_i + \theta_j\varkappa_{i3}$; $\varkappa_i, \tau_i, \varkappa_{i3}$ are expressed according to (1.7). The values C_{kl}, D_{kl} are the membrane and bending stiffnesses of an orthotropic material. For laminates treated within the framework of the kinematic model of a straight normal they are calculated according to well-known formulas.[21] In the case of a laminated package with a non-symmetric layup of layers through the thickness, the formulas (1.69) and (1.70) can be easily generalized in analogy with the linear theory of laminated shells.[21]

Further we shall substitute (1.69) and (1.70) into the formulas for generalized tractions and moments. In the expressions for T_{ii}^*, T_{ij}^*, as earlier, we shall restrict ourselves to the first degrees of ε_i, γ_i, as well as the squares and mutual derivations of θ_i, v_i^0. As a result we shall obtain

$$T^*{}_{ii}=C_{ii}\left(\varepsilon_i+\frac{1}{2}\theta_i{}^2\right)+C_{ij}\left(\varepsilon_j+\frac{1}{2}\theta_j{}^2\right)+C_{i3}v_i{}^0(v_i{}^0+\theta_i);$$
$$T^*{}_{ij}=C_{66}(\gamma_i+\gamma_j+\theta_i\theta_j)+C_{i3}v_j{}^0(v_i{}^0+\theta_i). \qquad (1.71)$$

In the expressions for T_{i3}^* the terms linear in respect to ε_i, γ_i and square to θ_i, v_i^0 are missing. Therefore it is necessary retain the terms of higher order of smallness: containing the products of the type $\varepsilon_i\theta_i$, $\gamma_i\theta_j$ and the third degrees of values θ_i, v_i^0. As a result we get the formula

$$T^*{}_{i3}=C_{i3}(v_i{}^0+\theta_i)+C_{ii}\theta_i\left(\varepsilon_i+\frac{1}{2}\theta_i{}^2\right)+C_{ij}\theta_i\left(\varepsilon_j+\frac{1}{2}\theta_j{}^2\right)+$$
$$+C_{66}\theta_j(\gamma_i+\gamma_j+\theta_i\theta_j). \qquad (1.72)$$

Further, by substituting (1.70) into (1.58) and confining ourselves to the linear relative to ϵ_i, γ_i and square relative to θ_i, v_i^0 terms, we find

$$M^*{}_{ii}=D_{ii}\left[\varkappa^*{}_i+k_j\left(\varepsilon_i+\frac{1}{2}\theta_i{}^2\right)\right]+D_{ij}\left[\varkappa^*{}_j+k_j\left(\varepsilon_j+\frac{1}{2}\theta_j{}^2\right)\right];$$
$$M^*{}_{ij}=D_{66}[\tau^*{}_i+\tau^*{}_j+k_j(\gamma_i+\gamma_j+\theta_i\theta_j)]. \qquad (1.73)$$

By substituting (1.69) and (1.70) into (1.59) and omitting the terms of the order $O(k_i^2h^2)$ compared to others, we find

$$\begin{gathered}T^{**}_{i3}=C_{i3}(\theta_i+v_i^0)+C_{j3}(\theta_j+v_j^0)(\gamma_j-v_i^0\theta_j)+\\+D_{i3}(k_i+k_j)(\theta_i+v_i^0)(\varkappa_i+\varkappa_j)+D_{j3}(k_i+k_j)(\theta_j+v_j^0)\tau_j-\\-k_jv_i^0(D_{jj}\varkappa_j+D_{ij}\varkappa_i)+k_jD_{66}v_j^0(\tau_i+\tau_j)+(k_i-k_j)D_{66}\theta_j(\tau_i+\tau_j).\end{gathered} \tag{1.74}$$

In the components proportional to D_{ii}, D_{ij}, D_{66}, the quadratic terms relative to θ_j, v_i^0 have been maintained in accordance with the accuracy assumed in deducing the formulas (1.73).

The formulas for generalized moments (1.73) can be somewhat simplified by assuming the linear expressions for $\varkappa_i^*$, τ_i^*:

$$\begin{aligned}\varkappa^*_i&=\frac{1}{A_i}\frac{\partial v_i^0}{\partial\alpha_i}+\frac{1}{A_iA_j}\frac{\partial A_i}{\partial\alpha_j}v_j^0;\\ \tau^*_i&=\frac{1}{A_i}\frac{\partial v_j^0}{\partial\alpha_i}-\frac{1}{A_iA_j}\frac{\partial A_i}{\partial\alpha_j}v_i^0.\end{aligned} \tag{1.75}$$

Besides, let us disregard the effect of membrane deformations on the moments. The first of these assumptions is correct, provided the ratios of variability of the strained state in a shell satisfy the conditions $\lambda_{ii}/A_ik_i \gg 1$. The second one is analogous to the assumption being used in the linear theory of shallow shells. Thus, each of them can essentially misrepresent the results in one and the same calculation case (a typical example is stability of a long cylindrical shell under longitudinal compression), which cannot be treated from the standpoint of the theory of shallow shells. It should be emphasized that the form of elasticity relationships, complying with this theory, has been assumed earlier (in Section 1.3). Besides, one of the basic assumptions — about the smallness of elongations and shears compared to the angles of rotation for normal sections — also loses its cogency beyond the limits of this theory. Introduction of the two indicated simplifications seems to be admissible due to that mentioned above. So we shall put the expressions for generalized moments down as follows:

$$\begin{aligned}M^*_{ii}&=D_{ii}\varkappa^*_i+D_{ij}\varkappa^*_j;\\ M^*_{ij}&=D_{66}(\tau^*_i+\tau^*_j).\end{aligned} \tag{1.76}$$

Accordingly, we have to also simplify formula (1.74) by omitting the non-linear terms. As a result we obtain

$$T^{**}_{i3}=C_{i3}(\theta_i+v_i^0). \tag{1.77}$$

Let us further treat the expressions for generalized tractions and moments by using the kinematic model of Kirchhoff-Love. By substituting (1.17) into (1.71) we find

$$T^*_{ii}=T_{ii}=C_{ii}\left(\varepsilon_i+\frac{1}{2}\theta_i^2\right)+C_{ij}\left(\varepsilon_j+\frac{1}{2}\theta_j^2\right);$$
$$T^*_{ij}=T_{ij}=C_{66}(\gamma_i+\gamma_j+\theta_i\theta_j), \tag{1.78}$$

but in accordance with (1.47) and (1.69) we obtain

$$T^*_{i3}=T_{i3}+C_{ii}\theta_i\left(\varepsilon_i+\frac{1}{2}\theta_i^2\right)+C_{ij}\theta_i\left(\varepsilon_j+\frac{1}{2}\theta_j^2\right)+$$
$$+C_{66}\theta_j(\gamma_i+\gamma_j+\theta_i\theta_j). \tag{1.79}$$

It has been taken into account in deducing the expression (1.79) that $\varepsilon_i \ll 1$, $\theta_i^2 \ll 1$, $v_3^0 \ll 1$.

The formulas for M_{ii}^* and M_{ij}^* have maintained the form (1.73) where

$$\varkappa^*_i=-\frac{1}{A_i}\frac{\partial\theta_i}{\partial\alpha_i}-\frac{1}{A_iA_j}\frac{\partial A_i}{\partial\alpha_j}\theta_j+k_iv_3{}^0+\frac{1}{A_i}\theta_i\frac{\partial v_3{}^0}{\partial\alpha_i}+k_i\theta_i^2;$$
$$\tau^*_i=-\frac{1}{A_i}\frac{\partial\theta_j}{\partial\alpha_i}+\frac{1}{A_iA_j}\frac{\partial A_i}{\partial\alpha_j}\theta_i+\frac{1}{A_i}\theta_j\frac{\partial v_3{}^0}{\partial\alpha_i}+k_i\theta_i\theta_j.$$

It is easy to obtain the expression for T_{i3}^{**} from (1.59) taking into consideration (1.69) and (1.70). Let us note that T_{i3} are the unknown functions in this case which are to be determined from the equation of equilibrium. For the case when in M_{ii}^*, M_{ij}^*, T_{i3}^{**} the non-linear terms are preserved, this problem is non-trivial.

When using the described simplified approach, there appear the formulas (1.76) for M_{ii}^*, M_{ij}^*, where

$$\varkappa_i=-\frac{1}{A_i}\frac{\partial\theta_i}{\partial\alpha_i}-\frac{1}{A_iA_j}\frac{\partial A_i}{\partial\alpha_j}\theta_j;$$
$$\tau_i=-\frac{1}{A_i}\frac{\partial\theta_j}{\partial\alpha_i}+\frac{1}{A_iA_j}\frac{\partial A_i}{\partial\alpha_j}\theta_i \tag{1.80}$$

and $T_{i3}^{**}=T_{i3}$. Considering this, the equations of equilibrium (1.57) take the form

$$\frac{\partial}{\partial\alpha_i}(A_jM^*_{ii})+\frac{\partial}{\partial\alpha_j}(A_iM^*_{ji})+\frac{\partial A_i}{\partial\alpha_j}M^*_{ij}-\frac{\partial A_j}{\partial\alpha_i}M^*_{jj}-$$
$$-A_iA_jT_{i3}+A_iA_j(F^*_j+\Phi^*_j)=0,$$

From this equation we obtain

$$T_{i3}=\frac{1}{A_iA_j}\left[\frac{\partial}{\partial\alpha_i}(A_jM^*_{ii})+\frac{\partial}{\partial\alpha_j}(A_iM^*_{ji})+\frac{\partial A_i}{\partial\alpha_j}M^*_{ij}-\right.$$
$$\left.-\frac{\partial A_j}{\partial\alpha_i}M^*_{jj}\right]+F^*_j+\Phi^*_j.$$

By substituting (1.76) into this formula, and considering that within the limits of the assumed accuracy $\tau_1=\tau_2=\tau$ and using then Kodatsi formulas

$$\frac{\partial}{\partial\alpha_i}(A_j\varkappa_j)-\frac{\partial A_j}{\partial\alpha_i}\varkappa_i=\frac{\partial}{\partial\alpha_j}(A_i\tau)+\frac{\partial A_i}{\partial\alpha_j}\tau,$$

we arrive at the following expression

$$T_{i3}=\frac{1}{A_iA_j}\frac{\partial}{\partial\alpha_i}\{A_j[D_{ii}\varkappa_i+(D_{12}+2D_{66})\varkappa_j]\}-$$
$$-\frac{1}{A_iA_j}\frac{\partial A_j}{\partial\alpha_i}[(D_{12}+2D_{66})\varkappa_i+D_{jj}\varkappa_j]+F^*_j+\Phi^*_j. \qquad (1.81)$$

In the case of an isotropic material

$$\left(D_{ii}=D_{12}+2D_{66}=D=\frac{h^3}{12}\frac{E}{1-\nu^2}\right)$$

at $F_j^*=0$, $\Phi_j^*=0$ we obtain the known formula:[94]

$$T_{i3}=\frac{D}{A_i}\frac{\partial}{\partial\alpha_i}(\varkappa_i+\varkappa_j). \qquad (1.82)$$

Substitution of (1.81) into (1.79) results in

$$T^*_{i3}=\frac{1}{A_iA_j}\frac{\partial}{\partial\alpha_i}\{A_j[D_{ii}\varkappa_i+(D_{12}+2D_{66})\varkappa_j]\}-\frac{1}{A_iA_j}\frac{\partial A_j}{\partial\alpha_i}\times$$
$$\times[(D_{12}+2D_{66})\varkappa_i+D_{jj}\varkappa_j]+C_{ii}\theta_i\left(\varepsilon_i+\frac{1}{2}\theta_i^2\right)+C_{ij}\theta_i\left(\varepsilon_j+\frac{1}{2}\theta_j^2\right)+$$
$$+C_{66}\theta_j(\gamma_i+\gamma_j+\theta_i\theta_j)+F^*_j+\Phi^*_j. \qquad (1.83)$$

Let us sum up the situation. By employing the kinematic model with allowance for the transverse shear deformations, we obtained five equations of equilibrium (1.42) and (1.57). The generalized tractions and moments are expressed by formulas (1.71), (1.72), (1.76) and (1.77), but the deformations entering these formulas are determined according to (1.6) and (1.75). The components of surface loads are specified by expressions (1.46), mass forces — (1.43), moments of mass and surface forces — (1.60) and (1.61). The use of the indicated formulas enable us to write down the system of equilibrium equations in terms of five unknown functions u_i^0, u_3^0, v_i^0.

In the case of employing Kirchhoff-Love's model, we have three equations of equilibrium (1.42). The generalized tractions are expressed in accordance with (1.78) and (1.83). The deformations are determined by the formulas (1.6) and (1.80). For components of surface loads and mass forces the expressions (1.48) and (1.43) are valid, for moments of mass and surface forces — eqns (1.65) and (1.66). Through subsequent substitution, the system of equilibrium equations can be written down in terms of three unknown functions u_i^0, u_3^0.

1.6 EQUATIONS OF THE TECHNICAL THEORY FOR ORTHOTROPIC SHELLS

The non-linear theory of shallow isotropic shells (henceforth we shall use the term 'the technical theory of shells'), based on Kirchhoff-Love hypotheses, was outlined in Ref. 196. The following additional simplifications are introduced. Firstly, in the formula for θ_i (1.6) the tangential displacement u_i^0 is disregarded. Accordingly, formulas (1.80) take the form

$$\begin{aligned} \varkappa_i &= -\frac{1}{A_i}\frac{\partial}{\partial \alpha_i}\left(\frac{1}{A_i}\frac{\partial u_3{}^0}{\partial \alpha_i}\right) - \frac{1}{A_i A_j{}^2}\frac{\partial A_i}{\partial \alpha_j}\frac{\partial u_3{}^0}{\partial \alpha_j}; \\ \tau_i &= -\frac{1}{A_i}\frac{\partial}{\partial \alpha_i}\left(\frac{1}{A_j}\frac{\partial u_3{}^0}{\partial \alpha_j}\right) + \frac{1}{A_i{}^2 A_j}\frac{\partial A_i}{\partial \alpha_j}\frac{\partial u_3{}^0}{\partial \alpha_i}. \end{aligned} \tag{1.84}$$

Secondly, in the first two equations of equilibrium (1.42) the shear tractions T_{i3}^* are disregarded and that results in

$$\frac{\partial}{\partial \alpha_i}(A_j T_{ii}) + \frac{\partial}{\partial \alpha_j}(A_i T_{ji}) + \frac{\partial A_i}{\partial \alpha_j} T_{ij} - \frac{\partial A_j}{\partial \alpha_i} T_{jj} + A_i A_j (X^*{}_i + P^*{}_i) = 0 \tag{1.85}$$

(here, it has been taken into account according to (1.78) that $T_{ii}^* = T_{ii}$, $T_{ij}^* = T_{ij}$).

If in eqn (1.85) we put $X_i^* = 0$, $P_i^* = 0$ and according to Ref. 94 introduce the traction function F in the form

$$\begin{aligned} T_{ii} &= \frac{1}{A_j}\frac{\partial}{\partial \alpha_j}\left(\frac{1}{A_j}\frac{\partial F}{\partial \alpha_j}\right) + \frac{1}{A_i{}^2 A_j}\frac{\partial A_j}{\partial \alpha_i}\frac{\partial F}{\partial \alpha_i}; \\ T_{ij} &= -\frac{1}{A_i A_j}\frac{\partial^2 F}{\partial \alpha_i \partial \alpha_j} + \frac{1}{A_i A_j{}^2}\frac{\partial A_j}{\partial \alpha_i}\frac{\partial F}{\partial \alpha_j} + \frac{1}{A_i{}^2 A_j}\frac{\partial A_i}{\partial \alpha_j}\frac{\partial F}{\partial \alpha_i}, \end{aligned} \tag{1.86}$$

then in the approximation of the technical theory, eqn (1.85) is satisfied identically. The expression of T^*_{i3} takes the form

$$T^*_{i3}=T_{i3}+T_{ii}\theta_i+T_{ij}\theta_j=\frac{1}{A_iA_j}\frac{\partial}{\partial\alpha_i}\{A_j[D_{ii}\varkappa_i+(D_{12}+2D_{66})\varkappa_j]\}-$$
$$-\frac{1}{A_iA_j}\frac{\partial A_j}{\partial\alpha_i}[(D_{12}+2D_{66})\varkappa_i+D_{jj}\varkappa_j]+\theta_i\left[\frac{1}{A_j}\frac{\partial}{\partial\alpha_j}\left(\frac{1}{A_j}\frac{\partial F}{\partial\alpha_j}\right)+\right.$$
$$\left.+\frac{1}{A_i^2A_j}\frac{\partial A_j}{\partial\alpha_i}\frac{\partial F}{\partial\alpha_i}\right]+\theta_j\left[-\frac{1}{A_iA_j}\frac{\partial^2F}{\partial\alpha_i\partial\alpha_j}+\frac{1}{A_iA_j^2}\frac{\partial A_j}{\partial\alpha_i}\frac{\partial F}{\partial\alpha_j}+\right.$$
$$\left.+\frac{1}{A_i^2A_j}\frac{\partial A_i}{\partial\alpha_j}\frac{\partial F}{\partial\alpha_i}\right]+F^*_j+\Phi^*_j. \tag{1.87}$$

As a result the system of equilibrium equations is reduced to one equation

$$\frac{\partial}{\partial\alpha_1}\left\{\frac{1}{A_1}\frac{\partial}{\partial\alpha_1}\{A_2[D_{11}\varkappa_1+(D_{12}+2D_{66})\varkappa_2]\}-\frac{1}{A_1}\frac{\partial A_2}{\partial\alpha_1}[(D_{12}+\right.$$
$$+2D_{66})\varkappa_1+D_{22}\varkappa_2]+\theta_1\left[\frac{\partial}{\partial\alpha_2}\left(\frac{1}{A_2}\frac{\partial F}{\partial\alpha_2}\right)+\frac{1}{A_1^2}\frac{\partial A_2}{\partial\alpha_1}\frac{\partial F}{\partial\alpha_1}\right]+$$
$$\left.+\theta_2\left[-\frac{1}{A_1}\frac{\partial^2F}{\partial\alpha_1\partial\alpha_2}+\frac{1}{A_1A_2}\frac{\partial A_2}{\partial\alpha_1}\frac{\partial F}{\partial\alpha_2}+\frac{1}{A_1^2}\frac{\partial A_1}{\partial\alpha_2}\frac{\partial F}{\partial\alpha_1}\right]\right\}+$$
$$+\frac{\partial}{\partial\alpha_2}\left\{\frac{1}{A_2}\frac{\partial}{\partial\alpha_2}\{A_1[D_{22}\varkappa_2+(D_{12}+2D_{66})\varkappa_1]\}-\right. \tag{1.88}$$
$$-\frac{1}{A_2}\frac{\partial A_1}{\partial\alpha_2}[(D_{12}+2D_{66})\varkappa_2+D_{11}\varkappa_1]+\theta_2\left[\frac{\partial}{\partial\alpha_1}\left(\frac{1}{A_1}\frac{\partial F}{\partial\alpha_1}\right)+\right.$$
$$\left.+\frac{1}{A_2^2}\frac{\partial A_1}{\partial\alpha_2}\frac{\partial F}{\partial\alpha_2}\right]+\theta_1\left[-\frac{1}{A_2}\frac{\partial^2F}{\partial\alpha_2\partial\alpha_1}+\frac{1}{A_2A_1}\frac{\partial A_1}{\partial\alpha_2}\frac{\partial F}{\partial\alpha_1}+\right.$$
$$\left.\left.+\frac{1}{A_2^2}\frac{\partial A_2}{\partial\alpha_1}\frac{\partial F}{\partial\alpha_2}\right]\right\}-k_1\left[A_1\frac{\partial}{\partial\alpha_2}\left(\frac{1}{A_2}\frac{\partial F}{\partial\alpha_2}+\frac{1}{A_1}\frac{\partial A_2}{\partial\alpha_1}\frac{\partial F}{\partial\alpha_1}\right)\right]-$$
$$-k_2\left[A_2\frac{\partial}{\partial\alpha_1}\left(\frac{1}{A_1}\frac{\partial F}{\partial\alpha_1}\right)+\frac{1}{A_2}\frac{\partial A_1}{\partial\alpha_2}\frac{\partial F}{\partial\alpha_2}\right]+\frac{\partial}{\partial\alpha_1}[A_2(F^*_2+\Phi^*_2)]+$$
$$+\frac{\partial}{\partial\alpha_2}[A_1(F^*_1+\Phi^*_1)]+A_1A_2(X^*_3+P^*_3)=0,$$

containing the unknown functions u_3^0 and F. The second equation can be derived from the Gaussian eqn (3.32), presented in ref. 196:

$$\frac{\partial}{\partial\alpha_1}\left\{\frac{1}{A_1}\left[\frac{\partial}{\partial\alpha_1}(A_2\hat{\varepsilon}_{22})-\frac{\partial A_1}{\partial\alpha_2}\hat{\varepsilon}_{12}-\frac{1}{2}A_1\frac{\partial\hat{\varepsilon}_{12}}{\partial\alpha_2}-\hat{\varepsilon}_{11}\frac{\partial A_2}{\partial\alpha_1}\right]\right\}+$$
$$+\frac{\partial}{\partial\alpha_2}\left\{\frac{1}{A_2}\left[\frac{\partial}{\partial\alpha_2}(A_1\hat{\varepsilon}_{11})-\frac{\partial A_2}{\partial\alpha_1}\hat{\varepsilon}_{12}-\frac{1}{2}A_2\frac{\partial\hat{\varepsilon}_{12}}{\partial\alpha_1}-\hat{\varepsilon}_{22}\frac{\partial A_1}{\partial\alpha_2}\right]\right\}=$$
$$=A_1A_2(\tau^2-\varkappa_1\varkappa_2-k_2\varkappa_1-k_1\varkappa_2), \tag{1.89}$$

where $\hat{\varepsilon}_{ii}=\varepsilon_i+\frac{1}{2}\theta_i^2$; $\hat{\varepsilon}_{12}=\gamma_1+\gamma_2+\theta_1\theta_2$. On expressing these values in terms of tractions in accordance with (1.78)

$$\hat{\varepsilon}_{ii}=\frac{C_{jj}T_{ii}-C_{ij}T_{jj}}{C_{11}C_{22}-C_{12}^2}\;;\quad \hat{\varepsilon}_{12}=\frac{T_{12}}{C_{66}}, \tag{1.90}$$

and inserting them into (1.89) and using the first two equations of equilibrium, we arrive at the following equation

$$\left(A_{22}-A_{12}-\frac{A_{66}}{2}\right)\left[\frac{\partial}{\partial\alpha_1}\left(\frac{1}{A_1}\frac{\partial A_2}{\partial\alpha_1}T_{22}\right)-\right.$$
$$\left.-\frac{\partial}{\partial\alpha_2}\left(\frac{1}{A_2}\frac{\partial A_1}{\partial\alpha_2}T_{22}\right)\right]+\left(A_{11}-A_{12}-\frac{A_{66}}{2}\right)\times$$
$$\times\left[\frac{\partial}{\partial\alpha_2}\left(\frac{1}{A_2}\frac{\partial A_1}{\partial\alpha_2}T_{11}\right)-\frac{\partial}{\partial\alpha_1}\left(\frac{1}{A_1}\frac{\partial A_2}{\partial\alpha_1}T_{11}\right)\right]+$$
$$+\frac{\partial}{\partial\alpha_1}\left\{\frac{A_2}{A_1}\frac{\partial}{\partial\alpha_1}\left[A_{22}T_{22}+\left(A_{12}+\frac{A_{66}}{2}\right)T_{11}\right]\right\}+ \tag{1.91}$$
$$+\frac{\partial}{\partial\alpha_2}\left\{\frac{A_1}{A_2}\frac{\partial}{\partial\alpha_2}\left[A_{11}T_{11}+\left(A_{12}+\frac{A_{66}}{2}\right)T_{22}\right]\right\}=$$
$$=A_1A_2(\tau^2-\varkappa_1\varkappa_2-k_2\varkappa_1-k_1\varkappa_2),$$

where

$$A_{ii}=\frac{C_{jj}}{C_{11}C_{22}-C_{12}^2}\;;\quad A_{ij}=-\frac{C_{ij}}{C_{11}C_{22}-C_{12}^2}\;;\quad A_{66}=\frac{1}{C_{66}}. \tag{1.92}$$

A substitution of expressions (1.86) and (1.84) into (1.91) results in an equation in respect to functions $\overset{0}{u}_3$, F, however, we failed to write it down in such a compact form as for an isotropic material. Let us note that in the latter case $A_{22}-A_{12}-(A_{66}/2)=0$, $A_{11}-A_{12}-(A_{66}/2)=0$, $A_{12}+(A_{66}/2)=A_{11}=A_{22}=1/Eh$ and by introducing an operator

$$\nabla^2F=\frac{1}{A_1A_2}\left\{\frac{\partial}{\partial\alpha_1}\left[\frac{A_2}{A_1}\frac{\partial F}{\partial\alpha_1}\right]+\frac{\partial}{\partial\alpha_2}\left[\frac{A_1}{A_2}\frac{\partial F}{\partial\alpha_2}\right]\right\}$$

(1.91) is reduced to

$$\nabla^2\nabla^2F=Eh(\tau^2-\varkappa_1\varkappa_2-k_2\varkappa_1-k_1\varkappa_2). \tag{1.93}$$

For a 'rather shallow' orthotropic shell, when A_1 and A_2 behave as constants under the differentiation,[21] it follows from (1.84) and (1.86) that

$$\varkappa_i = -\frac{1}{A_i^2}\frac{\partial^2 u_3^0}{\partial \alpha_i^2}; \quad \tau = -\frac{1}{A_1 A_2}\frac{\partial^2 u_3^0}{\partial \alpha_1 \partial \alpha_2};$$

$$T_{ii} = \frac{1}{A_j^2}\frac{\partial^2 F}{\partial \alpha_j^2}; \quad T_{12} = T_{21} = -\frac{1}{A_1 A_2}\frac{\partial^2 F}{\partial \alpha_1 \partial \alpha_2}, \tag{1.94}$$

but the eqn (1.88) assumes the form

$$\mathcal{L}_1 u_3^0 + \nabla_R F - \mathcal{L}(u_3^0, F) = Z, \tag{1.95}$$

where

$$\mathcal{L}_1 = \frac{D_{11}}{A_1^4}\frac{\partial^4}{\partial \alpha_1^4} + \frac{2(D_{12}+2D_{66})}{A_1^2 A_2^2}\frac{\partial^4}{\partial \alpha_1^2 \partial \alpha_2^2} + \frac{D_{22}}{A_2^4}\frac{\partial^4}{\partial \alpha_2^4};$$

$$\nabla_R = \frac{k_2}{A_1^2}\frac{\partial^2}{\partial \alpha_1^2} + \frac{k_1}{A_2^2}\frac{\partial^2}{\partial \alpha_2^2};$$

$$\mathcal{L}(f_1, f_2) = \frac{1}{A_1^2 A_2^2}\left(\frac{\partial^2 f_1}{\partial \alpha_1^2}\frac{\partial^2 f_2}{\partial \alpha_2^2} + \frac{\partial^2 f_2}{\partial \alpha_1^2}\frac{\partial^2 f_1}{\partial \alpha_2^2} - 2\frac{\partial^2 f_1}{\partial \alpha_1 \partial \alpha_2}\frac{\partial^2 f_2}{\partial \alpha_1 \partial \alpha_2}\right);$$

$$Z = X^*_3 + P^*_3 + \frac{1}{A_1}\frac{\partial}{\partial \alpha_1}(F^*_2 + \Phi^*_2) + \frac{1}{A_2}\frac{\partial}{\partial \alpha_2}(F^*_1 + \Phi^*_1).$$

Accordingly, the eqn (1.91) in the treated case is written in the form

$$\mathcal{L}_2 F - \nabla_R u_3^0 + \frac{1}{2}\mathcal{L}(u_3^0, u_3^0) = 0, \tag{1.96}$$

where

$$\mathcal{L}_2 = \frac{A_{22}}{A_1^4}\frac{\partial^4}{\partial \alpha_1^4} + \frac{2A_{12}+A_{66}}{A_1^2 A_2^2}\frac{\partial^4}{\partial \alpha_1^2 \partial \alpha_2^2} + \frac{A_{11}}{A_2^4}\frac{\partial^4}{\partial \alpha_2^4}$$

The eqns (1.95) and (1.96) were obtained in Ref. 24.

1.7 EFFECT OF INITIAL IMPERFECTIONS IN THE MIDDLE SURFACE OF A SHELL

In the process of fabrication or installation of structures, comprising shell elements, the overall or local technological imperfections inevitably arise, leading to deviation of the mid-surface shape from the 'ideal' one. The presence of such imperfections can be generally described through introduction of initial deformations and, as a consequence, initial stresses. The non-linear shell theory taking into account the stress–strain state was evolved by H. M. Mushtari.[194,195] It is also outlined in the monograph by H. M. Mushtari and K. Z. Galimov.[196]

Thereby, the initial deformations ε_{ii}^0, ε_{12}^0 are determined in terms of initial displacements u_i^0, u_3^0 according to the formulas of the theory of shallow shells.

Henceforth we shall treat a particular case, when the initial imperfections are of purely flexural character, i.e. $u_i^0 = 0$ over the whole surface of a shell. Divergence of the mid-surface shape from the 'ideal' one is completely specified by the field of normal displacement $w^0(\alpha_1, \alpha_2)$. Consequently, the initial deformations are evaluated within the limits of the technical theory according to the formulas

$$\begin{aligned}
\hat{\varepsilon}_{ii}{}^0 &= \varepsilon_i{}^0 + \frac{1}{2}\theta_i{}^{02} = k_i w^0 + \frac{1}{2}\left(\frac{1}{A_i}\frac{\partial w^0}{\partial \alpha_i}\right)^2;\\
\hat{\varepsilon}_{12}{}^0 &= \gamma_1{}^0 + \gamma_2{}^0 + \theta_1{}^0\theta_2{}^0 = \frac{1}{A_i A_j}\frac{\partial w^0}{\partial \alpha_i}\frac{\partial w^0}{\partial \alpha_j};\\
\varkappa_i{}^0 &= -\frac{1}{A_i}\frac{\partial}{\partial \alpha_i}\left(\frac{1}{A_i}\frac{\partial w^0}{\partial \alpha_i}\right) - \frac{1}{A_i A_j{}^2}\frac{\partial A_i}{\partial \alpha_j}\frac{\partial w^0}{\partial \alpha_j}; \qquad (1.97)\\
\tau^0 &= -\frac{1}{A_i}\frac{\partial}{\partial \alpha_i}\left(\frac{1}{A_i}\frac{\partial w^0}{\partial \alpha_j}\right) + \frac{1}{A_i{}^2 A_j}\frac{\partial A_i}{\partial \alpha_j}\frac{\partial w^0}{\partial \alpha_i}.
\end{aligned}$$

Let us assume that the initial 'distortions' of the shell shape do not give initial stresses. Therefore, under initial conditions on displacements $u_i^0|_{t=0} = 0$, $u_3^0|_{t=0} = w^0$ and in the absence of external effects on the shell at the moment $t = 0$, all the stresses at $t = 0$ should be zero. The equations of equilibrium thereby should be satisfied identically.

In such a way, to allow for initial geometric imperfections, we should either substitute $\hat{\varepsilon}_{ii}$ by $(\hat{\varepsilon}_{ii} - \hat{\varepsilon}_{ii}^0)$, $\hat{\varepsilon}_{12}$ — by $(\hat{\varepsilon}_{12} - \hat{\varepsilon}_{12}^0)$, $\varkappa_i$ — by $(\varkappa_i - \varkappa_i^0)$, τ_i — by $(\tau_i - \tau_i^0)$ or leave them in their previous form, but conveive $\hat{\varepsilon}_{ii}$, $\hat{\varepsilon}_{12}$, $\varkappa_i$, τ_i as 'additional' values having arisen under loading.

1.8 LINEARIZATION OF EQUILIBRIUM EQUATIONS

The problem of deducing the linearized stability equations in a shell theory has been treated by many authors. A short survey of the approaches and bibliography on the subject are given by V. V. Bolotin,[70] who has indicated a number of different versions for 'parametric' ('loading') terms, which can be attributed to a variety of ways for obtaining stability equations. In this context, we shall particularly mention the works[148, 196] where linearization is realized on the basis of non-linear equilibrium equations for a shell. An analogous approach was used in deducing the equations of stability in Refs. 134, 135.

To obtain linearized equations we shall proceed from equations of equilibrium (1.42) and expressions for generalized tractions (1.41). Let us consider a particular case when only normal forces T_{ii}^0 arise in the pre-critical state, which is characteristic when the longitudinal forces and external pressure are applied. We shall disregard all the deformations in the pre-critical state. Let us designate by T'_{ii}, T'_{ij}, T'_{i3} the 'additional' tractions arising in the shell upon derivation from the pre-critical equilibrium state. According to the method described by E. I. Grigolyuk and V. V. Kabanov,[134, 135] let us substitute T_{ii} by $T_{ii}^0 + T'_{ii}$ in (1.41), and then insert the obtained expressions for generalized tractions into (1.42), subtract the equation of equilibrium of the pre-critical state and perform linearization. The folllowing equations of stability will be finally obtained

$$\frac{\partial}{\partial\alpha_i}(A_jT'_{ii})+\frac{\partial}{\partial\alpha_j}(A_iT'_{ji})+\frac{\partial A_i}{\partial\alpha_j}T'_{ij}-\frac{\partial A_j}{\partial\alpha_i}T'_{jj}+$$
$$+A_iA_jk_iT'_{i3}+A_iA_jP'_i+X'_i=0; \tag{1.98}$$
$$\frac{\partial}{\partial\alpha_i}(A_jT'_{i3})+\frac{\partial}{\partial\alpha_j}(A_iT'_{j3})-A_iA_j(k_iT'_{ii}+k_jT'_{jj})+A_iA_jP'_3+X'_3=0$$

where

$$X'_i=\frac{\partial}{\partial\alpha_i}(A_jT_{ii}{}^0\varepsilon'_j)+\frac{\partial}{\partial\alpha_j}(A_iT_{jj}{}^0\gamma'_j)+\frac{\partial A_i}{\partial\alpha_j}T_{ii}{}^0\gamma'_i-$$
$$-\frac{\partial A_j}{\partial\alpha_i}T_{jj}{}^0\varepsilon'_i+A_iA_jk_iT_{ii}{}^0\theta'_i+A_iA_j[q_i(\varepsilon'_i+\varepsilon'_j)+\delta_c(q_j\gamma'_j+q_3v_i{}^{0'})]; \tag{1.99}$$
$$X'_3=\frac{\partial}{\partial\alpha_1}(A_2T_{11}{}^0\theta'_1)+\frac{\partial}{\partial\alpha_2}(A_1T_{22}{}^0\theta'_2)-A_1A_2(k_1T_{11}{}^0\varepsilon'_2+k_2T_{22}{}^0\varepsilon'_1)+$$
$$+A_1A_2[q_3(\varepsilon'_1+\varepsilon'_2)+\delta_c(q_1\theta'_1+q_2\theta'_2)];$$

$q_i = X_i^+ - X_i^-$; $q_3 = X_3^+ - X_3^-$; $\delta_e = 1$ and 0 at following and conservative loads, respectively. The eqns (1.98), as is seen, differ from the usual equations of equilibrium of the linear theory of shells only by the presence of 'loading' terms X'_i, X'_3. The formulas (1.99) are the particular case of the expressions obtained in Refs. 134, 135 and 148. The magnitudes X'_i, X'_3, P'_i, P'_3 represent the 'additional' surface and mass forces, acting on the shell only upon its deviation from the pre-critical equilibrium state.

The linear equations for moments in accordance with (1.57) have the form

$$\frac{\partial}{\partial \alpha_i}(A_j M'_{ii})+\frac{\partial}{\partial \alpha_j}(A_i M'_{ji})+\frac{\partial A_i}{\partial \alpha_j}M'_{ij}-\frac{\partial A_j}{\partial \alpha_i}M'_{jj}-$$
$$-A_iA_jT'_{i3}+A_iA_j(\Phi'_j+F'_j)=0, \qquad (1.100)$$

where Φ'_j and F'_j are the 'additional' moments of surface and mass forces, arising upon deviation from the pre-critical equilibrium state. By restricting ourselves to the linear terms in (1.60) and (1.61), we find

$$F'_i=\int_{-h/2}^{h/2} F_j[1+\alpha_3(k_i+k_j)]\alpha_3 d\alpha_3; \quad \Phi'_i=\frac{h}{2}(X_j^+ + X_j^-). \qquad (1.101)$$

The tractions T'_{ii}, T'_{ij}, T'_{i3} and moments M'_{ii}, M'_{ij} are expressed in terms of 'additional' deformations according to the formulas of the linear theory of shells:

$$T'_{ii}=C_{ii}\varepsilon'_i+C_{ij}\varepsilon'_j; \quad T'_{ij}=C_{66}(\gamma'_i+\gamma'_j);$$
$$T'_{i3}=C_{i3}(\theta'_i+v_i^{0\prime}); \quad M'_{ii}=D_{ii}\varkappa'_i+D_{ij}\varkappa'_j; \qquad (1.102)$$
$$M'_{ij}=D_{66}(\tau'_i+\tau'_j).$$

By employing the Kirchhoff-Love hypotheses the tractions T'_{i3} are determined from eqn (1.100).

To solve the stability problems on the base of equations, represented in a mixed form, it is necessary to put $P'_i=0$, $X'_i=0$ and express the tractions T'_{ii}, T'_{ij} in terms of the function F according to the formulas (1.86) and then perform analogous manipulations to those in Section (1.6) by omitting non-linear terms. As a result (1.88) takes the form

$$\frac{\partial}{\partial \alpha_1}\left\{\frac{1}{A_1}\frac{\partial}{\partial \alpha_1}\{A_2[D_{11}\varkappa'_1+(D_{12}+2D_{66})\varkappa'_2]\}-\right.$$
$$\left.-\frac{1}{A_1}\frac{\partial A_2}{\partial \alpha_1}[(D_{12}+2D_{66})\varkappa'_1+D_{22}\varkappa'_2]\right\}+\frac{\partial}{\partial \alpha_2}\left\{\frac{1}{A_2}\frac{\partial}{\partial \alpha_2}\times\right.$$
$$\times\{A_1[D_{22}\varkappa'_2+(D_{12}+2D_{66})\varkappa'_1]\}-\frac{1}{A_2}\frac{\partial A_1}{\partial \alpha_2}[(D_{12}+2D_{66})\varkappa'_2+$$
$$\left.+D_{11}\varkappa'_1]\right\}-k_1\left[A_1\frac{\partial}{\partial \alpha_2}\left(\frac{1}{A_2}\frac{\partial F}{\partial \alpha_2}\right)+\frac{1}{A_1}\frac{\partial A_2}{\partial \alpha_1}\frac{\partial F}{\partial \alpha_1}\right]-$$
$$-k_2\left[A_2\frac{\partial}{\partial \alpha_1}\left(\frac{1}{A_1}\frac{\partial F}{\partial \alpha_1}\right)+\frac{1}{A_2}\frac{\partial A_1}{\partial \alpha_2}\frac{\partial F}{\partial \alpha_2}\right]+$$
$$+\frac{\partial}{\partial \alpha_1}[A_2(F'_2+\Phi'_2)]+\frac{\partial}{\partial \alpha_2}[A_1(F'_1+\Phi'_1)]+A_1A_2P'_3+X'_3=0. \qquad (1.103)$$

In the equation of compatibility (1.91) non-linear terms in the right-hand part should be omitted.

When solving the problems of dynamic stability the tractions T_{ii}^0 are functions of time. Taking into account that the inertia forces

$$F_i' = -\rho \frac{\partial^2 u_i'}{\partial t^2}, \quad F_3' = -\rho \frac{\partial^2 u_3'}{\partial t^2}$$

(ρ is the density of the material; u_i', u_3' are 'additional' displacements), according to formulas (1.1), (1.43) and (1.101) we find

$$P'_i = -\rho h \left[\frac{\partial^2 u_i^{0\prime}}{\partial t^2} + \frac{h^2}{12}(k_i + k_j)\frac{\partial^2 v_i^{0\prime}}{\partial t^2} \right]; \qquad P'_3 = -\rho h \frac{\partial^2 u_3^{0\prime}}{\partial t^2}$$

$$F'_i = -\rho \frac{h^3}{12} \left[\frac{\partial^2 v_j^{0\prime}}{\partial t^2} + (k_i + k_j)\frac{\partial^2 u_j^{0\prime}}{\partial t^2} \right]. \qquad (1.104)$$

The equations and formulas presented in this chapter permit to solve non-linear problems of deformation, problems of static and dynamic stability for thin-wall orthotropic shells both with allowance for transverse shear deformations and on the basis of the Kirchhoff-Love kinematic model.

2

Equations of the Non-linear Theory for Orthotropic Cylindrical Shells

In Chapter 1, the non-linear and linearized equations of motion for orthotropic shells, written in an arbitrary curvilinear orthogonal system of coordinates, have been obtained. Their deduction was accomplished on the basis of the non-linear theory of elasticity under the assumption of the linear-elastic behavior of a material, thin-wallness of a shell, smallness of deformations in comparison to a unity, elongations and shears in comparison to the angles of rotation of normal sections. On the basis of these equations the non-linear and linearized equations of motion for an orthotropic circular cylindrical shell have been deduced in this chapter. The versions corresponding to Timoshenko and Kirchhoff-Love's kinematic models are treated. The linearized equations of motion for an orthotropic cylindrical shell have also been obtained, without the simplifying assumptions of the technical theory (a theory of shallow shells). The present chapter contains all the necessary basic formulas and equations, which will find its use throughout the monograph.

The theory of thin-wall cylindrical shells has been evolved by many authors. Detailed studies in the linear theory are presented in monographs.[21,24,94,123,143,209,254,322,441] The non-linear theory of isotropic cylindrical shells is outlined in Ref. 196. The equations of the non-linear theory for the case of an anisotropic material are treated in monographs.[92,224] Without dwelling on the historical aspect of the development of the theory of cylindrical shells, we shall indicate merely some review papers,[10,13,107,115,134,136,138,154,199,203,204,291] containing rich bibliography on this problem.

2.1 BASIC EQUATIONS OF THE NON-LINEAR THEORY FOR THIN-WALLED ORTHOTROPIC SHELLS

Let α_1, α_2, α_3 be the coordinates of a point in an arbitrary curvilinear orthogonal system of coordinates. Let us superpose the coordinate surface $\alpha_3 = 0$ on the mid-surface of a shell. The coordinate lines α_1, α_2 are oriented along the lines of principal curvatures of this surface. Let us denote the respective unit vectors by $\mathbf{e}_1$, $\mathbf{e}_2$ and assume that the unit vector $\mathbf{e}_3$ coincides with the direction of the outer normal to the mid-surface of a shell.

Let us designate the projections of the displacement vector at an arbitrary point of a shell on the directions $\mathbf{e}_1$, $\mathbf{e}_2$, $\mathbf{e}_3$ by $u_1(\alpha_1, \alpha_2, \alpha_3)$, $u_2(\alpha_1, \alpha_2, \alpha_3)$, $u_3(\alpha_1, \alpha_2, \alpha_3)$ and assume that they are the linear functions of the α_3 coordinate:[208]

$$u_i = u_i{}^0 + \alpha_3 v_i{}^0; \quad u_3 = u_3{}^0 + \alpha_3 v_3{}^0; \quad i = 1, 2. \tag{2.1}$$

Here, $u_1^0(\alpha_1, \alpha_2)$, $u_2^0(\alpha_1, \alpha_2)$, $u_3^0(\alpha_1, \alpha_2)$ are the projections of the displacement vector for an arbitrary point $\{\alpha_1, \alpha_2\}$ of the mid-surface of a shell on the directions $\mathbf{e}_1$, $\mathbf{e}_2$, $\mathbf{e}_3$; $v_1^0(\alpha_1, \alpha_2)$, $v_2^0(\alpha_1, \alpha_2)$ are the functions characterizing rotation of sections, passing through the normal to the coordinate surface, during the deformation of a shell.

The components of deformation tensor in accordance with Ref. 210 are defined by formulas (1.2) and (1.3). By substituting (2.1) into these formulas and neglecting the magnitudes of the order $\alpha_3 k_i$, compared to a unity, the expressions for deformations in a shell (1.12) have been obtained. Further simplifications are based on the following assumptions:

(1) on the indicators of variability along the coordinate lines α_1, α_2 of the characteristics in deformed state and the coefficients of the mid-surface first square form A_i are put upper bounds according to the inequalities in (1.13);
(2) elongations and shears are the values of higher order of smallness than angles of rotation of normal sections. In compliance with this assumption their squares, mutual products and their products by other unknown functions have been omitted.

As a result the formulas (1.15) have been obtained, accordingly

$$\varepsilon_{ii} = \hat{\varepsilon}_{ii} + \alpha_3 \varkappa^*{}_i; \quad \varepsilon_{12} = \hat{\varepsilon}_{12} + \alpha_3 (\tau^*{}_1 + \tau^*{}_2); \tag{2.2}$$

$$\varepsilon_{i3} = v_i{}^0 + \theta_i, \tag{2.3}$$

where

$$\hat{\varepsilon}_{ii}=\varepsilon_i+\frac{1}{2}\theta_i^2;\quad \hat{\varepsilon}_{12}=\gamma_1+\gamma_2+\theta_1\theta_2; \tag{2.4}$$

$$\varkappa^*_i=\varkappa_i+\theta_i\varkappa_{i3};\quad \tau^*_i=\tau_i+\theta_j\varkappa_{i3}; \tag{2.5}$$

$$\begin{aligned}\varepsilon_i&=\frac{1}{A_i}\frac{\partial u_i^0}{\partial\alpha_i}+\frac{1}{A_iA_j}\frac{\partial A_i}{\partial\alpha_j}u_j^0+k_iu_3^0;\\ \gamma_i&=\frac{1}{A_i}\frac{\partial u_j^0}{\partial\alpha_i}-\frac{1}{A_iA_j}\frac{\partial A_i}{\partial\alpha_j}u_i^0;\\ \theta_i&=\frac{1}{A_i}\frac{\partial u_3^0}{\partial\alpha_i}-k_iu_i^0;\end{aligned} \tag{2.6}$$

$$\varkappa_i=\frac{1}{A_i}\frac{\partial v_i^0}{\partial\alpha_i}+\frac{1}{A_iA_j}\frac{\partial A_i}{\partial\alpha_j}v_j^0+k_iv_3^0; \tag{2.7}$$

$$\tau_i=\frac{1}{A_i}\frac{\partial v_j^0}{\partial\alpha_i}-\frac{1}{A_iA_j}\frac{\partial A_i}{\partial\alpha_j}v_i^0; \tag{2.8}$$

$$\varkappa_{i3}=\frac{1}{A_i}\frac{\partial v_3^0}{\partial\alpha_i}-k_iv_i^0.$$

In these formulas, the index $j=2$ at $i=1$ and $j=1$ at $i=2$. The deformation ε_{33} according to (1.11) is expressed in the form

$$\varepsilon_{33}=v_3^0+\frac{1}{2}(v_1^{02}+v_2^{02}+v_3^{02}). \tag{2.9}$$

In Chapter 1, a variant of the non-linear theory of orthotropic shells based on Kirchhoff-Love's kinematic model was treated. Thereby in compliance with V. V. Novozhilov's interpretation[208] it has been assumed that in determining the change in unit vector $\mathbf{e}_3$ direction it is possible to disregard the values of transverse shears. This will lead to the condition (1.16) from which, by taking into account the smallness of v_3^0 compared to a unity, the relationship (1.17) is obtained:

$$v_i^0\approx-\theta_i. \tag{2.10}$$

Besides, the expression for the function v_3^0 (1.27) is obtained as

$$v_3^0\approx-(a_1\varepsilon_1+a_2\varepsilon_2)-\frac{1}{2}\theta_1^2(1+a_1)-\frac{1}{2}\theta_2^2(1+a_2), \tag{2.11}$$

which contains a combination of Poisson ratios for an orthotropic material (1.24). In consideration of highly anisotropic materials for which

$\nu_{13} \ll 1$, $\nu_{23} \ll 1$, the expression (1.27) differs slightly from V. V. Novozhilov's[208] formulas (1.19).

Taking into consideration (2.10) and (2.11) and restricting ourselves to the linear terms in respect to elongations and square terms — to angles of rotation, the formulas for $\varkappa_i^*$ and τ_i^* can be transformed into:

$$\varkappa^*_i = -\frac{1}{A_i}\frac{\partial \theta_i}{\partial \alpha_i} - \frac{1}{A_i A_j}\frac{\partial A_i}{\partial \alpha_j}\theta_j + k_i(\theta_i^2 + v_3^0);$$
$$\tau^*_i = -\frac{1}{A_i}\frac{\partial \theta_j}{\partial \alpha_i} + \frac{1}{A_i A_j}\frac{\partial A_i}{\partial \alpha_j}\theta_i + k_i\theta_i\theta_j. \qquad (2.12)$$

In the majority of solutions for linear problems of the shell theory the simplified expressions for $\varkappa_i^*$, τ_i^* are used:

$$\varkappa^*_i = \varkappa_i = \frac{1}{A_i}\frac{\partial v_i^0}{\partial \alpha_i} + \frac{1}{A_i A_j}\frac{\partial A_i}{\partial \alpha_j} v_j^0;$$
$$\tau^*_i = \tau_i, \qquad (2.13)$$

which correspond to Timoshenko type model and

$$\varkappa^*_i = -\frac{1}{A_i}\frac{\partial \theta_i}{\partial \alpha_i} - \frac{1}{A_i A_j}\frac{\partial A_i}{\partial \alpha_j}\theta_j;$$
$$\tau^*_i = -\frac{1}{A_i}\frac{\partial \theta_j}{\partial \alpha_i} + \frac{1}{A_i A_j}\frac{\partial A_i}{\partial \alpha_j}\theta_i, \qquad (2.14)$$

which correspond to Kirchhoff-Love's model. The possibility of disregarding the non-linear terms in the formulas for $\varkappa_i^*$, τ_i^* can be proven by setting a lower bound on the indicators of variability of the deformed state. Thus, under conditions of

$$\lambda_{ii} \geqslant k_i A_i; \quad \lambda_{ij} \geqslant k_i A_j \qquad (2.15)$$

(in accordance with definition, $\lambda_{ii} = (1/v_i^0)(\partial v_i^0/\partial \alpha_i)$, $\lambda_{ij} = (1/v_i^0)(\partial v_i^0/\partial a_j)$ the relative correction applied by the non-linear terms into the magnitude $\varkappa_i^*$ (2.5) is of the order $O(v_i^0/2) + O(v_j^{02}/2v_i^0) + O(\theta_i)$, accordingly, into the value $\tau^* = \tau_1^* + \tau_2^* - O(\theta_1) + O(\theta_2)$, but into a value $\varkappa_i^*$ (2.12) — $O(\theta_j^2/2\theta_i)$. In such a way, the accuracy of the formulas (2.13), (2.14) falls with the increase in the angles of rotation and the decrease in indicators of variability λ_{ii}, λ_{ij}. Let us note that the restrictions of the type (2.15) are usually imposed in the technical theory (theory of shallow shells).

The equations of the generalized Hooke's law upon disregarding the squares of the deformation tensor ε_{ij} components and the transverse normal deformation are written for the elastic orthotropic shell in the form:

$$\sigma_{ii}=B_{ii}\varepsilon_{ii}+B_{ij}\varepsilon_{jj};\quad \sigma_{ij}=B_{66}\varepsilon_{ij};\quad \sigma_{i3}=B_{i3}\varepsilon_{i3}. \tag{2.16}$$

The equations of equilibrium for a shell in terms of generalized tractions and moments assume the form (1.42) and (1.57):

$$\frac{\partial}{\partial\alpha_i}(A_jT^*_{ii})+\frac{\partial}{\partial\alpha_j}\cdot(A_iT^*_{ji})+\frac{\partial A_i}{\partial\alpha_j}T^*_{ij}-\frac{\partial A_j}{\partial\alpha_i}T^*_{jj}+$$
$$+A_iA_jk_iT^*_{i3}+A_iA_j(X^*_i+P^*_i)=0; \tag{2.17}$$

$$\frac{\partial}{\partial\alpha_1}(A_2T^*_{13})+\frac{\partial}{\partial\alpha_2}(A_1T^*_{23})-A_1A_2(k_1T^*_{11}+k_2T^*_{22})+$$
$$+A_1A_2(X^*_3+P^*_3)=0; \tag{2.18}$$

$$\frac{\partial}{\partial\alpha_i}(A_jM^*_{ii})+\frac{\partial}{\partial\alpha_j}(A_iM^*_{ji})+\frac{\partial A_i}{\partial\alpha_j}M^*_{ij}-\frac{\partial A_j}{\partial\alpha_i}M^*_{jj}-$$
$$-A_iA_jT^{**}_{i3}+A_iA_j(F^*_j+\Phi^*_j)=0. \tag{2.19}$$

Here, in accordance with (1.71)–(1.74) we have:

$$T^*_{ii}=C_{ii}\left(\varepsilon_i+\frac{1}{2}\theta_i^2\right)+C_{ij}\left(\varepsilon_j+\frac{1}{2}\theta_j^2\right)+C_{i3}v_i^0(v_i^0+\theta_i);$$
$$T^*_{ij}=C_{66}(\gamma_i+\gamma_j+\theta_i\theta_j)+C_{i3}v_j^0(v_i^0+\theta_i); \tag{2.20}$$

$$T^*_{i3}=C_{i3}(v_i^0+\theta_i)+C_{ii}\theta_i\left(\varepsilon_i+\frac{1}{2}\theta_i^2\right)+C_{ij}\theta_i\left(\varepsilon_j+\frac{1}{2}\theta_j^2\right)+$$
$$+C_{66}\theta_j(\gamma_i+\gamma_j+\theta_i\theta_j); \tag{2.21}$$

$$M^*_{ii}=D_{ii}\left[\varkappa^*_i+k_j\left(\varepsilon_i+\frac{1}{2}\theta_i^2\right)\right]+D_{ij}\left[\varkappa^*_j+k_j\left(\varepsilon_j+\frac{1}{2}\theta_j^2\right)\right];$$
$$M^*_{ij}=D_{66}[\tau^*_i+\tau^*_j+k_j(\gamma_i+\gamma_j+\theta_i\theta_j)]; \tag{2.22}$$

$$T^{**}_{i3}=C_{i3}(\theta_i+v_i^0)+C_{j3}(\theta_j+v_j^0)(\gamma_j-v_i^0\theta_j)+D_{i3}(k_i+k_j)\times$$
$$\times(\theta_i+v_i^0)(\varkappa_i+\varkappa_j)+D_{j3}(k_i+k_j)(\theta_j+v_j^0)\tau_j-k_jv_i^0(D_{jj}\varkappa_j+D_{ij}\varkappa_i)+$$
$$+k_jD_{66}v_j^0(\tau_i+\tau_j)+(k_i-k_j)D_{66}\theta_j(\tau_i+\tau_j). \tag{2.23}$$

The values C_{kl}, D_{kl} are the membrane and flexural stiffnesses of an orthotropic homogeneous or multilayer (with a symmetric layer layup through the thickness) package which can be calculated according to the formulas from Refs. 21, 24 and 180.

The expressions (2.22) and (2.23) can be simplified by the approach used in the technical theory. By disregarding in (2.22) the membrane deformations, and using the formulas (2.13) for $\varkappa_i^*$, τ_i^* we have

$$M^{**}_{ii}=D_{ii}\varkappa_i+D_{ij}\varkappa_j;\quad M^*_{ij}=D_{66}(\tau_i+\tau_j). \tag{2.24}$$

Besides, on the basis of smallness of the values γ_j, $v_i^0\theta_j$ compared to a unity and taking into account the introduced restrictions for the indicators of variability of deformed state, it is possible to simplify essentially the expression for T_{i3}^{**} (2.23) by omitting the non-linear terms. In such a way,

$$T^{**}_{i3}=C_{i3}(\theta_i+v_i{}^0). \tag{2.25}$$

The surface loads X_i^*, X_3^*, mass forces P_i^*, P_3^*, moments of surface and mass forces Φ_i^*, F_i^*, entering equations (2.17)–(2.19), taking into account the formulas (1.46), (1.43), (1.61), (1.60) and considering smallness of ε_i, γ_i, θ_i^2, $v_i^{0^2}$ in comparison to a unity, can be written in the form

$$X^*_i=(X_i^+-X_i^-)+(X_3^+-X_3^-)v_i{}^0; \tag{2.26}$$

$$X^*_3=(X_3^+-X_3^-)+(X_1^+-X_1^-)\theta_1+(X_2^+-X_2^-)\theta_2; \tag{2.27}$$

$$P^*_i=\int_{-h/2}^{h/2}[1+\alpha_3(k_i+k_j)]F_i d\alpha_3;\quad P^*_3=\int_{-h/2}^{h/2}[1+\alpha_3(k_1+k_2)]F_3 d\alpha_3; \tag{2.28}$$

$$\Phi^*_i=\frac{h}{2}\left\{(X_j^++X_j^-)+\frac{h}{2}\left[(X_j^+-X_j^-)(\varkappa_i+\varkappa_j)+(X_i^+-X_i^-)\tau_i\right]\right\}; \tag{2.29}$$

$$F^*_i=\int_{-h/2}^{h/2}\{(F_j-F_3v_j{}^0)[1+\alpha_3(k_i+k_j)]+\alpha_3F_j(\varkappa_i+\varkappa_j)\}\alpha_3 d\alpha_3. \tag{2.30}$$

As a result of substituting the formulas (2.6)–(2.8), (2.13), (2.20), (2.21), (2.24), (2.25), (2.26)–(2.30) into the system (2.17)–(2.19) the latter can be written in respect to five unknown functions u_i^0, u_3^0, v_i^0.

On employing Kirchhoff-Love's kinematic model we have three equations of equilibrium (2.17), (2.18). The expressions for generalized tractions in approximation of the technical shell theory are written as follows (1.78), (1.83)

$$\begin{aligned} T^*_{ii}&=C_{ii}\left(\varepsilon_i+\frac{1}{2}\theta_i{}^2\right)+C_{ij}\left(\varepsilon_j+\frac{1}{2}\theta_j{}^2\right);\\ T^*_{ij}&=C_{66}(\gamma_i+\gamma_j+\theta_i\theta_j); \end{aligned} \tag{2.31}$$

$$T^*_{i3}=\frac{1}{A_iA_j}\frac{\partial}{\partial\alpha_i}\{A_j[D_{ii}\varkappa_i+(D_{12}+2D_{66})\varkappa_j]\}-$$

$$-\frac{1}{A_iA_j}\frac{\partial A_j}{\partial\alpha_i}[(D_{12}+2D_{66})\varkappa_i+D_{jj}\varkappa_j]+C_{ii}\theta_i\left(\varepsilon_i+\frac{1}{2}\theta_i^2\right)+$$

$$+C_{ij}\theta_i\left(\varepsilon_j+\frac{1}{2}\theta_j^2\right)+C_{66}\theta_j(\gamma_i+\gamma_j+\theta_i\theta_j)+F^*_j+\Phi^*_j. \qquad (2.32)$$

The surface loads X_i^*, X_3^* in compliance with (1.48) and considering smallness of ε_i, γ_i, θ_i^2 compared to a unity, are specified by the formulas

$$X^*_i=(X_i^+-X_i^-)-(X_3^+-X_3^-)\theta_i \qquad (2.33)$$

and (2.27). For mass forces P_i^*, P_3^* the expressions (2.28) are maintained. For moments of surface forces, taking into account the condition (1.13) which have been used earlier, instead of (2.29) the simplified formula

$$\Phi^*_i=\frac{h}{2}(X_j^++X_j^-). \qquad (2.34)$$

can be used. The moments of mass forces F_i^* according to (1.65) in the mentioned approach have the form

$$F^*_i=\int_{-h/2}^{h/2}\{(F_j+F_3\theta_j)[1+\alpha_3(k_i+k_j)]+\alpha_3F_j(\varkappa_i+\varkappa_j)\}\alpha_3d\alpha_3. \qquad (2.35)$$

Through subsequent substitution of (2.31), (2.32), (2.27), (2.28), (2.33)–(2.35), (2.6), (2.14) into the system (2.17), (2.18) the latter can be written in respect to the three unknown functions $\overset{0}{u}_i$, $\overset{0}{u}_3$.

2.2 NON-LINEAR EQUATIONS OF THE TECHNICAL THEORY FOR ORTHOTROPIC CIRCULAR CYLINDRICAL SHELLS

Let us consider a particular case of a thin-walled shell — a closed circular orthotropic cylindrical shell of radius R. Let us introduce the coordinates: $\alpha_1=x$ along the generating line; $\alpha_2=y$ along the circumference; $\alpha_3=z$ along the normal to the mid-surface (the coordinate z is positive in the direction of the outer normal). Thereby $A_1=A_2=1$, $k_1=0$, $k_2=1/R$. Henceforth we shall use the notations: $\overset{0}{u}_1=u$, $\overset{0}{u}_2=v$, $\overset{0}{u}_3=w$, $\overset{0}{v}_1=\gamma_x$, $\overset{0}{v}_2=\gamma_y$.

The equations of equilibrium for a shell within the framework of the non-linear technical theory with consideration of transverse shear deformations, obtained from (2.17) to (2.19), have the form

$$\frac{\partial T^*_{11}}{\partial x}+\frac{\partial T^*_{21}}{\partial y}+X^*_1+P^*_1=0;$$
$$\frac{\partial T^*_{22}}{\partial y}+\frac{\partial T^*_{12}}{\partial x}+X^*_2+P^*_2=0;$$
$$\frac{\partial T^*_{13}}{\partial x}+\frac{\partial T^*_{23}}{\partial y}-\frac{1}{R}T^*_{22}+X^*_3+P^*_3=0; \tag{2.36}$$
$$\frac{\partial M^*_{11}}{\partial x}+\frac{\partial M^*_{21}}{\partial y}-T^{**}_{13}+F^*_2+\Phi^*_2=0;$$
$$\frac{\partial M^*_{22}}{\partial y}+\frac{\partial M^*_{12}}{\partial x}-T^{**}_{23}+F^*_1+\Phi^*_1=0.$$

The generalized tractions and moments are determined according to the formulas (2.20), (2.21), (2.24) and (2.25) in which in accordance with (2.6), (2.8) and (2.13)

$$\varepsilon_1=\frac{\partial u}{\partial x};\quad \varepsilon_2=\frac{\partial v}{\partial y}+\frac{w}{R};\quad \gamma_1=\frac{\partial v}{\partial x};\quad \gamma_2=\frac{\partial u}{\partial y};$$
$$\theta_1=\frac{\partial w}{\partial x};\quad \theta_2=\frac{\partial w}{\partial y}; \tag{2.37}$$

$$\varkappa_1=\frac{\partial \gamma_x}{\partial x};\quad \varkappa_2=\frac{\partial \gamma_y}{\partial y};\quad \tau_1=\frac{\partial \gamma_y}{\partial x};\quad \tau_2=\frac{\partial \gamma_x}{\partial y}. \tag{2.38}$$

In compliance with the usual assumptions of the technical theory the term $(1/R)T^*_{23}$ in the second equation of the system (2.36) and the term v/R in the formula for θ_2 have been omitted. The values X^*_i, X^*_3, Φ^*_i are maintained in the form (2.26), (2.27), (2.34). For P^*_i, P^*_3, F^*_i according to (2.28), (2.35) we have

$$P^*_i=\int_{-h/2}^{h/2}\left(1+\frac{z}{R}\right)F_i dz;\quad P^*_3=\int_{-h/2}^{h/2}\left(1+\frac{z}{R}\right)F_3 dz; \tag{2.39}$$

$$F^*_i=\int_{-h/2}^{h/2}\left[(F_j+F_3\theta_j)\left(1+\frac{z}{R}\right)+zF_j(\varkappa_i+\varkappa_j)\right]z\,dz. \tag{2.40}$$

On substituting expressions (2.20), (2.21), (2.24)–(2.27), (2.34), (2.37)–(2.40) into the system (2.36), we can reduce it to five non-linear equations in partial derivatives in respect to functions u, v, w, γ_x, γ_y (we shall not put down this system in explicit form).

Furthermore, we shall treat the equations for an orthotropic cylindrical shell in the version of technical theory and Kirchhoff-Love's kinematic model. The equations of equilibrium coincide with the first three equations of the system (2.36), where T^*_{ii}, T^*_{ij}, T^*_{i3}, in accordance with (2.31) and (2.32), are written as follows

$$T^*_{11}=C_{11}\hat{\varepsilon}_{11}+C_{12}\hat{\varepsilon}_{22};\quad T^*_{22}=C_{12}\hat{\varepsilon}_{11}+C_{22}\hat{\varepsilon}_{22};\quad T^*_{12}=T^*_{21}=C_{66}\hat{\varepsilon}_{12};$$

$$T^*_{13}=\frac{\partial}{\partial x}\left[D_{11}\varkappa_1+(D_{12}+2D_{66})\varkappa_2\right]+\theta_1(C_{11}\hat{\varepsilon}_{11}+C_{12}\hat{\varepsilon}_{22})+$$
$$+C_{66}\theta_2\hat{\varepsilon}_{12}+F^*_2+\Phi^*_2; \qquad (2.41)$$

$$T^*_{23}=\frac{\partial}{\partial y}\left[D_{22}\varkappa_2+(D_{12}+2D_{66})\varkappa_1\right]+\theta_2(C_{22}\hat{\varepsilon}_{22}+C_{12}\hat{\varepsilon}_{11})+$$
$$+C_{66}\theta_1\hat{\varepsilon}_{12}+F^*_1+\Phi^*_1.$$

Here,

$$\hat{\varepsilon}_{11}=\varepsilon_1+\frac{1}{2}\theta_1^2;\quad \hat{\varepsilon}_{22}=\varepsilon_2+\frac{1}{2}\theta_2^2;\quad \hat{\varepsilon}_{12}=\gamma_1+\gamma_2+\theta_1\theta_2. \qquad (2.42)$$

The values ε_1, ε_2, γ_1, γ_2, θ_1, θ_2 are determined from the formulas (2.37). Besides, according to (2.14), (2.12)

$$\varkappa_1=-\frac{\partial\theta_1}{\partial x};\quad \varkappa_2=-\frac{\partial\theta_2}{\partial y};\quad \tau_1=-\frac{\partial\theta_2}{\partial x};\quad \tau_2=-\frac{\partial\theta_1}{\partial y}. \qquad (2.43)$$

Upon substituting (2.37) into (2.42) and (2.43) we find

$$\hat{\varepsilon}_{11}=\frac{\partial u}{\partial x}+\frac{1}{2}\left(\frac{\partial w}{\partial x}\right)^2;\quad \hat{\varepsilon}_{22}=\frac{\partial v}{\partial y}+\frac{w}{R}+\frac{1}{2}\left(\frac{\partial w}{\partial y}\right)^2;$$
$$\hat{\varepsilon}_{12}=\frac{\partial u}{\partial y}+\frac{\partial v}{\partial x}+\frac{\partial w}{\partial x}\frac{\partial w}{\partial y}; \qquad (2.44)$$

$$\varkappa_1=-\frac{\partial^2 w}{\partial x^2};\quad \varkappa_2=-\frac{\partial^2 w}{\partial y^2};\quad \tau_1=\tau_2=-\frac{\partial^2 w}{\partial x\partial y}.$$

In treatment of a shell with initial deflection w^0, according to the approach outlined in (1.7), the formulas (2.44) are replaced by

$$\hat{\varepsilon}_{11}=\frac{\partial u}{\partial x}+\frac{1}{2}\left[\left(\frac{\partial w}{\partial x}\right)^2-\left(\frac{\partial w^0}{\partial x}\right)^2\right];$$
$$\hat{\varepsilon}_{22}=\frac{\partial v}{\partial y}+\frac{w-w^0}{R}+\frac{1}{2}\left[\left(\frac{\partial w}{\partial y}\right)^2-\left(\frac{\partial w^0}{\partial y}\right)^2\right]; \qquad (2.45)$$
$$\hat{\varepsilon}_{12}=\frac{\partial u}{\partial y}+\frac{\partial v}{\partial x}+\frac{\partial w}{\partial x}\frac{\partial w}{\partial y}-\frac{\partial w^0}{\partial x}\frac{\partial w^0}{\partial y};$$
$$\varkappa_1=-\frac{\partial^2(w-w^0)}{\partial x^2}; \quad \varkappa_2=-\frac{\partial^2(w-w^0)}{\partial y^2};$$
$$\tau_1=\tau_2=-\frac{\partial^2(w-w^0)}{\partial x\partial y}.$$

By using (2.2) and (2.45) the deformations are expressed as follows

$$\varepsilon_{11}=\frac{\partial u}{\partial x}+\frac{1}{2}\left[\left(\frac{\partial w}{\partial x}\right)^2-\left(\frac{\partial w^0}{\partial x}\right)^2\right]-z\frac{\partial^2(w-w^0)}{\partial x^2};$$
$$\varepsilon_{22}=\frac{\partial v}{\partial y}+\frac{w-w^0}{R}+\frac{1}{2}\left[\left(\frac{\partial w}{\partial y}\right)^2-\left(\frac{\partial w^0}{\partial y}\right)^2\right]-$$
$$-z\frac{\partial^2(w-w^0)}{\partial y^2}; \qquad (2.46)$$
$$\varepsilon_{12}=\frac{\partial u}{\partial y}+\frac{\partial v}{\partial x}+\frac{\partial w}{\partial x}\frac{\partial w}{\partial y}-\frac{\partial w^0}{\partial x}\frac{\partial w^0}{\partial y}-2z\frac{\partial^2(w-w^0)}{\partial x\partial y}.$$

Proceeding from the equations and formulas presented above, it is easy to deduce the equations of equilibrium for a shell in terms of displacements. Upon substituting sequentially (2.41) and (2.45) into the first three equations of the system (2.36) we obtain

$$C_{11}\left(\frac{\partial^2 u}{\partial x^2}+\frac{\partial w}{\partial x}\frac{\partial^2 w}{\partial x^2}-\frac{\partial w^0}{\partial x}\frac{\partial^2 w^0}{\partial x^2}\right)+$$
$$+C_{12}\left[\frac{\partial^2 v}{\partial x\partial y}+\frac{1}{R}\frac{\partial(w-w^0)}{\partial x}+\frac{\partial w}{\partial y}\frac{\partial^2 w}{\partial x\partial y}-\right.$$
$$\left.-\frac{\partial w^0}{\partial y}\frac{\partial^2 w^0}{\partial x\partial y}\right]+C_{66}\left(\frac{\partial^2 u}{\partial y^2}+\frac{\partial^2 v}{\partial x\partial y}+\frac{\partial^2 w}{\partial x\partial y}\frac{\partial w}{\partial y}-\right.$$
$$\left.-\frac{\partial^2 w^0}{\partial x\partial y}\frac{\partial w^0}{\partial y}+\frac{\partial w}{\partial x}\frac{\partial^2 w}{\partial y^2}-\frac{\partial w^0}{\partial x}\frac{\partial^2 w^0}{\partial y^2}\right)+$$
$$+X^*_1+P^*_1=0; \qquad (2.47)$$
$$C_{12}\left(\frac{\partial^2 u}{\partial x\partial y}+\frac{\partial w}{\partial x}\frac{\partial^2 w}{\partial x\partial y}-\frac{\partial w^0}{\partial x}\frac{\partial^2 w^0}{\partial x\partial y}\right)+C_{22}\times$$
$$\times\left[\frac{\partial^2 v}{\partial y^2}+\frac{1}{R}\frac{\partial(w-w^0)}{\partial y}+\frac{\partial w}{\partial y}\frac{\partial^2 w}{\partial y^2}-\frac{\partial w^0}{\partial y}\frac{\partial^2 w^0}{\partial y^2}\right]+$$

$$+C_{66}\left(\frac{\partial^2 u}{\partial x\partial y}+\frac{\partial^2 v}{\partial x^2}+\frac{\partial^2 w}{\partial x^2}\frac{\partial w}{\partial y}-\frac{\partial^2 w^0}{\partial x^2}\frac{\partial w^0}{\partial y}+\right.$$

$$\left.+\frac{\partial w}{\partial x}\frac{\partial^2 w}{\partial x\partial y}-\frac{\partial w^0}{\partial x}\frac{\partial^2 w^0}{\partial x\partial y}\right)+X^*_2+P^*_2=0; \qquad (2.48)$$

$$D_{11}\frac{\partial^4(w-w^0)}{\partial x^4}+2(D_{12}+2D_{66})\frac{\partial^4(w-w^0)}{\partial x^2\partial y^2}+D_{22}\frac{\partial^4(w-w^0)}{\partial y^4}+$$

$$+\frac{C_{12}}{R}\frac{\partial u}{\partial x}+\frac{C_{22}}{R}\left(\frac{\partial v}{\partial y}+\frac{w-w^0}{R}\right)+\frac{C_{12}}{2R}\times$$

$$\times\left[\left(\frac{\partial w}{\partial x}\right)^2-\left(\frac{\partial w^0}{\partial x}\right)^2\right]+\frac{C_{22}}{2R}\left[\left(\frac{\partial w}{\partial y}\right)^2-\left(\frac{\partial w^0}{\partial y}\right)^2\right]-$$

$$-\frac{\partial}{\partial x}\left\{C_{11}\frac{\partial w}{\partial x}\left[\frac{\partial u}{\partial x}+\frac{1}{2}\left(\frac{\partial w}{\partial x}\right)^2-\frac{1}{2}\left(\frac{\partial w^0}{\partial x}\right)^2\right]+\right.$$

$$+C_{12}\frac{\partial w}{\partial x}\left[\frac{\partial v}{\partial y}+\frac{w-w^0}{R}+\frac{1}{2}\left(\frac{\partial w}{\partial y}\right)^2-\frac{1}{2}\left(\frac{\partial w^0}{\partial y}\right)^2\right]+$$

$$\left.+C_{66}\frac{\partial w}{\partial y}\left[\frac{\partial u}{\partial y}+\frac{\partial v}{\partial x}+\frac{\partial w}{\partial x}\frac{\partial w}{\partial y}-\frac{\partial w^0}{\partial x}\frac{\partial w^0}{\partial y}\right]\right\}-$$

$$-\frac{\partial}{\partial y}\left\{C_{22}\frac{\partial w}{\partial y}\left[\frac{\partial v}{\partial y}+\frac{w-w^0}{R}+\frac{1}{2}\left(\frac{\partial w}{\partial y}\right)^2-\right.\right.$$

$$\left.-\frac{1}{2}\left(\frac{\partial w^0}{\partial y}\right)^2\right]+C_{12}\frac{\partial w}{\partial y}\left[\frac{\partial u}{\partial x}+\frac{1}{2}\left(\frac{\partial w}{\partial x}\right)^2-\right.$$

$$\left.-\frac{1}{2}\left(\frac{\partial w^0}{\partial x}\right)^2\right]+C_{66}\frac{\partial w}{\partial x}\left[\frac{\partial u}{\partial y}+\frac{\partial v}{\partial x}+\frac{\partial w}{\partial x}\frac{\partial w}{\partial y}-\right.$$

$$\left.\left.-\frac{\partial w^0}{\partial x}\frac{\partial w^0}{\partial y}\right]\right\}=X^*_3+P^*_3+\frac{\partial}{\partial x}(F^*_2+\Phi^*_2)+\frac{\partial}{\partial y}(F^*_1+\Phi^*_1). \qquad (2.49)$$

The equations of equilibrium in the mixed form can directly be obtained from (1.88) and (1.91). Taking into account that according to (1.86) we have for a cylindrical shell the relations

$$T_{11}=\frac{\partial^2 F}{\partial y^2}; \quad T_{22}=\frac{\partial^2 F}{\partial x^2}; \quad T_{12}=T_{21}=-\frac{\partial^2 F}{\partial x\partial y}, \qquad (2.50)$$

and using the formulas (2.45) we arrive at the well-known system of two equations:

$$D_{11}\frac{\partial^4(w-w^0)}{\partial x^4}+2(D_{12}+2D_{66})\frac{\partial^4(w-w^0)}{\partial x^2\partial y^2}+D_{22}\frac{\partial^4(w-w^0)}{\partial y^4}+$$
$$+\frac{1}{R}\frac{\partial^2 F}{\partial x^2}-\frac{\partial^2 w}{\partial x^2}\frac{\partial^2 F}{\partial y^2}-\frac{\partial^2 w}{\partial y^2}\frac{\partial^2 F}{\partial x^2}+2\frac{\partial^2 w}{\partial x\partial y}\frac{\partial^2 F}{\partial x\partial y}=$$
$$=X^*_3+P^*_3+\frac{\partial}{\partial x}(F^*_2+\Phi^*_2)+\frac{\partial}{\partial y}(F^*_1+\Phi^*_1); \qquad (2.51)$$

$$A_{22}\frac{\partial^4 F}{\partial x^4}+(2A_{12}+A_{66})\frac{\partial^4 F}{\partial x^2\partial y^2}+A_{11}\frac{\partial^4 F}{\partial y^4}=\frac{1}{R}\frac{\partial^2(w-w^0)}{\partial x^2}+$$
$$+\left(\frac{\partial^2 w}{\partial x\partial y}\right)^2-\left(\frac{\partial^2 w^0}{\partial x\partial y}\right)^2-\frac{\partial^2 w}{\partial x^2}\frac{\partial^2 w}{\partial y^2}+\frac{\partial^2 w^0}{\partial x^2}\frac{\partial^2 w^0}{\partial y^2}. \qquad (2.52)$$

In deduction of (2.52) it was kept in mind that in the initial stressed state the equation of the compatibility of deformations assumed the form

$$\tau^{0^2}-\varkappa_1{}^0\varkappa_2{}^0-\frac{\varkappa_1{}^0}{R}=0.$$

On the basis of systems (2.47)–(2.49) and (2.51) and (2.52) the solutions of various non-linear dynamic problems for imperfect orthotropic cylindrical shells will be constructed in subsequent chapters.

2.3 LINEARIZED EQUATIONS FOR AN ORTHOTROPIC CIRCULAR CYLINDRICAL SHELL

A procedure of deducing the linearized equations of stability for thin-walled orthotropic shells was treated in (1.8). For a particular cylindrical shell we have the following equations in accordance with (1.98) and (1.100):

$$\frac{\partial T'_{11}}{\partial x}+\frac{\partial T'_{21}}{\partial y}+P'_1+X'_1=0; \quad \frac{\partial T'_{22}}{\partial y}+\frac{\partial T'_{12}}{\partial x}+\frac{T'_{23}}{R}+P'_2+X'_2=0;$$
$$\frac{\partial T'_{13}}{\partial x}+\frac{\partial T'_{23}}{\partial y}-\frac{T'_{22}}{R}+P'_3+X'_3=0; \qquad (2.53)$$
$$\frac{\partial M'_{11}}{\partial x}+\frac{\partial M'_{21}}{\partial y}-T'_{13}+\Phi'_2+F'_2=0;$$
$$\frac{\partial M'_{22}}{\partial y}+\frac{\partial M'_{12}}{\partial x}-T'_{23}+\Phi'_1+F'_1=0.$$

Here, T'_{ii}, T'_{ij}, T'_{i3} are 'additional' tractions, having arisen in the shell upon deviating from the pre-critical equilibrium state. Within the framework of the technical theory these equations have the form (1.102)

$$T'_{11}=C_{11}\varepsilon'_1+C_{12}\varepsilon'_2;\quad T'_{22}=C_{22}\varepsilon'_2+C_{12}\varepsilon'_1;\quad T'_{12}=T'_{21}=C_{66}(\gamma'_1+\gamma'_2);$$
$$T'_{13}=C_{13}(\theta'_1+\gamma'_x);\quad T'_{23}=C_{23}(\theta'_2+\gamma'_y);\qquad (2.54)$$
$$M'_{11}=D_{11}\varkappa'_1+D_{12}\varkappa'_2;\quad M'_{22}=D_{22}\varkappa'_2+D_{12}\varkappa'_1;\quad M'_{12}=M'_{21}=D_{66}(\tau'_1+\tau'_2).$$

The values X'_i and X'_3 represent themselves as the 'additional' surface forces, acting on a shell only upon deviating from the pre-critical equilibrium state. In accordance with (1.99), they can be written down for a cylindrical shell as follows

$$X'_1=\frac{\partial}{\partial x}(T_{11}{}^0\varepsilon'_2)+\frac{\partial}{\partial y}(T_{22}{}^0\gamma'_2)+q_1(\varepsilon'_1+\varepsilon'_2)+\delta_c(q_2\gamma'_2+q_3\gamma'_x);$$
$$X'_2=\frac{\partial}{\partial y}(T_{22}{}^0\varepsilon'_1)+\frac{\partial}{\partial x}(T_{11}{}^0\gamma'_1)+\frac{T_{22}{}^0}{R}\theta'_2+q_2(\varepsilon'_1+\varepsilon'_2)+$$
$$+\delta_c(q_1\gamma'_1+q_3\gamma'_y);\qquad (2.55)$$
$$X'_3=\frac{\partial}{\partial x}(T_{11}{}^0\theta'_1)+\frac{\partial}{\partial y}(T_{22}{}^0\theta'_2)-\frac{T_{22}{}^0}{R}\varepsilon'_1+q_3(\varepsilon'_1+\varepsilon'_2)+$$
$$+\delta_c(q_1\theta'_1+q_2\theta'_2).$$

The moments of surface forces Φ'_i are expressed in the form (1.101).

The values ε'_i, γ'_i, θ'_i, $\varkappa'_i$, τ'_i entering (2.54) and (2.55), within the framework of the technical theory, are determined by the formulas (2.37) and (2.38).

If we assume that the only mass forces acting on a shell are the inertial forces, then according to (1.104) we have for a cylindrical shell

$$P'_1=-\rho h\left(\frac{\partial^2 u'}{\partial t^2}+\frac{h^2}{12R}\frac{\partial^2\gamma_x}{\partial t^2}\right);$$
$$P'_2=-\rho h\left(\frac{\partial^2 v'}{\partial t^2}+\frac{h^2}{12R}\frac{\partial^2\gamma_y}{\partial t^2}\right);\qquad (2.56)$$
$$P'_3=-\rho h\frac{\partial^2 w'}{\partial t^2};$$

$$F'_1=-\rho\frac{h^3}{12}\left(\frac{\partial^2\gamma'_y}{\partial t^2}+\frac{1}{R}\frac{\partial^2 v'}{\partial t^2}\right);$$
$$F'_2=-\rho\frac{h^3}{12}\left(\frac{\partial^2\gamma'_x}{\partial t^2}+\frac{1}{R}\frac{\partial^2 u'}{\partial t^2}\right),\qquad (2.57)$$

where ρ is the density of material.

On employing Kirchhoff-Love's model after eliminating T'_{13} and T'_{23} from the last two equations of the system (2.53) we obtain

$$\begin{gathered}
\frac{\partial T'_{11}}{\partial x}+\frac{\partial T'_{21}}{\partial y}+P'_1+X'_1=0;\\
\frac{\partial T'_{22}}{\partial y}+\frac{\partial T'_{12}}{\partial x}+\frac{1}{R}\left(\frac{\partial M'_{22}}{\partial y}+\frac{\partial M'_{12}}{\partial x}\right)+P'_2+X'_2+\\
+\frac{1}{R}(\Phi'_1+F'_1)=0;\\
\frac{T'_{22}}{R}-\frac{\partial^2 M'_{11}}{\partial x^2}-\frac{\partial^2 M'_{22}}{\partial y^2}-\frac{\partial^2(M'_{12}+M'_{21})}{\partial x\partial y}=P'_3+X'_3+\\
+\frac{\partial}{\partial x}(\Phi'_2+F'_2)+\frac{\partial}{\partial y}(\Phi'_1+F'_1).
\end{gathered} \tag{2.58}$$

At last let us consider a more general expression for deformations, tractions and moments than that referred to in the technical theory. From formulas (1.2) and (1.3) for the linear case it follows (dashes are omitted):

$$\varepsilon_{ii}=\frac{\varepsilon_i+\alpha_3\varkappa_i}{1+\alpha_3 k_i};\quad \varepsilon_{ij}=\frac{\gamma_i+\alpha_3\tau_i}{1+\alpha_3 k_i}+\frac{\gamma_j+\alpha_3\tau_j}{1+\alpha_3 k_j}. \tag{2.59}$$

By expanding these values into power series and retaining all the terms up to the first degree of α_3 also, we obtain for a cylindrical shell

$$\varepsilon_{11}=\varepsilon_1+\alpha_3\chi_1;\quad \varepsilon_{22}=\varepsilon_2+\alpha_3\chi_2;\quad \varepsilon_{12}=\gamma+\alpha_3\xi, \tag{2.60}$$

where

$$\begin{gathered}
\chi_1=\varkappa_1=-\frac{\partial^2 w}{\partial x^2};\quad \chi_2=\varkappa_2-\frac{\varepsilon_2}{R}=-\left(\frac{\partial^2 w}{\partial y^2}+\frac{w}{R^2}\right);\\
\xi=\tau_1+\tau_2-\frac{\gamma_2}{R}=-2\frac{\partial^2 w}{\partial x\partial y}+\frac{1}{R}\left(\frac{\partial v}{\partial x}-\frac{\partial u}{\partial y}\right),
\end{gathered} \tag{2.61}$$

The values ε_1, ε_2, $\gamma=\gamma_1+\gamma_2$ are determined by the formulas (2.37).

Furthermore, by substituting (2.59) into the expressions for tractions (1.40) and moments (1.62) and considering (1.38), we find

$$\begin{gathered}
T_{ii}=\int_{-h/2}^{h/2}\left[B_{ii}(\varepsilon_i+\alpha_3\varkappa_i)\frac{1+\alpha_3 k_j}{1+\alpha_3 k_i}+B_{ij}(\varepsilon_j+\alpha_3\varkappa_j)\right]d\alpha_3;\\
T_{ij}=\int_{-h/2}^{h/2}B_{66}\left[(\gamma_i+\alpha_3\tau_i)\frac{1+\alpha_3 k_j}{1+\alpha_3 k_i}+(\gamma_j+\alpha_3\tau_j)\right]d\alpha_3;
\end{gathered} \tag{2.62}$$

$$M_{ii}=\int_{-h/2}^{h/2}\left[B_{ii}(\varepsilon_i+\alpha_3\varkappa_i)\frac{1+\alpha_3k_j}{1+\alpha_3k_i}+B_{ij}(\varepsilon_j+\alpha_3\varkappa_j)\right]\alpha_3d\alpha_3;$$

$$M_{ij}=\int_{-h/2}^{h/2}B_{66}\left[(\gamma_i+\alpha_3\tau_i)\frac{1+\alpha_3k_j}{1+\alpha_3k_i}+(\gamma_j+\alpha_3\tau_j)\right]\alpha_3d\alpha_3.$$

Further, the expansion of subintegral functions into power series with respect to α_3 and retention of all the terms up to α_3^2 also, yield the following expressions for tractions and moments in an orthotropic cylindrical shell:

$$\begin{aligned}
T_{11}&=C_{11}\frac{\partial u}{\partial x}+C_{12}\left(\frac{\partial v}{\partial y}+\frac{w}{R}\right)-\frac{D_{11}}{R}\frac{\partial^2 w}{\partial x^2};\\
T_{22}&=C_{22}\left(\frac{\partial v}{\partial y}+\frac{w}{R}\right)+C_{12}\frac{\partial u}{\partial x}+\frac{D_{22}}{R}\left(\frac{\partial^2 w}{\partial y^2}+\frac{w}{R^2}\right);\\
T_{12}&=C_{66}\left(\frac{\partial v}{\partial x}+\frac{\partial u}{\partial y}\right)+\frac{D_{66}}{R}\left(\frac{1}{R}\frac{\partial v}{\partial x}-\frac{\partial^2 w}{\partial x\partial y}\right);\\
T_{21}&=C_{66}\left(\frac{\partial v}{\partial x}+\frac{\partial u}{\partial y}\right)+\frac{D_{66}}{R}\left(\frac{1}{R}\frac{\partial u}{\partial y}+\frac{\partial^2 w}{\partial x\partial y}\right);\\
M_{11}&=-D_{11}\frac{\partial^2 w}{\partial x^2}-D_{12}\left(\frac{\partial^2 w}{\partial y^2}-\frac{1}{R}\frac{\partial w}{\partial y}\right)+\frac{D_{11}}{R}\frac{\partial u}{\partial x};\\
M_{22}&=-D_{22}\left(\frac{\partial^2 w}{\partial y^2}+\frac{w}{R^2}\right)-D_{12}\frac{\partial^2 w}{\partial x^2};\\
M_{12}&=-2D_{66}\left(\frac{\partial^2 w}{\partial x\partial y}-\frac{1}{R}\frac{\partial v}{\partial x}\right);\\
M_{21}&=-2D_{66}\left(\frac{\partial^2 w}{\partial x\partial y}+\frac{1}{2R}\frac{\partial u}{\partial y}-\frac{1}{2R}\frac{\partial v}{\partial x}\right).
\end{aligned}\tag{2.63}$$

The procedure of deducing the relationships of elasticity, used above, was first employed by W. Flügge[321] for isotropic and structurally-orthotropic cylindrical shells. The respective theory of cylindrical shells was also outlined in monographs.[254,322] A generalization of this approach for shells of arbitrary mid-surface shape was given by A. I. Luriye.[181]

The system (2.58) together with the formulas (2.63) permits to solve a wider class of problems than the equations of the technical theory of cylindrical shells. Among them are the problems of deformation, vibration and stability for comparatively long shells as well as for shells of medium length at small indicators of variability of stress–strain state in the circumferential direction. In Chapter 3, this version of the theory will be employed for finding out the limits of applicability of the technical theory to the problems of vibrations and stability of orthotropic cylindrical shells.

3

Free Vibrations of Orthotropic Cylindrical Shells

The investigation of the applicability of the equations of technical theory (theory of shallow shells) based on the Kirchhoff-Love model to the calculation of natural frequencies of orthotropic cylindrical shells is underway. It is realized by comparing with the calculated results received by the use of two more perfect versions of the theory — Flügge's type theory and Timoshenko's type improved theory. The first one renders more precisely the equations of equilibrium, elasticity relations and expressions for the changes in curvatures and torsion, but disregards the transverse shear deformations. The second one, in its turn, takes into account transverse shears, but in all other respects it is equivalent to the technical theory. An independent comparison with these two versions of the theory of cylindrical shells allows one to establish the applicability limits for the technical theory, based on the Kirchhoff-Love model, for all parameters entering the problem: length-to-radius and radius-to-thickness ratios, the ratios of the moduli of transverse shears to in-plane moduli, as well as the modes of vibration in longitudinal and circumferential directions.

The solution of natural vibration problems for structural elements is of paramount significance for developing all the problems of dynamics. In studying the forced and parametric vibrations, flutter, transient wave problems, dynamic instability, impact and other processes the information about a large number of natural frequencies and modes of vibrations is directly or indirectly to be put into the calculations. Thereby, it should be noted that in many problems of dynamics, comparatively high spatial forms of vibrations play an important role. This attaches particular importance to the problem of careful controlling of the basic calculation model, since even the high accuracy of the model in solving the statics and low-frequency vibrations problems does not guarantee its applicability to the analysis of high-frequency or high-velocity deformation processes.

The problem of natural vibrations of thin-wall shells was first formulated by A. Love,[370] who also obtained the frequency equation for the case of small vibrations. Furthermore, he investigated the equations and found out the form of solution for the case of flexural vibrations of cylindrical shells.[182] In 1894, J. W. S. Rayleigh obtained the formula for natural frequencies in the case of purely flexural vibrations of a cylindrical shell.[407] Calculations of natural frequencies of a freely supported cylindrical shell have been first performed by W. Flügge,[321] who established the existence of a group of three natural frequencies for each flexural form of vibrations. The problem of calculating the natural vibrations of a cylindrical shell with restrained end faces was first treated by A. P. Filippov.[252]

The works of R. N. Arnold and G. B. Warburton[285,286] contributed greatly to the theory of vibrations in cylindrical shells. In particular, a conclusion was drawn about the existence of a minimum of deformation energy, and, as a consequence, also a minimum of natural frequency for a rigorously definite number of circumferential harmonics depending on the geometric parameters of a shell. In these works, the results of comprehensive experimental studies of vibrations of steel cylindrical shells were presented.

The experimental work of S. A. Tobias[442] should especially be pointed out because of the effects of initial imperfections on vibrations of the cylindrical shells that were revealed. In these papers it was shown particularly that:

(1) in presence of imperfections the location of nodal planes becomes rather fixed;
(2) for each form of vibrations there exist two preferable nodal configurations having, in a general case, different natural frequencies; and
(3) the difference between these frequencies can be considered to be the measure of imperfections present in the body.

The works[182,254,285,286,442,462] developed on the basis of further theoretical and experimental studies in the field of vibrations of cylindrical shells.

Since the mid-50s the problem of harmonic wave propagation in infinitely long cylindrical shells, which is relative to the problem of natural vibrations, has undergone a theoretical development. The first investigations of P. W. Smith,[421,422] P. M. Naghdi and R. M. Cooper,[388] T. Lin and C. Morgan,[366] G. Herrmann and I. Mirsky,[344] were further developed by R. M. Cooper and P. M. Naghdi,[304] I. Mirsky and G. Herrmann,[382,383] Yi-Yuan Yu,[463] J. Greenspon,[334–337] and D. C. Gazis.[326] These works were devoted to calculation of dispersion relations and to studying the effect of transverse shear deformations and inertia of rotation on the dispersion curves (for plates, these questions were first investigated by R. D. Mindlin).[379]

Since the phase velocity of an harmonic wave is expressed in terms of frequency of natural vibrations of a freely supported shell, the results and conclusions of the mentioned works are connected directly with the problem of natural vibrations of cylindrical shells having finite length. The following basic conclusion can be formulated. Consideration of transverse shears and inertia of rotation widens essentially (moving down the limiting value of wavelength) the area of applicability of the shell theory. Though the authors, in the majority of cases did not distinguish the effects, ascribed to consideration of transverse shears and inertia of rotation, some results (for instance, in Ref. 344) show that the effect of taking into account the transverse shears in quantitative relation greatly excels the effect of taking into account the inertia of rotation. This branch of dynamics for cylindrical shells was further developed[352,353,355,390,455,456,464] where homogeneous orthotropic, two-ply, sandwich and later on also multilayer shells of a general type structure were considered. Yu. N. Novichkov[206] obtained a solution for a multi-layer cylindrical shell of regular structure.

The results of investigations on the problem of harmonic wave propagation[206,337,344,366] are of fundamental significance for selection, verification and substantiation of calculation models in dynamics of cylindrical shells. Thus, in Ref. 206 a conclusion was drawn that taking into account the shear effects in rigid layers and inertia terms, associated with rotation of normal sections, is essential only for dynamic processes with sufficiently high indicators of variability. In wave processes it is necessary only for such wave lengths which are comparable with the thickness of rigid layers. It should be added that in the design of composite shells having comparatively low transverse shear stiffness, it is necessary to pay attention to one more factor — the magnitude of transverse shear moduli. If these moduli are sufficiently small compared to the in-plane moduli of elasticity, the classical Kirchhoff-Love theory can turn out to be invalid even for wave lengths considerably exceeding the thickness of a shell (or that of an individual rigid layer).

The investigations accomplished in the 60–70s within the framework of the classical theory for thin cylindrical shells, were devoted mainly to studying the effect of end conditions on frequencies and forms of natural vibrations. A general solution of the problem of natural vibrations for an isotropic cylindrical shell was suggested by K. Forsberg[323] and G. B. Warburton.[450] The results obtained by the use of finite-difference methods were described in Ref. 163. Let us point out also the works of I. I. Vorovich and M. A. Shlenev[109] (a theory of dynamic edge effect, evolved by V. V. Bolotin was used) and V. S. Gontkevich,[127] Yu. Yu. Sheiko *et al.*,[263] and C. B. Sharma[417] (for boundary conditions other than simple supporting, various approximate methods were used). The investigations of natural vibrations of cylindrical shells by asymptotical methods were thoroughly treated in a monograph by A. L. Gol'denveiser *et al.*[124]

The problem of natural vibrations of a cylindrical shell was treated also on the basis of equations of a three-dimensional theory of elasticity.[278,292,381] Besides, an analysis of the effects associated with static pre-loading[205] and with tangential inertia terms[338] was performed. A large number of works was devoted to the effects of initial shape imperfections. Comprehensive experimental investigations[329,454] were also conducted.

Since the beginning of the 60s a theory of natural vibrations of orthotropic cylindrical shells was under development. Among the first works were those of V. S. Gontkevich,[126] I. Mirsky,[381] and Y. C. Das.[307] Vibrations in structurally-orthotropic shells were investigated by M. V. Nikulin.[205] In R. S. Sabirova's work[228] a shell of anisotropic material of general type was considered. On the whole, a way of developing the theory of isotropic shells was repeated: from boundary conditions of free supporting to approximate approaches to satisfying more complex types of boundary conditions and further to derivation of a general equation of frequency, allowing to consider any type of boundary conditions.

In the last years, great attention was focussed on the problems of natural vibrations of reinforced and multi-layer cylindrical shells (both orthotropic and those pertaining to more general types of anisotropy). Calculation of natural frequencies for a multi-layer shell, symmetrically located in respect to the mid-surface isotropic layers, was performed by V. I. Weingarten.[453] In many subsequent works a more general case of shell structures formed of orthotropic, symmetrically laid in respect to the mid-surface, layers has been treated. Within the framework of Kirchhoff-Love and Timoshenko's models, applied to a whole package of layers, such shells, as is known, are reduced to homogeneous ones. As a result, the respective problems of vibrations are of interest mainly from the viewpoint of management of the frequency characteristics by means of the change in stiffness and layup sequence. A comparatively small number of studies were devoted to the problems of vibrations of multi-layer shells with an unsymmetric layup of orthotropic layers through the thickness.[52,307,354,425] In these works, the effects ascribed to coupling of membrane deformations with moments, and flexural deformations with tangential tractions were studied. And, finally, more general types of multi-layer anisotropic shells in which the axes of orthotropy do not coincide with the directions of coordinate lines of the shell, have been treated in a paper by C. W. Bert *et al.*,[290] later by J. B. Greenberg and Y. Stavsky,[332] K. P. Soldatos[424] and a number of other works. This problem contains certain principal difficulties and up to now it has been the least studied.

In addition to what has been said before let us emphasize that a number of interesting results on problems of natural vibrations in anisotropic, reinforced and multi-layer shells were obtained in other works.[24,176,177,274,294,308–310,330,331,347,405,406,420,432,433] On the whole, the problem of natural vibrations in cylindrical shells from the theoretical point of view is sufficiently well developed. However, our viewpoint is that on the three essential applied problems — assessment of the limits of applicability of the technical theory (a theory of shallow shells), the range of applicability of the Kirchhoff-Love model and admissibility of disregard for tangential inertial terms and inertia of rotation for orthotropic shells, there are only fragmental quantitative results. For the sake of filling this gap a comprehensive study of these problems is carried out in this chapter.

3.1 CALCULATION OF NATURAL FREQUENCIES AND INVESTIGATION OF THE RANGE OF APPLICABILITY OF THE TECHNICAL THEORY

As is known, the applicability of the technical theory of cylindrical shells is limited (in foreign literature, this theory is usually called Donnell's theory and which was first formulated and used in stability problems[311]). One of the requirements of this theory is that in the case of vibrations and loss of stability each of the regions enclosed by neighbouring nodal lines might be considered as a shallow shell. In axisymmetric problems this requirement imposes a restriction from above on the width of ring belts encompassed between two neighbouring nodal lines, in non-axisymmetric ones — a restriction from above on the size of formed dents and bulges in the circumferential direction. The question of applicability of Donnell's theory to calculation of cylindrical shells under static loads was treated in Refs 97, 115, 134, 135, 254, 351, 369, etc.

In the given paragraph the range of applicability of the technical theory for calculation of natural frequencies of orthotropic cylindrical shells by means of comparison with the results calculated from the equations based on Flügge's model is established. The solution of the problem of natural vibrations of isotropic cylindrical shells on the basis of Flügge's equations was first presented by W. Flügge[321] in 1933 (the respective results are also outlined in Ref. 254). In subsequent years, Flügge's equations were utilized for calculation of frequencies of natural vibrations by many authors (for instance, in Refs 314, 323, 450). A variant of Timoshenko's theory,[441] simpler than that of Flügge, was used in Refs 285 and 286. Sanders' theory[297] was made use of by C. B. Sharma.[417] A comparative analysis of the results of calculating the natural frequencies according to Sanders, Timoshenko and Flügge's equations performed in Ref. 417 has shown that the differences are of the order of fractions of a per cent. Such a conclusion seems to be evident, since all the mentioned variants of the theory of thin shells have been obtained within the framework of Kirchhoff-Love kinematic model and are based on identical (or very close in essence) equations of equilibrium and relations between deformations and displacements.

As to the comparison of the results of calculation of natural frequencies according to Donnell's equations and more exact (though remaining within the framework of Kirchhoff-Love model) variants of equations, then, according to C. L. Dym[314] and M. El-Raheb and C. D. Babcock Jr.,[315] the differences can turn out to be considerable (up to several tenths of a percent) and increase with increasing length of half-waves in the axial direction and the thickness-to-radius ratio of the shell. It is worth noting that in all the mentioned works the isotropic shells have been treated.

A thorough comparative analysis of the calculation results of frequencies of natural vibrations for orthotropic shells on the basis of Love[182] and Donnell theories was performed by Y. Stavsky and R. Loewy.[425] Two-layer and three-layer shells with orthotropic layers were considered. The effect of the relations of the length L to radius R and radius R to thickness h on the difference in results for the lowest frequency of vibrations was investigated. As the numerical data show, Donnell's equations yield overestimated magnitudes as against Love's equations. At $R/h = 20$, the lowest frequency for the two-layer shell, treated in Ref. 425, differs by 0.8%, 4.5%, 24.6% at $L/R = 0.5$, 2 and 10, respectively. At $L/R = 50$, the lowest frequency according to Donnell's equations is 3.1 times greater. For three-layer shells with $R/h = 100$, treated in Ref. 425 the values of the lowest frequencies according to Donnell's equations become overestimated (depending on the structure of the shell) by 0.5–1.0% at $L/R = 0.5$; 2.1–3.7% at $L/R = 2$, 14.0–22.2% at $L/R = 10$ and up to 40% at $L/R - 30$. The relation between layer thicknesses in a three-layer package appreciably affects the value of difference in natural frequencies. Obviously, the difference in the calculated natural frequencies, introduced by the assumptions of the technical theory, would depend on the character of material orthotropy in orthotropic shells as well. Henceforth, we shall give special attention to this aspect of the problem.

Let us consider the solution of linearized equations of motion for an orthotropic cylindrical shell (2.58) with tractions and moments, evaluated in accordance with (2.63). The inertia terms in compliance with (2.56), (2.57) and relationships $\gamma_x = -(\partial w/\partial x)$, $\gamma_y = -[(\partial w/\partial y) - (v/R)]$ have the form (primes in subsequent formulas are omitted):

$$P_1 = -\rho h\left(\frac{\partial^2 u}{\partial t^2} - \frac{h^2}{12R}\frac{\partial^3 w}{\partial x \partial t^2}\right);$$

$$P_2 = -\rho h\left(\frac{\partial^2 v}{\partial t^2} - \frac{h^2}{12R}\frac{\partial^3 w}{\partial y \partial t^2}\right);$$

$$P_3 = -\rho h\frac{\partial^2 w}{\partial t^2}; \quad F_1 = -\frac{\rho h^3}{12}\left(-\frac{\partial^3 w}{\partial y \partial t^2} + \frac{2}{R}\frac{\partial^2 v}{\partial t^2}\right); \tag{3.1}$$

$$F_2 = -\rho\frac{h^3}{12}\left(-\frac{\partial^3 w}{\partial x \partial t^2} + \frac{1}{R}\frac{\partial^2 u}{\partial t^2}\right).$$

Assume that each end face of the shell is subjected to a uniformly distributed normal compressive force $P(t)$ which generates the uniform stress state, defined by the traction $T_{11}^0 = -P$. Further, let us assume that the outer side surface of the shell is subjected to a uniformly distributed normal following load-pressure $q(t)$, generating a uniform precritical stress state which is defined by the traction $T_{22}^0 = -qR$. Then, in accordance with the formulas (2.55), we obtain the expressions for 'loading' terms:

$$X_1 = T_{11}^0\left(\frac{\partial^2 v}{\partial x \partial y} + \frac{1}{R}\frac{\partial w}{\partial x}\right) + T_{22}^0\left(\frac{\partial^2 u}{\partial y^2} - \frac{1}{R}\frac{\partial w}{\partial x}\right);$$

$$X_2 = T_{11}^0\frac{\partial^2 v}{\partial x^2} + T_{22}^0\frac{\partial^2 u}{\partial x \partial y}; \tag{3.2}$$

$$X_3 = T_{11}^0\frac{\partial^2 w}{\partial x^2} + T_{22}^0\left(\frac{\partial^2 w}{\partial y^2} + \frac{w}{R^2}\right).$$

Substitution of (2.63), (3.1), (3.2) into (2.58) results in the system of equation:

$$C_{11}\frac{\partial^2 u}{\partial x^2} + \left(C_{66} + \frac{D_{66}}{R^2}\right)\frac{\partial^2 u}{\partial y^2} + (C_{12} + C_{66})\frac{\partial^2 v}{\partial x \partial y} + \frac{C_{12}}{R}\frac{\partial w}{\partial x} -$$

$$-\frac{D_{11}}{R}\frac{\partial^3 w}{\partial x^3} + \frac{D_{66}}{R}\frac{\partial^3 w}{\partial x \partial y^2} + T_{11}^0\left(\frac{\partial^2 v}{\partial x \partial y} + \frac{1}{R}\frac{\partial w}{\partial x}\right) +$$

$$+T_{22}^0\left(\frac{\partial^2 u}{\partial y^2} - \frac{1}{R}\frac{\partial w}{\partial x}\right) = \rho h\left(\frac{\partial^2 u}{\partial t^2} - \frac{h^2}{12R}\frac{\partial^3 w}{\partial x \partial t^2}\right); \tag{3.3}$$

$$C_{22}\frac{\partial^2 v}{\partial y^2}+\left(C_{66}+\frac{3D_{66}}{R^2}\right)\frac{\partial^2 v}{\partial x^2}+(C_{12}+C_{66})\frac{\partial^2 u}{\partial x\partial y}+\frac{C_{22}}{R}\frac{\partial w}{\partial y}-$$

$$-\frac{D_{12}+3D_{66}}{R}\frac{\partial^3 w}{\partial x^2\partial y}+T_{11}{}^0\frac{\partial^2 v}{\partial x^2}+T_{22}{}^0\frac{\partial^2 u}{\partial x\partial y}=\rho h\left(\frac{\partial^2 v}{\partial t^2}-\right.$$

$$\left.-\frac{h^2}{6R}\frac{\partial^3 w}{\partial y\partial t^2}\right);$$

$$D_{11}\frac{\partial^4 w}{\partial x^4}+2(D_{12}+2D_{66})\frac{\partial^4 w}{\partial x^2\partial y^2}+D_{22}\frac{\partial^4 w}{\partial y^4}+\frac{C_{12}}{R}\frac{\partial u}{\partial x}+$$

$$+\frac{C_{22}}{R}\frac{\partial v}{\partial y}+\frac{C_{22}}{R^2}w+\frac{2D_{22}}{R^2}\frac{\partial^2 w}{\partial y^2}+\frac{D_{22}}{R^4}w-\frac{D_{11}}{R}\frac{\partial^3 u}{\partial x^3}+$$

$$+\frac{D_{66}}{R}\frac{\partial^3 u}{\partial x\partial y^2}-\frac{D_{12}+3D_{66}}{R}\frac{\partial^3 v}{\partial x^2\partial y}-T_{11}{}^0\frac{\partial^2 w}{\partial x^2}-T_{22}{}^0\left(\frac{\partial^2 w}{\partial y^2}+\right.$$

$$\left.+\frac{w}{R^2}\right)=-\rho h\left[\frac{\partial^2 w}{\partial t^2}-\frac{h^2}{12}\left(\frac{\partial^4 w}{\partial x^2\partial t^2}+\frac{\partial^4 w}{\partial y^2\partial t^2}-\right.\right.$$

$$\left.\left.-\frac{1}{R}\frac{\partial^3 u}{\partial x\partial t^2}-\frac{2}{R}\frac{\partial^3 v}{\partial y\partial t^2}\right)\right].$$

At the end faces of the shell $x=0$ and $x=L$ we assume the end conditions of free support. Since for 'additional' non-axisymmetric components of stress–strain state they are homogeneous, we have

$$T_{11}\Big|_{x=0,L}=0;\quad v\Big|_{x=0,L}=0;\quad w\Big|_{x=0,L}=0;\quad M_{11}\Big|_{x=0,L}=0. \quad (3.4)$$

The following series satisfy term-by-term the eqns (3.3) and boundary conditions (3.4)

$$u(x,y,t)=\sum_{m=1}^{\infty}\sum_{n=1}^{\infty}[U_{mn}(t)\cos\beta_n y+\tilde{U}_{mn}(t)\sin\beta_n y]\cos\alpha_m x;$$

$$v(x,y,t)=\sum_{m=1}^{\infty}\sum_{n=1}^{\infty}[V_{mn}(t)\sin\beta_n y-\tilde{V}_{mn}(t)\cos\beta_n y]\sin\alpha_m x; \quad (3.5)$$

$$w(x,y,t)=\sum_{m=1}^{\infty}\sum_{n=1}^{\infty}[W_{mn}(t)\cos\beta_n y+\tilde{W}_{mn}(t)\sin\beta_n y]\sin\alpha_m x,$$

where

$$\alpha_m=\frac{\pi m}{L},\quad \beta_n=\frac{n}{R}.$$

The substitution of (3.5) into (3.3) at fixed pair of $\{m, n\}$ gives two independent and equivalent in form systems of equations in respect to U_{mn}, V_{mn}, W_{mn} and $\tilde{U}_{mn}$, $\tilde{V}_{mn}$, $\tilde{W}_{mn}$. Let us present the first system as follows:

$$\left[\alpha_m^2 C_{11}+\beta_n^2\left(C_{66}+\frac{D_{66}}{R^2}+T_{22}^0\right)\right] U_{mn}-\alpha_m\beta_n(C_{12}+C_{66}+T_{11}^0)\times$$
$$\times V_{mn}=\frac{\alpha_m}{R}(C_{12}+\alpha_m^2 D_{11}-\beta_n^2 D_{66}+T_{11}^0-T_{22}^0)W_{mn}-$$
$$-\rho h\left(\frac{d^2U_{mn}}{dt^2}-\frac{h^2\alpha_m}{12R}\frac{d^2W_{mn}}{dt^2}\right); \tag{3.6}$$

$$\left[\beta_n^2 C_{22}+\alpha_m^2\left(C_{66}+\frac{3D_{66}}{R^2}+T_{11}^0\right)\right] V_{mn}-\alpha_m\beta_n\times$$
$$\times(C_{12}+C_{66}+T_{22}^0)U_{mn}=-\frac{\beta_n}{R}[C_{22}+\alpha_m^2(D_{12}+3D_{66})]W_{mn}-$$
$$-\rho h\left(\frac{d^2V_{mn}}{dt^2}+\frac{h^2\beta_n}{6R}\frac{d^2W_{mn}}{dt^2}\right); \tag{3.7}$$

$$\left[D_{11}\alpha_m^4+2(D_{12}+2D_{66})\alpha_m^2\beta_n^2+D_{22}\beta_n^4-\frac{2D_{22}}{R^2}\beta_n^2+\frac{D_{22}}{R^4}+\right.$$
$$\left.+\frac{C_{22}}{R^2}+T_{11}^0\alpha_m^2+T_{22}^0\left(\beta_n^2-\frac{1}{R^2}\right)\right]W_{mn}-\frac{\alpha_m}{R}\times$$
$$\times(C_{12}+D_{11}\alpha_m^2-D_{66}\beta_n^2)U_{mn}+\frac{\beta_n}{R}[C_{22}+(D_{12}+3D_{66})\alpha_m^2]V_{mn}=$$
$$=-\rho h\left[\frac{d^2W_{mn}}{dt^2}+\frac{h^2}{12}(\alpha_m^2+\beta_n^2)\frac{d^2W_{mn}}{dt^2}-\frac{h^2}{12}\frac{\alpha_m}{R}\times\right.$$
$$\left.\times\frac{d^2U_{mn}}{dt^2}+\frac{h^2}{12}\frac{2\beta_n}{R}\frac{d^2V_{mn}}{dt^2}\right]. \tag{3.8}$$

It is not difficult to make sure that a group of inertia terms

$$-\frac{\rho h^3}{12}\left[(\alpha_m^2+\beta_n^2)\frac{d^2W_{mn}}{dt^2}-\frac{\alpha_m}{R}\frac{d^2U_{mn}}{dt^2}+\frac{2\beta_n}{R}\frac{d^2V_{mn}}{dt^2}\right]$$

in (3.8), reflecting jointly the effect of rotation inertia of elements, normal to the mid-surface, is comparable as to its value with $\rho h(d^2W_{mn}/dt^2)$ only under the condition that $m\approx(\sqrt{12}/\pi)(L/h)$ or $n\approx\sqrt{12}(R/h)$. For very thin shells this corresponds to very high wave numbers which are not of practical interest. It can be shown (for instance, by studying the expressions for kinetic energy of a shell) that the inertia terms

$$\frac{\rho h^3}{12}\frac{\alpha_m}{R}\frac{d^2W_{mn}}{dt^2} \quad \text{and} \quad -\frac{\rho h^3}{6}\frac{\beta_n}{R}\frac{d^2W_{mn}}{dt^2}$$

in (3.6) and (3.7) are the values of the same order as the above indicated group of inertia terms in (3.8). Let us add that the quantitative study of the effects associated with taking into account inertia of rotation, will be accomplished on the basis of Timoshenko type model in paragraph (3.4), where the effects which may be attributed to tangential inertia terms

$$-\rho h\frac{d^2U_{mn}}{dt^2} \quad \text{and} \quad -\rho h\frac{d^2V_{mn}}{dt^2}$$

will be investigated in all the details.

Taking into account the above we shall use the following system of equations for calculation of the frequencies of natural flexural vibrations of cylindrical shells (assuming $T_{11}^0 = 0$, $T_{22}^0 = 0$):

$$\begin{aligned}
&\left[\alpha_m^2C_{11}+\beta_n^2\left(C_{66}+\frac{D_{66}}{R^2}\right)\right]U_{mn}-\alpha_m\beta_n(C_{12}+C_{66})V_{mn}=\\
&\qquad=\frac{\alpha_m}{R}(C_{12}+\alpha_m^2D_{11}-\beta_n^2D_{66})W_{mn};\\
&\left[\beta_n^2C_{22}+\alpha_m^2\left(C_{66}+\frac{3D_{66}}{R^2}\right)\right]V_{mn}-\alpha_m\beta_n(C_{12}+C_{66})U_{mn}=\\
&\qquad=-\frac{\beta_n}{R}[C_{22}+\alpha_m^2(D_{12}+3D_{66})]W_{mn};\\
&\left[D_{11}\alpha_m^4+2(D_{12}+2D_{66})\alpha_m^2\beta_n^2+D_{22}\beta_n^4-\frac{2D_{22}}{R^2}\beta_n^2+\right.\\
&\left.+\frac{D_{22}}{R^4}+\frac{C_{22}}{R^2}\right]W_{mn}-\frac{\alpha_m}{R}(C_{12}+D_{11}\alpha_m^2-D_{66}\beta_n^2)U_{mn}+\\
&+\frac{\beta_n}{R}[C_{22}+(D_{12}+3D_{66})\alpha_m^2]V_{mn}=-\rho h\frac{d^2W_{mn}}{dt^2}.
\end{aligned} \tag{3.9}$$

In accordance with the technical theory of cylindrical shells, instead of (2.58) and (2.63) we have

$$\frac{\partial T_{11}}{\partial x}+\frac{\partial T_{21}}{\partial y}+P_1+X_1=0; \quad \frac{\partial T_{22}}{\partial y}+\frac{T_{12}}{\partial x}+P_2+X_2=0;$$

$$\frac{T_{22}}{R}-\frac{\partial^2M_{11}}{\partial x^2}-\frac{\partial^2M_{22}}{\partial y^2}-\frac{\partial^2(M_{12}+M_{21})}{\partial x\partial y}=$$

$$=P_3+X_3+\frac{\partial}{\partial x}(\Phi_2+F_2)+\frac{\partial}{\partial y}(\Phi_1+F_1);$$

$$T_{11}=C_{11}\frac{\partial u}{\partial x}+C_{12}\left(\frac{\partial v}{\partial y}+\frac{w}{R}\right); \quad T_{22}=C_{22}\left(\frac{\partial v}{\partial y}+\frac{w}{R}\right)+C_{12}\frac{\partial u}{\partial x};$$

$$T_{12}=T_{21}=C_{66}\left(\frac{\partial v}{\partial x}+\frac{\partial u}{\partial y}\right); \quad M_{12}=M_{21}=-2D_{66}\frac{\partial^2 w}{\partial x \partial y};$$

$$M_{11}=-D_{11}\frac{\partial^2 w}{\partial x^2}-D_{12}\frac{\partial^2 w}{\partial y^2}; \quad M_{22}=-D_{22}\frac{\partial^2 w}{\partial y^2}-D_{12}\frac{\partial^2 w}{\partial x^2}.$$

By substituting the expressions of tractions and moments into equations of motion and repeating manipulations of the given paragraph, we obtain under the end conditions (3.4) the following system:

$$\begin{gathered}
(\alpha_m^2 C_{11}+B_n^2 C_{66})\,U_{mn}-\alpha_m\beta_n\,(C_{12}+C_{66})\,V_{mn}=\frac{\alpha_m}{R}C_{12}W_{mn};\\
(\beta_n^2 C_{22}+\alpha_m^2 C_{66})\,V_{mn}-\alpha_m\beta_n\,(C_{12}+C_{66})\,U_{mn}=-\frac{\beta_n}{R}C_{22}W_{mn};\\
\left[D_{11}\alpha_m^4+2\,(D_{12}+2D_{66})\,\alpha_m^2\beta_n^2+D_{22}\beta_n^4+\frac{C_{22}}{R^2}\right]W_{mn}-\\
-\frac{\alpha_m}{R}C_{12}U_{mn}+\frac{\beta_n}{R}C_{22}V_{mn}=-\rho h\frac{d^2 W_{mn}}{dt^2}.
\end{gathered} \tag{3.10}$$

The solution of (3.9) and (3.10) is carried out in an obvious manner. From the first two equations U_{mn} and V_{mn} are expressed; then they are substituted into the third one, which ultimately is written in the form

$$\frac{d^2 W_{mn}}{dt^2}+\omega_{mn}^2 W_{mn}=0. \tag{3.11}$$

The expression of ω_{mn}^2, obtained from the system (3.10) is well-known:

$$\begin{gathered}
\omega_{mn}^2=\frac{1}{\rho h}\left[\alpha_m^4 D_{11}+2\alpha_m^2\beta_n^2\,(D_{12}+2D_{66})+\beta_n^4 D_{22}+\right.\\
\left.+\frac{\alpha_m^4 C_{66}\,(C_{11}C_{22}-C_{12}^2)}{\Delta_{mn}}\right],
\end{gathered} \tag{3.12}$$

There

$$\Delta_{mn}=R^2[\alpha_m^4 C_{11}C_{66}+\beta_n^4 C_{22}C_{66}+\alpha_m^2\beta_n^2\,(C_{11}C_{22}-C_{12}^2-2C_{12}C_{66})]. \tag{3.13}$$

The formula for ω_{mn}^2 being obtained from the system (3.9) contains many terms of the order (h^2/R^2) in comparison to unity. By omitting them, we arrive at the following final expression

$$\begin{gathered}
\omega_{mn}^2=\frac{1}{\rho h}\left\{\alpha_m^4 D_{11}+2\alpha_m^2\beta_n^2\,(D_{12}+2D_{66})+\beta_n^4 D_{22}+\right.\\
+\frac{1}{\Delta_{mn}}\left\{\alpha_m^4 C_{66}\,(C_{11}C_{22}-C_{12}^2)+\frac{\beta_n^4}{R^2}C_{66}C_{22}D_{22}+\frac{\alpha_m^2\beta_n^2}{R^2}[(C_{11}C_{22}-\right.
\end{gathered} \tag{3.14}$$

$$-C_{12}^2-2C_{12}C_{66})D_{22}+4C_{22}C_{66}D_{66}]-2\alpha_m^4\beta_n^2[C_{66}(C_{11}D_{22}-C_{22}D_{11})+$$
$$+(3C_{11}C_{22}-3C_{12}^2-4C_{12}C_{66})D_{66}+(C_{11}C_{22}-C_{12}^2-C_{12}C_{66})D_{12}]-$$
$$-2\beta_n^6C_{66}C_{22}D_{22}-2\alpha_m^2\beta_n^4[(C_{11}C_{22}-C_{12}^2-2C_{12}C_{66})D_{22}+$$
$$+C_{22}C_{66}(D_{12}+4D_{66})]-2\alpha_m^6C_{12}C_{66}D_{11}\Big\}\Big\}.$$

Further we shall investigate the formula (3.14) in several ultimate cases.

1. Let us suppose the condition $\alpha_m \ll (1/R)$ which corresponds to low values of m and long shells $[(L/\pi R) \gg 1]$. From (3.14) we obtain the expression

$$\omega_{mn}^2=\frac{1}{\rho h}\left[\frac{\alpha_m^4}{R^2\beta_n^4}\cdot\frac{C_{11}C_{22}-C_{12}^2}{C_{22}}+\left(\beta_n^2-\frac{1}{R^2}\right)^2D_{22}\right]. \quad (3.15)$$

At $n = 1$, the formula yields

$$\omega_{m1}^2=\frac{1}{\rho h}\,\alpha_m^4R^2\cdot\frac{C_{11}C_{22}-C_{12}^2}{C_{22}}.$$

For an isotropic material with elastic modulus E from (3.15) we find

$$\omega_{m1}^2=\frac{\alpha_m^4R^2E}{\rho}.$$

The given value is twice the square of flexural vibration frequency of a beam with an annular cross section. The fact that the asymptotic formula (3.15) does not allow at $n = 1$ the limiting transition to the formula for a beam, may be attributed to disregard of circumferential inertia of the shell made earlier. If we maintain the inertia term $-\rho h(d^2V_{mn}/dt^2)$ in eqn (3.7), then in the asymptotic case $\alpha_m \ll (1/R)$ under consideration the following expression for the lowest frequency of natural vibrations is obtained:

$$\omega_{mn}^2=\frac{1}{\rho h}\left[\frac{\alpha_m^4}{R^2\beta_n^4}\cdot\frac{C_{11}C_{22}-C_{12}^2}{C_{22}}+\right.$$
$$\left.+\left(\beta_n^2-\frac{1}{R^2}\right)^2D_{22}\right]\frac{\beta_n^2R^2}{1+\beta_n^2R^2}. \quad (3.16)$$

In the case of an isotropic material (3.16) coincides with the asymptotic formula (4.7) presented in Ref. 124, Chapter 6. At $n = 1$, (3.16) yields the square of natural vibration frequency for a beam with an annular cross section. As is seen from comparison of (3.15) and (3.16), the effect of circumferential inertia on the lowest frequency of natural vibrations of the shell diminishes with increasing n.

From expression (3.12) at $\alpha_m \ll (1/R)$, in its turn, we find

$$\omega_{mn}^2 = \frac{1}{\rho h}\left[\frac{\alpha_m^4}{R^2\beta_n^4}\frac{C_{11}C_{22}-C_{12}^2}{C_{22}} + \beta_n^4 D_{22}\right]. \tag{3.17}$$

A comparison of (3.17) and (3.15) shows that the error introduced due to the assumptions of the technical theory in the case of vibrations of long shells reaches its maximum for the wave number $n = 1$; it increases sharply with increasing h/R. In the case when the second term, which takes into account the flexural stiffness of the shell is the dominating one in (3.15) and (3.17), the formula (3.17) yields a highly over-rated magnitude of natural vibration frequency at low n.

2. Let $\alpha_m \sim (1/R)$, i.e. $(\pi mR/L) \sim 1$. This condition is satisfied for shells of medium length at low wave numbers along the generatrix and for long shells at high m. Let us consider separately three cases.

For circumferential wave numbers satisfying the condition $n \gg [(12R^2/h^2)]^{1/8}$, from formulas (3.12) and (3.14) we obtain one and the same asymptotic expression

$$\omega_n^2 = \frac{n^4}{R^4}\frac{D_{22}}{\rho h}. \tag{3.18}$$

Under the condition $n \sim [(12R^2/h^2)]^{1/8}$ the fomula (3.12) yields

$$\omega_{mn}^2 = \frac{1}{\rho h[\beta_n^4 C_{22}C_{66} + \alpha_m^2\beta_n^2(C_{11}C_{22} - C_{12}^2 - 2C_{12}C_{66})]} \times$$
$$\times \left\{\frac{\alpha_m^4}{R^2} C_{66}(C_{11}C_{22} - C_{12}^2) + \beta_n^8 C_{22}C_{66}D_{22} + \alpha_m^2\beta_n^6[(C_{11}C_{22} - \right.$$
$$\left. - C_{12}^2 - 2C_{12}C_{66})D_{22} + 2C_{22}C_{66}(D_{12} + 2D_{66})]\right\}, \tag{3.19}$$

whereas from (3.14) we find

$$\omega_{mn}^2 = \frac{1}{\rho h[\beta_n^4 C_{22}C_{66} + \alpha_m^2\beta_n^2(C_{11}C_{22} - C_{12}^2 - 2C_{12}C_{66})]} \times$$
$$\times \left\{\frac{\alpha_m^4}{R^2} C_{66}(C_{11}C_{22} - C_{12}^2) + \beta_n^6\left(\beta_n^2 - \frac{2}{R^2}\right) C_{22}C_{66}D_{22} + \right.$$
$$\left. + \alpha_m^2\beta_n^6[(C_{11}C_{22} - C_{12}^2 - 2C_{12}C_{66})D_{22} + 2C_{22}C_{66}(D_{12} + 2D_{66})]\right\}. \tag{3.20}$$

As is seen, the expressions (3.19) and (3.20) differ by a sufficiently small value and this difference decreases with increasing n.

Within the range of low values $n \ll [(12R^2/h^2)]^{1/8}$ the 'membrane' term becomes the dominating one and we obtain one and the same expression from the formulas (3.12) and (3.14)

$$\omega_{mn}^2 = \frac{1}{\rho h} \frac{\alpha_m^4 C_{66}(C_{11}C_{22} - C_{12}^2)}{\Delta_{mn}}. \qquad (3.21)$$

In such a way, in the case $\alpha_m \sim (1/R)$ both for low and sufficiently high n the results of calculation of natural frequencies according to the technical theory and Flügge's theory should differ slightly. The technical theory yields a distinct error (toward overrated frequency) within the medium range of n, where 'membrane' and 'flexural' components in the formula for frequencies are close in their magnitudes.

3. Let $\alpha_m \gg (1/R)$, i.e. $(\pi mR/L) \gg 1$. This variant is realized either for short shells or shells of medium length at high axial wave numbers. It is, consequently, of particular interest for problems of longitudinal impact. As shown in the analysis of the expression (3.14), all the additional terms to formula (3.12) for any value of n yield correction of the order h^2/R^2.

Let us further consider the results of numerical calculations. As an example we shall consider shells made of two materials: steel with

$$E = 20 \cdot 10^{10}\ \text{Н/м}^2; \quad \nu = 0{,}3; \quad \rho = 7{,}8 \cdot 10^3\ \text{кг/м}^3, \qquad (3.22)$$

and a unidirectionally reinforced carbon plastic T 300/5208, having the following characteristics:[259]

$$E_1 = 15,4 \cdot 10^{10}\ \text{Н/м}^2; \quad E_2 = 1,08 \cdot 10^{10}\ \text{Н/м}^2; \quad G_{12} = 0,57 \cdot 10^{10}\ \text{Н/м}^2$$
$$\nu_{21} = 0{,}28; \quad \rho = 1{,}6 \cdot 10^3\ \text{кг/м}^3; \quad E_1\nu_{12} = E_2\nu_{21} \qquad (3.23)$$

(the axis 1 is oriented in the direction of reinforcing fibers). Further two layer layup variants are considered:

(a) the axis 1 of the material coincides with the x-axis of a shell (longitudinal reinforcement) and
(b) the axis 1 of the material coincides with the y-axis of a shell (circumferential reinforcement). In all calculational variants $R = 1$ m.

Let us denote the frequencies of natural vibrations calculated by the formulas (3.12), (3.14) and (3.20), respectively, by ω_{mn}^T, ω_{mn}^F and ω_{mn}^A. Henceforth, the following values of relative differences in frequency magnitudes will be considered:

$$\Delta_T^\omega(m,n) = \frac{\omega_{mn}^T}{\omega_{mn}^F} - 1; \quad \Delta_A^\omega(m,n) = \frac{\omega_{mn}^A}{\omega_{mn}^F} - 1. \qquad (3.24)$$

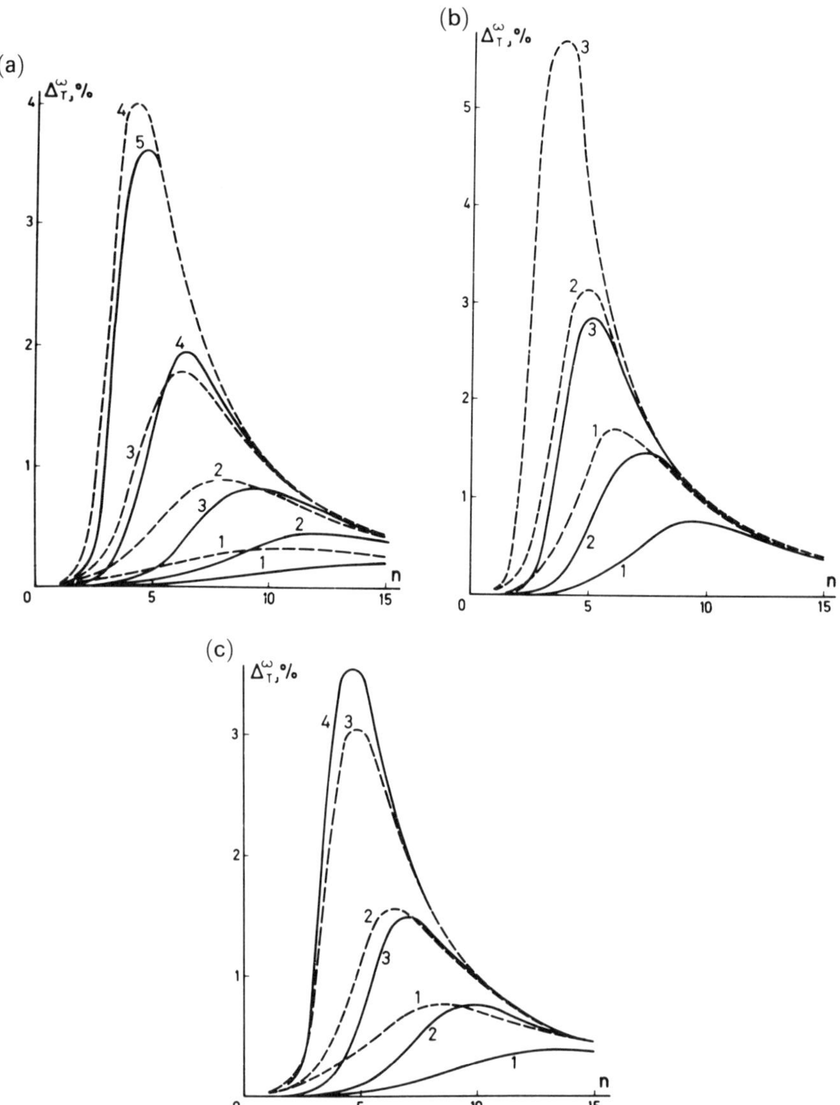

Fig. 3.1. The $\Delta_T^\omega(n)$ functions for shells with a longitudinal (a), circumferential (b) reinforcement and steel shell (c) at $m = 1$, $R/h = 200$ (———) and 50 (– – –). Numbers at the curves correspond to the magnitudes of L/R: 0.5 (1); 1 (2); 2 (3); 5 (4); and 10 (5).

In Fig. 3.1, the dependencies of Δ_T^ω on n are given for the carbon plastic shells with longitudinal and circumferential reinforcement and for the steel shell. The most interesting feature of these dependencies is the presence of its maximum at a definite value of n. This maximum is moving toward lower n at increasing L/R and h/R. As was expected from the qualitative analysis of the above formulas, the error given by the technical theory is rather small both for low and sufficiently high circumferential harmonics in the case of shells of medium length.

It is well-known that the dependency of frequency of natural vibrations on the circumferential wave number has a minimum for shells of medium length. The location of this minimum, as first shown in Ref. 285, is defined from the condition of equality of two components of deformation energy: membrane and flexural. The value of n, to which the maximum error corresponds, also obeys the condition of proximity of the values of 'membrane' and 'flexural' terms in the formula for ω_{mn}^2. As a result of the above said we can assume that a minimum frequency of natural vibrations and a maximum error in its calculation by the use of the technical theory for shells of medium length correspond to one and the same value n (or, at least, to very close n values). This is fully confirmed by

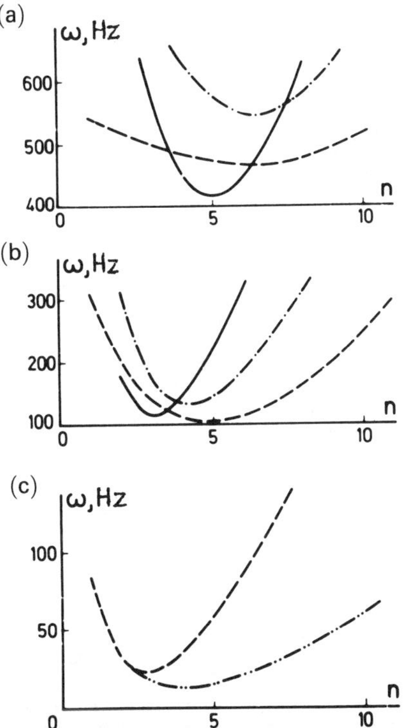

Fig. 3.2. The $\omega(n)$ functions for shells with circumferential (———) and longitudinal (– – –) reinforcement and steel shell (– · –) at $R/h = 50$, $L/R = 0.5$ (a), 2 (b), 10 (c); (– · · –) corresponds to $R/h = 200$.

comparing the results (see Fig. 3.1) with the calculated magnitudes of frequencies (Fig. 3.2).

In such a way, we can formulate the conclusion: the relative error in the calculation of natural vibration frequencies according to the equations of the technical theory for cylindrical shells of medium length reaches its maximum in the case of calculating the lowest frequencies.

The analysis of the data of Fig. 3.1 shows also that the maximum of the relative error increases with increasing L/R. Figure 3.3 is a convincing illustration of this (it should be kept in mind that the parameters L/R and m enter the solution only in combination $\pi mR/L$, so that the decrease in m corresponds to proportional growth of L/R). The results, presented in Table 3.1, reflect the following interesting feature of the effect of L/R variation: for each fixed n, other than $n = 1$, we can find such a value of $L = L_0$ that at $L > L_0$ the calculation error of frequencies remains constant. Thereby the 'ultimate' error decreases with growing n. As to the 'beam' form of vibration $n = 1$, the error increases sharply with the increase in the shell length.

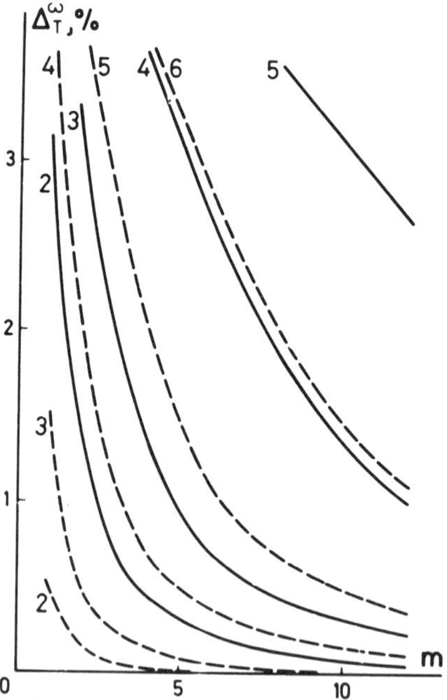

Fig. 3.3. The $\Delta_T^\omega(m)$ functions for shells with circumferential (———) and longitudinal (– – –) reinforcement; $n = 5$, $R/h = 50$. Notations are the same as in Fig. 3.1 (6 corresponds to $L/R = 20$).

TABLE 3.1

Relative Correction Δ_T^{ω} in Per Cents for a Shell with Circumferential Reinforcement at $R/h = 50$, $m = 1$

L/R	Number of circumferential harmonics n										
	1	*2*	*3*	*4*	*5*	*6*	*7*	*8*	*9*	*10*	*11*
0.25	0.021	0.06	0.14	0.27	0.45	0.63	0.76	0.81	0.80	0.74	0.66
0.5	0.015	0.08	0.30	0.76	1.34	1.66	1.60	1.39	1.16	0.96	0.81
1	0.017	0.21	1.07	2.64	3.13	2.60	2.01	1.56	1.24	1.00	0.83
2	0.035	1.00	5.02	5.70	4.03	2.83	2.08	1.58	1.25	1.01	0.83
5	0.29	12.7	11.8	6.63	4.16	2.86	2.08	1.59	1.25	1.01	0.83
10	2.92	29.8	12.5	6.66	4.17	2.86	2.08	1.59	1.25	1.01	0.83
20	35.1	33.1	12.5	6.67	4.17	2.86	2.08	1.59	1.25	1.01	0.83
50	465	33.3	12.5	6.67	4.17	2.86	2.08	1.59	1.25	1.01	0.83
100	2120	33.3	12.5	6.67	4.17	2.86	2.08	1.59	1.25	1.01	0.83

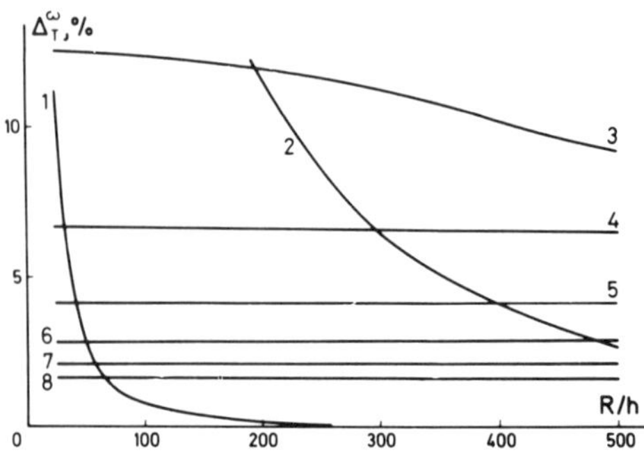

Fig. 3.4. The $\Delta_T^\omega(R/h)$ functions for a shell with circumferential reinforcement at $m = 1$, $L/R = 10$. Numbers at the curves correspond to the values of n.

The dependencies on R/h of the relative calculation error for a sufficiently long shell, as seen from Fig. 3.4, differ qualitatively at $n = 1$ and at $n \geq 4$. If at $n = 1$ the error increases sharply with an increase in thickness, then for higher, 'flexural' forms of vibrations, it practically does not depend on the value of R/h.

In Figure 3.5, the calculation error of frequencies according to an approximate formula (3.20) is presented. As is seen, at $n \geq 5$, this formula yields results that are fairly acceptable for practical use in all cases under consideration. The accuracy of this formula increases with increasing n, and that corresponds to the condition under which the formula has been developed.

Further we shall dwell on the problem about the effect of material characteristics on the magnitudes of the errors under consideration. On passing over from the longitudinal reinforcement to the circumferential one, the stiffness ratio in the axial and circumferential directions for a carbon plastic shell changes approximately from 14 to 1/14. The steel shell with this ratio equal to a unity, occupies the intermediate position. As follows from the comparison of the results of Fig. 3.1 as well as of Fig. 3.5, the calculation error of natural frequencies for a steel shell in all cases falls between respective errors for two carbon plastic shells. Thereby it should be emphasized that the difference between the errors is sufficiently vast. Thus, for instance, the error according to the technical theory at $m = 1$, $L/R = 2$, $R/h = 200$ is 0.8% for longitudinal and 2.8% for circumferential reinforcement, while at $R/h = 50$ accordingly 1.8% and 5.7%; at $m = 1$, $L/R = 20$, $R/h = 50$, the error is 0.34% and 35.1%, respectively.

The conclusion can be drawn from the given results that the range of applicability for the technical theory in the calculation of natural frequencies of anisotropic structures varies essentially at fixed geometric

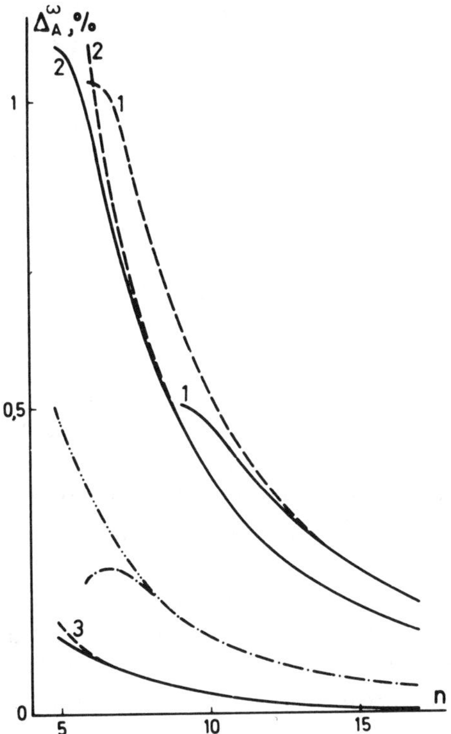

Fig. 3.5. The $\Delta_A^\omega(n)$ functions at $m = 1$ for shells with longitudinal (1), circumferential (3) reinforcement and steel shell (2); $L/R = 5$ and $R/h = 200$ (———); $L/R = 5$ and $R/h = 50$ (– – –); $L/R = 10$ and $R/h = 200$ (–·–); $L/R = 10$ and $R/h = 50$ (–··–).

parameters and modes of vibrations in dependence on the anisotropy degree of the material.

In the completion of considering the question about the accuracy of calculation results for natural frequencies let us point out that in the most unfavorable, from the viewpoint of the technical theory, situation (circumferential reinforcement), the relative difference in frequencies, calculated in accordance with two theories, at any value of m and n does not exceed 3%, if $L/R \leq 2$, $R/h \geq 200$; 6%, if $L/R \leq 2$, $R/h \geq 50$; 7% if $L/R \leq 5$, $R/h \geq 200$; 13%, if $L/R \leq 5$, $R/h \geq 50$. The obtained estimates of the range of applicability of this theory should necessarily be taken into account when solving the problems of dynamics for orthotropic cylindrical shells on the basis of the technical theory.

3.2 APPLICABILITY OF THE TECHNICAL THEORY FOR CALCULATION OF STATIC CRITICAL LOADS

The solution of problems of parametric vibrations, dynamic stability, deformation of shells under compressive loads involves, evidently, the values of critical static tractions, corresponding to a fixed pair of wave numbers $\{m, n\}$. It is natural that the accuracy of calculating the critical static tractions defines essentially the accuracy of solving these problems. Thereby it should be borne in mind that in many dynamic problems an important role is played by spatial modes of movement corresponding to the values of m and n which are much higher than those which yield a minimum of critical static loads. This circumstance makes it necessary to verify the applicability of simplified theories to calculation of critical static loads, within a sufficiently great range of variation of m and n.

Let us consider on the basis of eqns (3.3) the solution of static stability problem for an orthotropic cylindrical shell loaded on end faces by uniformly distributed compressive tractions P_0 and on side faces — by uniformly distributed external pressure q_0. It is assumed that in a momentless homogeneous pre-critical state the tractions $T_{11}^0 = -P_0$, $T_{22}^0 = -q_0 R$ arise. In the case of free support end conditions (3.4), the system (3.3) takes the form (3.6)–(3.8), but without inertia terms.

If we express the values U_{mn}, V_{mn} from eqn (3.6), (3.7) in terms of W_{mn} and substitute the obtained relations into (3.8), then after cumbersome transformations we obtain the equation:

$$\frac{F_{mn}(T_{11}{}^0, T_{22}{}^0)}{\Delta'_{mn}(T_{11}{}^0, T_{22}{}^0)} + \rho h \omega_{mn}{}^2 = 0, \tag{3.25}$$

where ω_{mn}^2 is determined according to the formula (3.14) while

$$\begin{aligned} F_{mn}(T_{11}{}^0, T_{22}{}^0) &= R^2\alpha_m{}^6(C_{66}+T_{11}{}^0)\left(T_{11}{}^0C_{11}+T_{22}{}^0\frac{D_{11}}{R^2}\right)+ \\ &+R^2\alpha_m{}^2\beta_n{}^4[T_{22}{}^0(C_{11}C_{22}-C_{12}{}^2)+(C_{66}+T_{22}{}^0)(T_{11}{}^0C_{22}-T_{22}{}^0C_{12})- \\ &-(C_{66}+T_{11}{}^0)T_{22}{}^0C_{12}]+R^2\alpha_m{}^4\beta_n{}^2\left[T_{11}{}^0(C_{11}C_{22}-C_{12}{}^2)+(C_{66}+\right. \\ &\left.+T_{11}{}^0)(T_{22}{}^0C_{11}-T_{11}{}^0C_{12})-(C_{66}+T_{22}{}^0)T_{11}{}^0C_{12}+2T_{22}{}^0C_{22}\frac{D_{11}}{R^2}\right]+ \\ &+R^2\beta_n{}^6T_{22}{}^0C_{22}\left[(C_{66}+T_{22}{}^0)-2\frac{D_{22}}{R^2}\right]+\alpha_m{}^4\{T_{11}{}^0(C_{11}C_{22}-C_{12}{}^2)- \\ &-(C_{66}+T_{11}{}^0)[T_{11}{}^0C_{12}+T_{22}{}^0(C_{11}-C_{12})]\}-\beta_n{}^4T_{22}{}^0C_{22}\left[(C_{66}+\right. \\ &\left.+T_{22}{}^0)-\frac{D_{22}}{R^2}\right]-\alpha_m{}^2\beta_n{}^2\{T_{22}{}^0(C_{11}C_{22}-C_{12}{}^2)-(C_{66}+T_{11}{}^0)T_{22}{}^0C_{12}- \end{aligned} \tag{3.26}$$

$$-(C_{66}+T_{22}^0)[T_{11}^0C_{22}-T_{22}^0(C_{22}-C_{12})]\};$$
$$\Delta'_{mn}(T_{11}^0, T_{22}^0)=R^2[\alpha_m^4(C_{66}+T_{11}^0)C_{11}+\beta_n^4(C_{66}+T_{22}^0)C_{22}+$$
$$+\alpha_m^2\beta_n^2(C_{11}C_{22}-C_{12}^2)-\alpha_m^2\beta_n^2(C_{66}+T_{11}^0)C_{12}-\alpha_m^2\beta_n^2(C_{66}+$$
$$+T_{22}^0)C_{12}]. \qquad (3.27)$$

As it follows from (3.25) and expressions (3.26) and (3.27), on the plane $\{T_{11}^0, T_{22}^0\}$ a critical combination of axial compression and external pressure loads yield a curve of second order.

Without dwelling at length on this equation, let us note that at $T_{22}^0 = 0$, for example, it has two actual negative roots, one of which (of the order $-C_{66}$) defines the value of critical compressive traction for a shear form of instability.

The possibility for the shear form of instability to appear in compression of thick-walled anisotropic structural elements was discussed in literature (for instance, see Ref. 14). For thin-walled shells at realistic values of shear modulus of composites, a shear form can manifest itself only if rather great values of m and n are considered, therefore, in the case when the thickness of the shell is comparable within the length of a buckling half-wave. In such a situation, the methodology for calculation of critical tractions to be treated further, becomes inapplicable. By excluding the indicated case from consideration, let us make an assumption

$$-T_{11}^0 \ll C_{66}, \quad -T_{22}^0 \ll C_{66}, \qquad (3.28)$$

which must be verified in the problems of stability and dynamics of anisotropic shells for all those values of m and n, which are involved in the solution. By disregarding the values T_{11}^0 and T_{22}^0 in (3.26) compared to C_{66}, we obtain a simplified formula:

$$F_{mn}(T_{11}^0, T_{22}^0)=R^2\alpha_m^6C_{66}\left(T_{11}^0C_{11}+T_{22}^0\frac{D_{11}}{R^2}\right)+R^2\alpha_m^2\beta_n^4\times$$
$$\times[T_{22}^0(C_{11}C_{22}-C_{12}^2-2C_{12}C_{66})+T_{11}^0C_{22}C_{66}]+R^2\alpha_m^4\beta_n^2\times$$
$$\times\left[T_{11}^0(C_{11}C_{22}-C_{12}^2-2C_{12}C_{66})+T_{22}^0C_{11}C_{66}+2T_{22}^0C_{22}\frac{D_{22}}{R^2}\right]+$$
$$+R^2\beta_n^6T_{22}^0C_{22}\left(C_{66}-2\frac{D_{22}}{R^2}\right)+\alpha_m^4[T_{11}^0(C_{11}C_{22}-C_{12}^2-C_{12}C_{66})-$$
$$-T_{22}^0C_{66}(C_{11}-C_{12})]-\beta_n^4T_{22}^0C_{22}\left(C_{66}-\frac{D_{22}}{R^2}\right)-\alpha_m^2\beta_n^2\times$$
$$\times[T_{22}^0(C_{11}C_{22}-C_{12}^2-2C_{12}C_{66})-C_{22}C_{66}(T_{11}^0-T_{22}^0)],$$

which with $\Delta'_{mn} = \Delta_{mn}$ taken into account, is transformed into the form

$$F_{mn}(T_{11}{}^0, T_{22}{}^0) = \Delta_{mn}\left[T_{11}{}^0\alpha_m{}^2 + T_{22}{}^0\left(\beta_n{}^2 - \frac{1}{R^2}\right)\right] + T_{11}{}^0\times$$
$$\times\left[\alpha_m{}^2(C_{11}C_{22} - C_{12}{}^2 - C_{12}C_{66}) + \beta_n{}^2C_{22}C_{66}\right]\alpha_m{}^2 + T_{22}{}^0\times$$
$$\times\left[\alpha_m{}^6C_{66}D_{11} + 2\alpha_m{}^4\beta_n{}^2C_{22}D_{11} + \alpha_m{}^4C_{12}C_{66} - \beta_n{}^4C_{22}D_{22}\times\right.$$
$$\left.\times\left(2\beta_n{}^2 - \frac{1}{R^2}\right) - \alpha_m{}^2\beta_n{}^2C_{22}C_{66}\right]. \tag{3.29}$$

On considering the cases of axial compression and external pressure separately, we obtain the following expressions for critical values of axial traction P^*_{mn} and external pressure q^*_{mn} from (3.25) and (3.29)

$$P^*{}_{mn} = \frac{\rho h\omega_{mn}{}^2}{\alpha_m{}^2(1+\delta_P)}; \quad q^*{}_{mn} = \frac{\rho h\omega_{mn}{}^2}{R\beta_n{}^2(1+\delta_q)}, \tag{3.30}$$

where

$$\delta_P = \frac{\alpha_m{}^2(C_{11}C_{22} - C_{12}{}^2 - C_{12}C_{66}) + \beta_n{}^2C_{22}C_{66}}{R^2[\alpha_m{}^4C_{11}C_{66} + \beta_n{}^4C_{22}C_{66} + \alpha_m{}^2\beta_n{}^2(C_{11}C_{22} - C_{12}{}^2 - 2C_{12}C_{66})]}; \tag{3.31}$$

$$\delta_q = \frac{1}{(\beta_nR)^2}\left[-1 + \frac{\alpha_m{}^6C_{66}D_{11} + 2\alpha_m{}^4\beta_n{}^2C_{22}D_{11} - \beta_n{}^4C_{22}D_{22}}{\alpha_m{}^4C_{11}C_{66} + \beta_n{}^4C_{22}C_{66} + \alpha_m{}^2\beta_n{}^2(C_{11}C_{22} -}\times\right.$$
$$\left.\times\frac{\left(2\beta_n{}^2 - \frac{1}{R^2}\right) + \alpha_m{}^4C_{12}C_{66} - \alpha_m{}^2\beta_n{}^2C_{22}C_{66}}{-C_{12}{}^2 - 2C_{12}C_{66})}\right], \tag{3.32}$$

whereas ω^2_{mn} is determined by the formula (3.14).

If we follow the technical theory, then for ω^2_{mn} we have the formula (3.12). Besides, in the simplest approach to the calculation of critical loads, which is usually employed within the framework of the technical theory, 'loading' terms X_1 and X_2 are disregarded and X_3 is written in a simplified form as

$$X_3 = T_{11}{}^0\frac{\partial^2 w}{\partial x^2} + T_{22}{}^0\frac{\partial^2 w}{\partial y^2}. \tag{3.33}$$

Then the expressions for critical loads take the form

$$P^*{}_{mn} = \frac{\rho h\omega_{mn}{}^2}{\alpha_m{}^2}; \quad q^*{}_{mn} = \frac{\rho h\omega_{mn}{}^2}{R\beta_n{}^2}. \tag{3.34}$$

The calculation error for critical loads according to the formulas (3.34) is defined by two factors: the error in the calculation of frequencies of natural vibrations when using (3.12) (this error was treated in detail in paragraph 3.1) and the values δ_P and δ_q. Let us consider these values in some characteristic cases. Henceforth we shall employ the condition $D_{22} \ll R^2 C_{66}$ which, apparently, does not require special comments.

1. Let $\alpha_m \ll (1/R)$. Then $\delta_P \sim (1/n^2)$ and under the condition

$$D_{22}\beta_n{}^2 \ll C_{66}, \quad \delta_q \sim -\frac{1+\dfrac{\alpha_m{}^2}{\beta_n{}^2}}{n^2}.$$

With the formula (3.15) taken into consideration we obtain the following expressions for critical loads (the designation $\lambda_m = (\pi m R/L)$ is introduced):

$$P^*_{mn} = \frac{1}{n^2+1}\left[\frac{\lambda_m{}^2}{n^2}\frac{C_{11}C_{22}-C_{12}{}^2}{C_{22}} + \frac{n^2(n^2-1)^2}{\lambda_m{}^2}\frac{D_{22}}{R^2}\right]; \tag{3.35}$$

$$q^*_{mn} = \frac{1}{R\left(n^2-1-\dfrac{\lambda_m{}^2}{n^2}\right)}\left[\frac{\lambda_m{}^4}{n^4}\frac{C_{11}C_{22}-C_{12}{}^2}{C_{22}} + (n^2-1)^2\frac{D_{22}}{R^2}\right]. \tag{3.36}$$

The formulas (3.35) and (3.36) also follow from the characteristic equation (24.23) from Ref. 134.

For the case of an isotropic material, from (3.35) and (3.36) we obtain

$$P^*_{mn} = \frac{Eh}{n^2+1}\left[\frac{\lambda_m{}^2}{n^2} + \frac{n^2}{\lambda_m{}^2}(n^2-1)^2\frac{h^2}{12R^2(1-\nu^2)}\right]; \tag{3.37}$$

$$q^*_{mn} = \frac{Eh}{R}\left[\frac{\lambda_m{}^4}{n^4\left(n^2-1-\dfrac{\lambda_m{}^2}{n^2}\right)} + (n^2-1)\frac{h^2}{12R^2(1-\nu^2)}\right]. \tag{3.38}$$

These expressions are true also at $n = 1$. For $n \geq 2$ instead of (3.38) an approximate expression can be used

$$q^*_{mn} = \frac{Eh}{R}\left[\frac{\lambda_m{}^4}{n^4(n^2-1)} + (n^2-1)\frac{h^2}{12R^2(1-\nu^2)}\right]. \tag{3.39}$$

The formulas (3.37) and (3.39) have been obtained in several works (for example, in Refs 97 and 135). If we put the condition $n \gg 1$ into (3.35) and (3.36), then

$$P^*_{mn} = \frac{\lambda_m{}^2}{n^4}\frac{C_{11}C_{22}-C_{12}{}^2}{C_{22}} + \frac{n^4}{\lambda_m{}^2}\frac{D_{22}}{R^2};$$
$$q^*_{mn} = \frac{1}{R}\left(\frac{\lambda_m{}^4}{n^6}\frac{C_{11}C_{22}-C_{12}{}^2}{C_{22}} + n^2\frac{D_{22}}{R^2}\right). \tag{3.40}$$

These formulas follow directly from the simplified expressions (3.34) and (3.17).

2. Let $\alpha_m \sim (1/R)$. Just as in 3.1, let us consider separately the three cases. For high circumferential harmonics, $n \gg [(12R^2/h^2)]^{1/8}$ we have the estimates $\delta_P \sim (1/n^2)$, $\delta_q \sim -(1/n^2)$. Thereby, both (3.30) and (3.34) with (3.18) taken into account result in equivalent asymptotic formulas:

$$P^*_{mn} = \frac{D_{22}}{R}\frac{n^4}{\lambda_m{}^2}; \quad q^*_{mn} = \frac{D_{22}}{R^2}\frac{n^2}{R}. \tag{3.41}$$

Within the medium range of circumferential modes, $n \sim [(12R^2/h^2)]^{1/8}$ (at $R/h = 200$, this corresponds to $n = 5$), δ_P and δ_q, as follows from (3.31) and (3.32), are also of the order $1/n^2$. In this case, the error of application of the technical theory together with the formulas (3.34) is determined by two factors: by the calculation error in ω^2_{mn} and the error introduced by disregarding δ_P and δ_q. The first error, as shown in paragraph 3.1, is very small for shells of medium length (it is of the order of several per cent). The second error, as it follows from the above presented analysis, is of the same order at $n \geq 5$. Consequently, one can conclude that for shells of medium length at $n \geq 5$ the formulas (3.34) and (3.12) ensure sufficient accuracy. This result is in complete agreement with the estimation of the range of applicability of the technical theory (Donnell theory) for the design of cylindrical shells of medium length presented earlier in Refs 97, 345 and 384. By the way, it should be taken into account that the criterion for applicability of the technical theory '$n \geq 5$' is only of a reference character, and for orthotropic shells, the value of error at the given geometric parameters and n-values can considerably vary depending on the moduli ratio E_1/E_2.

In calculation of the critical loads, corresponding to low wave numbers over the circumference, the error in determination of ω^2_{mn} according to the formula (3.12) is small for shells of medium length. At the same time, the error due to disregard of values δ_P, δ_q in (3.30) is of the order $1/n^2$. In this case, it is reasonable to use the formulas (3.30) and (3.12) together with (3.31) and with a simplified expression for δ_q:

$$\delta_q = \frac{1}{(\beta_n R)^2}\left[-1 + \frac{\alpha_m{}^2 C_{66}(\alpha_m{}^2 C_{12} - \beta_n{}^2 C_{22})}{\alpha_m{}^4 C_{11} C_{66} + \beta_n{}^4 C_{22} C_{66} + \alpha_m{}^2 \beta_n{}^2 (C_{11} C_{22} - C_{12}{}^2 - 2C_{12} C_{66})}\right]. \tag{3.42}$$

3. Let us assume $\alpha_m \gg (1/R)$. As is shown in paragraph 3.1, the error when using the formulas (3.12) is of the order h^2/R^2 for any n values. The

value δ_P according to (3.31) is of the order $1/(\alpha_m R)^2$. Consequently, the technical theory and expression (3.30) for P^*_{mn} are applicable to the calculation of the critical loads in uniaxial compression. For shells loaded with external pressure, the given limiting case is of no interest (except the shells with $(\pi R/L) \gg 1$, which will not be treated here).

In conclusion to the above analysis, we can state that the only exception, when the expressions for ω^2_{mn} together with the simplest formulas for critical loads (3.34), derived in accordance with the technical theory, do not ensure sufficient accuracy in calculating the critical static loads, is that of low circumferential modes. For this case, special formulas ensuring the necessary calculation accuracy have been derived in this paragraph.

3.3 EFFECT OF TRANSVERSE SHEARS ON NATURAL FREQUENCIES

One of the drawbacks of all shell theories, based on the kinematic Kirchhoff-Love model is the disregard of transverse shear deformations. This factor can essentially affect the accuracy of calculation of the frequencies of natural vibrations, critical static loads and, as a consequence, the accuracy of the results in a wide class of more complicated dynamic problems. In this paragraph, a thorough quantitative study of the range of applicability of the technical theory, based on the Kirchhoff-Love model, for calculation of the natural frequencies of orthotropic cylindrical shells, is given.

Up to now the simplest theories based on the hypothesis about the linear distribution of tangential displacement components through the thickness (2.1) (a kinematic Timoshenko type model) and theories based on approximation of the transverse tangential stresses through the thickness by quadratic parabolas (a kinematic Ambartsumian's model)[20,21] have been most widely used in solving different designing problems for plates and shells with consideration of transverse shears.

The methods of calculation for the frequencies of natural vibrations for an orthotropic cylindrical shell of finite length under boundary conditions of free support are outlined in Ref. 310. The deformations of transverse shears have been taken into account on the basis of Timoshenko type theory. All the inertia terms have been taken into account also. The results of frequency calculations, corresponding mainly to flexural, torsional and longitudinal vibrations for a three-layer shell with orthotropic layers, whose elastic constants look rather artificial, were treated. A comparison with the results obtained on the basis of equations of elasticity theory[390] and according to the classical theory of shells was presented.

The effect of magnitudes of the moduli of transverse shears on the dynamic characteristics of orthotropic cylindrical shells was first investigated by A. E. Bogdanovich,[45] where the main regions of dynamic instability (the lower points of these regions represent the twofold frequencies of natural vibrations) were calculated according to the classical theory and refined theory of Ambartsumian.[20] The numerical results showed that the relative correction introduced into the values of natural vibrations by taking into account the transverse shears at other fixed parameters is inversely proportional to the values of moduli of transverse shears. It was also established by A. E. Bogdanovich[45] that this correction is appreciably increased with the growth of the circumferential wave number. Later on the problem of natural vibrations with transverse shears and inertia of rotation being taken into account has been investigated both for isotropic[419,423] and orthotropic laminated[331,405] cylindrical shells. In Ref. 331 it was evidently mentioned for the first time in foreign literature about this problem that the importance of shear deformations being taken into account increases for anisotropic structures with very low shear moduli.

Let us pass over to expounding the methods and results of calculating the frequencies of natural vibrations for orthotropic cylindrical shells on the basis of equations obtained in paragraph 2.3 with transverse shear deformations taken into account. In conformity with the assumptions of the technical theory, we shall disregard the term T'_{23}/R in the second equation of equilibrium (2.53). By inserting (2.54), (2.37) and (2.38) into (2.53), we obtain the following equations of free vibrations for a shell (primes at unknown functions in subsequent formulas will be omitted):

$$C_{11}\frac{\partial^2 u}{\partial x^2}+C_{66}\frac{\partial^2 u}{\partial y^2}+(C_{12}+C_{66})\frac{\partial^2 v}{\partial x\partial y}+\frac{C_{12}}{R}\frac{\partial w}{\partial x}=$$
$$=\rho h\left(\frac{\partial^2 u}{\partial t^2}+\frac{h^2}{12R}\frac{\partial^2 \gamma_x}{\partial t^2}\right);$$
$$(C_{12}+C_{66})\frac{\partial^2 u}{\partial x\partial y}+C_{22}\frac{\partial^2 v}{\partial y^2}+C_{66}\frac{\partial^2 v}{\partial x^2}+\frac{C_{22}}{R}\frac{\partial w}{\partial y}=$$
$$=\rho h\left(\frac{\partial^2 v}{\partial t^2}+\frac{h^2}{12R}\frac{\partial^2 \gamma_y}{\partial t^2}\right);$$
$$C_{13}\frac{\partial^2 w}{\partial x^2}+C_{23}\frac{\partial^2 w}{\partial y^2}-\frac{C_{22}}{R^2}w-\frac{C_{12}}{R}\frac{\partial u}{\partial x}-\frac{C_{22}}{R}\frac{\partial v}{\partial y}+$$
$$+C_{13}\frac{\partial \gamma_x}{\partial x}+C_{23}\frac{\partial \gamma_y}{\partial y}=\rho h\frac{\partial^2 w}{\partial t^2}; \qquad (3.43)$$

$$D_{11}\frac{\partial^2\gamma_x}{\partial x^2}+D_{66}\frac{\partial^2\gamma_x}{\partial y^2}-C_{13}\gamma_x+(D_{12}+D_{66})\frac{\partial^2\gamma_y}{\partial x\partial y}-C_{13}\frac{\partial w}{\partial x}=$$
$$=\rho\frac{h^3}{12}\left(\frac{\partial^2\gamma_x}{\partial t^2}+\frac{1}{R}\frac{\partial^2 u}{\partial t^2}\right);$$
$$D_{22}\frac{\partial^2\gamma_y}{\partial y^2}+D_{66}\frac{\partial^2\gamma_y}{\partial x^2}-C_{23}\gamma_y(D_{12}+D_{66})\frac{\partial^2\gamma_x}{\partial x\partial y}-C_{23}\frac{\partial w}{\partial y}=$$
$$=\rho\frac{h^3}{12}\left(\frac{\partial^2\gamma_y}{\partial t^2}+\frac{1}{R}\frac{\partial^2 v}{\partial t^2}\right).$$

By assuming that the conditions of free support (3.4) are fulfilled at the end faces of a shell, let us expand the unknown functions u, v, w in Fourier series (3.5) and represent, in analogous form, the unknown functions γ_x, γ_y:

$$\gamma_x(x\ y,t)=\sum_{m=1}^{\infty}\sum_{n=1}^{\infty}[X_{mn}(t)\cos\beta_n y+\tilde{X}_{mn}(t)\sin\beta_n y]\cos\alpha_m x;$$
$$\gamma_y(x\ y,t)=\sum_{m=1}^{\infty}\sum_{n=1}^{\infty}[Y_{mn}(t)\sin\beta_n y-\tilde{Y}_{mn}(t)\cos\beta_n y]\sin\alpha_m x. \tag{3.44}$$

Substitution of (3.5) and (3.44) into (3.43) leads to two that are identical in their form systems of equations in respect to U_{mn}, V_{mn}, W_{mn}, X_{mn}, Y_{mn} and $\tilde{U}_{mn}$, $\tilde{V}_{mn}$, $\tilde{W}_{mn}$, $\tilde{X}_{mn}$, $\tilde{Y}_{mn}$. Let us write down the first one:

$$-(\alpha_m^2C_{11}+\beta_n^2C_{66})U_{mn}+\alpha_m\beta_n(C_{12}+C_{66})V_{mn}+\frac{C_{12}}{R}\alpha_m W_{mn}=$$
$$=\rho h\left(\frac{d^2U_{mn}}{dt^2}+\frac{h^2}{12R}\frac{d^2X_{mn}}{dt^2}\right);$$
$$\alpha_m\beta_n(C_{12}+C_{66})U_{mn}-(\alpha_m^2C_{66}+\beta_n^2C_{22})V_{mn}-\frac{C_{22}}{R}\beta_n W_{mn}=$$
$$=\rho h\left(\frac{d^2V_{mn}}{dt^2}+\frac{h^2}{12R}\frac{d^2Y_{mn}}{dt^2}\right); \tag{3.45}$$
$$-\left(\frac{C_{22}}{R^2}+C_{13}\alpha_m^2+C_{23}\beta_n^2\right)W_{mn}+\frac{C_{12}}{R}\alpha_m U_{mn}-\frac{C_{22}}{R}\beta_n V_{mn}-$$
$$-C_{13}\alpha_m X_{mn}+C_{23}\beta_n Y_{mn}=\rho h\frac{d^2W_{mn}}{dt^2};$$
$$-(\alpha_m^2D_{11}+\beta_n^2D_{66}+C_{13})X_{mn}+\alpha_m\beta_n(D_{12}+D_{66})Y_{mn}-C_{13}\alpha_m W_{mn}=$$
$$=\rho\frac{h^3}{12}\left(\frac{d^2X_{mn}}{dt^2}+\frac{1}{R}\frac{d^2U_{mn}}{dt^2}\right);$$
$$\alpha_m\beta_n(D_{12}+D_{66})X_{mn}-(\alpha_m^2D_{66}+\beta_n^2D_{22}+C_{23})Y_{mn}+C_{23}\beta_n W_{mn}=$$
$$=\rho\frac{h^3}{12}\left(\frac{d^2Y_{mn}}{dt^2}+\frac{1}{R}\frac{d^2V_{mn}}{dt^2}\right).$$

The system of five ordinary differential eqns (3.45) can be reduced to a matrix form

$$\hat{M}\frac{d^2\mathbf{f}_{mn}}{dt^2}+\hat{G}_{mn}\mathbf{f}_{mn}=0, \tag{3.46}$$

where $\mathbf{f}_{mn}=\{U_{mn}, V_{mn}, W_{mn}, X_{mn}, Y_{mn}\}$, while $\hat{G}_{mn}$, $\hat{M}$ are symmetrical matrices with the following non-zero elements:

$$g_{mn}{}^{11}=\alpha_m{}^2C_{11}+\beta_n{}^2C_{66};\quad g_{mn}{}^{12}=-\alpha_m\beta_n(C_{12}+C_{66});$$
$$g_{mn}{}^{13}=-\frac{C_{12}}{R}\alpha_m;\quad g_{mn}{}^{22}=\alpha_m{}^2C_{66}+\beta_n{}^2C_{22};\quad g_{mn}{}^{23}=\frac{C_{22}}{R}\beta_n;$$
$$g_{mn}{}^{33}=\frac{C_{22}}{R^2}+C_{13}\alpha_m{}^2+C_{23}\beta_n{}^2;\quad g_{mn}{}^{34}=C_{13}\alpha_m;\quad g_{mn}{}^{35}=-C_{23}\beta_n;$$
$$g_{mn}{}^{44}=\alpha_m{}^2D_{11}+\beta_n{}^2D_{66}+C_{13};\quad g_{mn}{}^{45}=-\alpha_m\beta_n(D_{12}+D_{66});$$
$$g_{mn}{}^{55}=\alpha_m{}^2D_{66}+\beta_n{}^2D_{22}+C_{23};\quad m_{11}=m_{22}=m_{33}=\rho h;$$
$$m_{44}=m_{55}=\frac{\rho h^3}{12};\quad m_{14}=m_{25}=\frac{\rho h^3}{12R}.$$

Principally, the problem of finding out the frequencies of natural vibrations from the system (3.46) does not represent difficulties. Various approbated algorithms and standard programs, described, for instance, in Refs 246, 247 and 378 have been used for its solution (some numerical results are treated in paragraph 3.4). Bearing in mind the studies of the effect of transverse shears on the frequencies of natural flexural vibrations of a shell, let us restrict ourselves here to only one inertia term, entering the third equation, in the system (3.45). This will allow us to reduce this system to one equation of the type (3.11) where

$$\omega_{mn}{}^2=\frac{1}{\rho h}\left[\frac{\alpha_m{}^4C_{66}(C_{11}C_{22}-C_{12}{}^2)}{\Delta_{mn}}+\frac{A_{mn}}{B_{mn}}\right]; \tag{3.47}$$

Δ_{mn} is expressed according to (3.13), whereas

$$\begin{aligned}A_{mn}&=\alpha_m{}^6C_{13}D_{11}D_{66}+\alpha_m{}^4\beta_n{}^2[C_{13}(D_{11}D_{22}-D_{12}{}^2-2D_{12}D_{66})+\\&+C_{23}D_{11}D_{66}]+\alpha_m{}^2\beta_n{}^4[C_{13}D_{22}D_{66}+C_{23}(D_{11}D_{22}-D_{12}{}^2-2D_{12}D_{66})]+\\&+\beta_n{}^6C_{23}D_{22}D_{66}+C_{13}C_{23}[\alpha_m{}^4D_{11}+\beta_n{}^4D_{22}+2\alpha_m{}^2\beta_n{}^2(D_{12}+2D_{66})];\end{aligned} \tag{3.48}$$
$$\begin{aligned}B_{mn}&=\alpha_m{}^4D_{11}D_{66}+\alpha_m{}^2\beta_n{}^2(D_{11}D_{22}-D_{12}{}^2-2D_{12}D_{66})+\beta_n{}^4D_{22}D_{66}+\\&+\alpha_m{}^2(C_{13}D_{66}+C_{23}D_{11})+\beta_n{}^2(C_{13}D_{22}+C_{23}D_{66})+C_{13}C_{23}.\end{aligned}$$

In the limiting case, $C_{13}\to\infty$, $C_{23}\to\infty$, the formula (3.47) is transformed into (3.12).

Further we shall dwell on the solution of an analogous problem in terms of the equations of Ambartsumian's refined theory. The following representations of transversal stresses are assumed:

$$\sigma_{13}=\frac{1}{2}\left(\frac{h^2}{4}-z^2\right)\varphi(x, y, t); \quad \sigma_{23}=\frac{1}{2}\left(\frac{h^2}{4}-z^2\right)\psi(x, y, t). \tag{3.49}$$

Taking into account the assumptions of the technical theory, the displacements and deformations are written in the form

$$\begin{gathered}
u_1=u+z\left[-\frac{\partial w}{\partial x}+a_{55}\varphi\left(\frac{h^2}{8}-\frac{z^2}{6}\right)\right]; \\
u_2=v+z\left[-\frac{\partial w}{\partial y}+a_{44}\psi\left(\frac{h^2}{8}-\frac{z^2}{6}\right)\right]; \\
u_3=w; \\
\varepsilon_{11}=\frac{\partial u}{\partial x}+z\left[-\frac{\partial^2 w}{\partial x^2}+a_{55}\frac{\partial \varphi}{\partial x}\left(\frac{h^2}{8}-\frac{z^2}{6}\right)\right]; \\
\varepsilon_{22}=\frac{\partial v}{\partial y}+\frac{w}{R}+z\left[-\frac{\partial^2 w}{\partial y^2}+a_{44}\frac{\partial \psi}{\partial y}\left(\frac{h^2}{8}-\frac{z^2}{6}\right)\right]; \\
\varepsilon_{12}=\frac{\partial u}{\partial y}+\frac{\partial v}{\partial x}+z\left[-2\frac{\partial^2 w}{\partial x\partial y}+\right. \\
\left.+\left(a_{55}\frac{\partial \varphi}{\partial y}+a_{44}\frac{\partial \psi}{\partial x}\right)\left(\frac{h^2}{8}-\frac{z^2}{6}\right)\right],
\end{gathered} \tag{3.50}$$

where

$$a_{55}=\frac{1}{G_{13}}; \quad a_{44}=\frac{1}{G_{23}};$$

G_{i3} are the moduli of transverse shears. The transversal tractions and the moments, in such a way, are equal to

$$\begin{gathered}
T_{13}=\frac{h^3}{12}\varphi; \quad T_{23}=\frac{h^3}{12}\psi; \\
M_{11}=D_{11}\frac{\partial}{\partial x}\left(-\frac{\partial w}{\partial x}+a_{55}\frac{h^2}{10}\varphi\right)+ \\
+D_{12}\frac{\partial}{\partial y}\left(-\frac{\partial w}{\partial y}+a_{44}\frac{h^2}{10}\psi\right); \\
M_{22}=D_{12}\frac{\partial}{\partial x}\left(-\frac{\partial w}{\partial x}+a_{55}\frac{h^2}{10}\varphi\right)+ \\
+D_{22}\frac{\partial}{\partial y}\left(-\frac{\partial w}{\partial y}+a_{44}\frac{h^2}{10}\psi\right);
\end{gathered} \tag{3.51}$$

$$M_{12}=M_{21}=D_{66}\left[\frac{\partial}{\partial x}\left(-\frac{\partial w}{\partial y}+a_{44}\frac{h^2}{10}\psi\right)+\right.$$
$$\left.+\frac{\partial}{\partial y}\left(-\frac{\partial w}{\partial x}+a_{55}\frac{h^2}{10}\varphi\right)\right].$$

It is not difficult to establish the correlation between (3.51) and expressions having been used above in Timoshenko type theory. The transversal tractions and the moments in both cases are equal, if

$$\gamma_x=-\frac{\partial w}{\partial x}+a_{55}\frac{h^2}{10}\varphi;\quad \gamma_y=-\frac{\partial w}{\partial y}+a_{44}\frac{h^2}{10}\psi \tag{3.52}$$

and, besides,

$$C_{13}=k'hG_{13};\quad C_{23}=k''hG_{23};\quad k'=k''=\frac{5}{6}. \tag{3.53}$$

The equations of motion for a shell in respect to the functions u, v, w, ϕ, ψ, obtained through substitution of (3.51) into (2.53), under the condition (3.52), coincide with the equations (3.43) in which the condition (3.53) has been taken into account.

In such a way, when employing the refined theory of Ambartsumian the frequency of natural vibrations of an orthotropic cylindrical shell is determined according to the formula (3.47) with $k'=k''=5/6$. Let us note that the coefficient of shear $k=5/6$ was first introduced by E. Reissner in solving the bending problem for a plate. Earlier S. P. Timoshenko suggested a value $k=8/9$ for a beam of rectangular cross section.[439,440] In a paper by I. Mirsky and G. Herrmann[382] the condition of equality between two magnitudes of shear wave velocity was employed for finding k' and k''. The first of these velocities was obtained from the equations of motion for a shell. The second one was an asymptotic velocity, obtained as a limiting value from the equations of the three-dimensional theory of elasticity in the case of infinitesimal wave lengths. It was established that the values k' and k'' depend on the circumferential wave number n. For not very high n and thin shells the value $k'=k''=(\pi/\sqrt{12})$ was proposed which differs slightly from 5/6.

Let us pass over to a comparison of the calculational results for the frequencies of natural vibrations of orthotropic shells, obtained by the formulas (3.48) and (3.12). Let us designate the values of frequencies corresponding to these formulas by ω_{mn}^C and ω_{mn}^T and introduce the relative difference in frequencies $\delta(m,n)=\omega_{mn}^T/\omega_{mn}^C-1$.

The numerical results have been obtained for shells of a unidirectionally reinforced carbon plastic having the characteristics (3.23). Then both cases

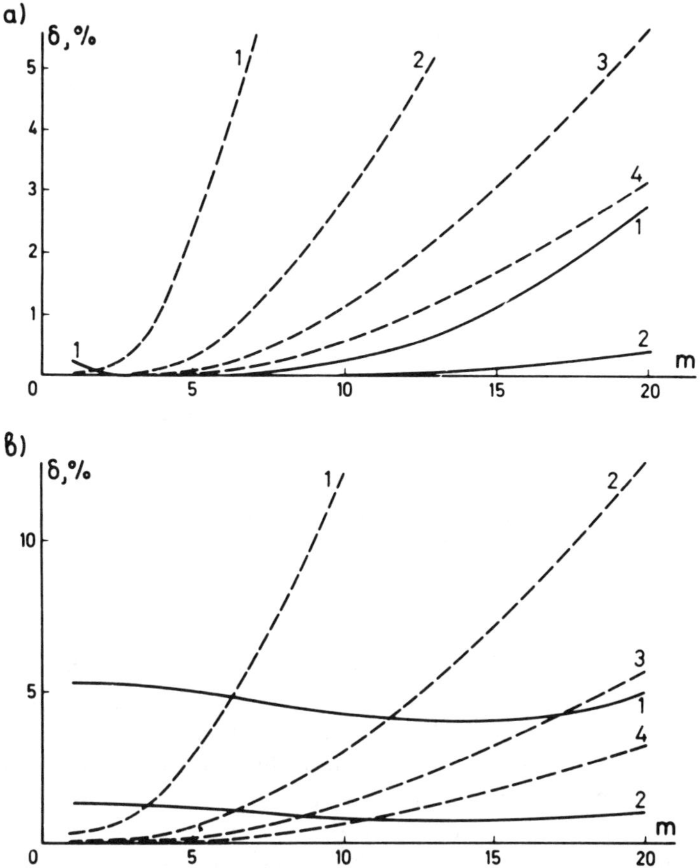

Fig. 3.6. The $\delta(m)$ functions for shells with circumferential (———) and longitudinal (– – –) reinforcement at $n = 3$ (a) and $n = 10$ (b); $R/h = 50$ (1); 100 (2); 150 (3); and 200 (4).

of reinforcing the shell along the generatrix and in the circumferential direction are considered. In all calculational versions $k' = k'' = 5/6$, $R = 1$ m, $L/R = 2$. Let us note that the parameters L and m enter the formulas for frequencies only in combination $\pi mR/L$, so that the dependencies on the value m at fixed L are equivalent to their dependencies on $1/L$ at fixed m.

In Fig. 3.6, the dependence of δ on the number of axial harmonic m are presented at $G_{12}/G_{i3} = 1$ and two values of n. As is shown, in the case of longitudinal reinforcement, δ increases monotonously with increasing m for all the variants considered. The character of these dependencies in the case of circumferential reinforcement is essentially different: δ changes

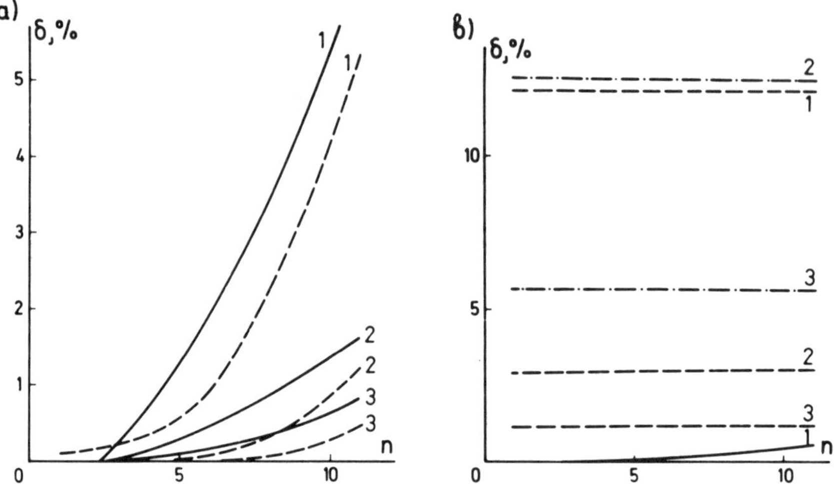

Fig. 3.7. The $\delta(n)$ functions for shells with circumferential (a) and longitudinal (b) reinforcement at $m = 1$ (———); 10 (– – –), 20 (–•–). The notations are the same as in Fig. 3.6.

non-monotonously and the functions presented here in all cases have a minimum. It can be also noted that at $n = 3$ the value of δ is considerably higher for the shell with longitudinal reinforcement, whereas at $n = 10$, up to $m = 6$, δ is higher in the case of circumferential reinforcement.

The dependence of δ on the number of circumferential harmonics for $G_{12}/G_{i3} = 1$ is presented in Fig. 3.7. As is seen, in the case of circumferential reinforcement, the dependencies are of monotonously increasing character, whereas in the case of longitudinal reinforcement δ does not practically change with increasing n.

The following general conclusion can be drawn from the results above according to the correction introduced by taking the transverse shears into account in calculating the frequency of natural vibrations. Firstly, the higher the elastic moduli ratio in the longitudinal and circumferential directions of a shell is, the faster this correction increases with increasing m. Secondly, contrary to the previous case, the higher the elastic moduli ratio in the circumferential and longitudinal direction is, the faster the correction increases with increasing n. This conclusion is confirmed by the analysis of the formula (3.48) in the limiting cases $C_{11} \gg C_{22}$ and $C_{22} \gg C_{11}$. Let us also note that for shells with comparatively slightly differing moduli E_1 and E_2 the dependencies $\delta(m)$ and $\delta(n)$ are qualitatively the same: they both increase monotonously. The GFRP shell of such type with $E_2/E_1 = 2$ was considered by A. E. Bogdanovich.[45]

Let us note that the numerical results obtained by S. B. Dong and F. K. W. Tso[310] at $R/h = 10$, are contradictory to the formulated conclusion: for the shell with $E_1/E_2 \approx 11$ the relative correction of the classical theory increases from 0.3% at $n = 2$ to 70.5% at $n = 10$ ($L/h = 100$) and from 3.1% at $n = 2$ to 50% at $n = 10$ ($L/h = 10$). Besides, in accordance with Ref. 310, the correction increases monotonously at $n = 2$ (this corresponds to the results of Fig. 3.6*a*) with decreasing L (in our case, this corresponds to increasing m), but has a minimum as to L at $n = 10$ (we obtained the relationships of such nature only for $E_2 \geq E_1$).

The complex of results, presented in Figs. 3.6 and 3.7 allows to evaluate the ranges of applicability of the classical theory for calculation of the frequencies of natural vibrations of orthotropic cylindrical shells with $G_{i3} = G_{12}$. Thus in the case of circumferential reinforcement within the range of wave numbers $m = 1$–20 and $n = 1$–10, the correction does not exceed 1.3% at $R/h = 100$ and 5.3% — at $R/h = 50$. In the case of longitudinal reinforcement the correction does not exceed 3% within the same range of variation of m and n at $R/h = 200$, but at $R/h = 100$ exceeds 5% at $m \geq 13$. Let us note that the length of axial half-wave $\lambda_m \approx 16\,\mathrm{h}$ corresponds to the value $m = 13$.

In Fig. 3.8, the dependencies of δ on the shell thickness for the case of circumferential and longitudinal reinforcements are presented. As is seen, their character in both cases is the same and corresponds to *a priori* considerations: the correction, introduced by taking the transverse shears into account, increases monotonously approximately as $(h/R)^2$ (such a result was stated by A. E. Bogdanovich[45]) with increasing thickness.

As it follows from Fig. 3.9, δ increases with decreasing moduli of transverse shears G_{i3} of the shell, thereby the dependence $\delta(G_{12}/G_{i3})$ is very close to the linear one (it was noted by A. E. Bogdanovich)[45]. The given circumstance allows to introduce a correlation of the calculated ranges of applicability of the classical theory in dependence on the values of transverse shear moduli, by using only the basic results (Figs 3.6–3.8) obtained for $G_{i3} = G_{12}$.

The main conclusion of the analysis presented above is that the correction in the results of calculation of the frequencies of natural vibrations depends for orthotropic shells on moduli ratios E_1/E_2 and G_{i3}/G_{12}. The estimated ranges of applicability of the classical theory, obtained on the basis of calculations made for isotropic shells and pertaining only to geometric parameters, longitudinal and circumferential wave numbers are inapplicable for shells of anisotropic materials.

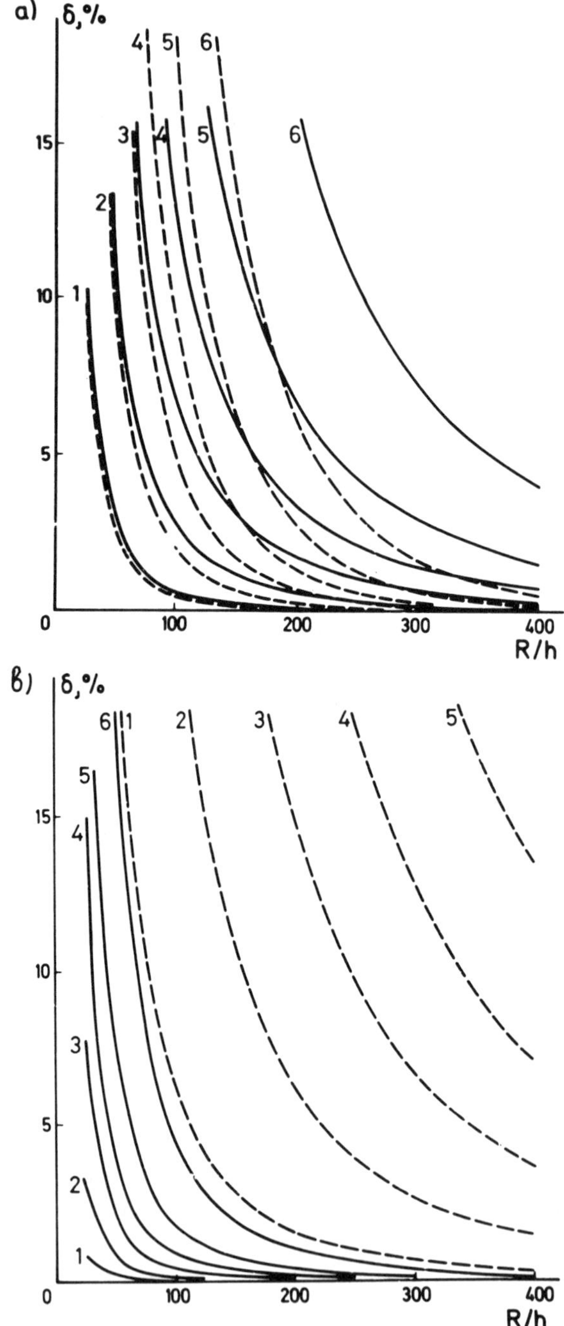

Fig. 3.8. The $\delta(R/h)$ functions for shells with circumferential (a) and longitudinal (b) reinforcement at $n = 10$; $m = 1$ (———) and 20 (– – –). Numbers at the curves correspond to the magnitudes of G_{12}/G_{i3}: 0.5 (1); 2 (2); 5 (3); 10 (4); 20 (5); and 50 (6).

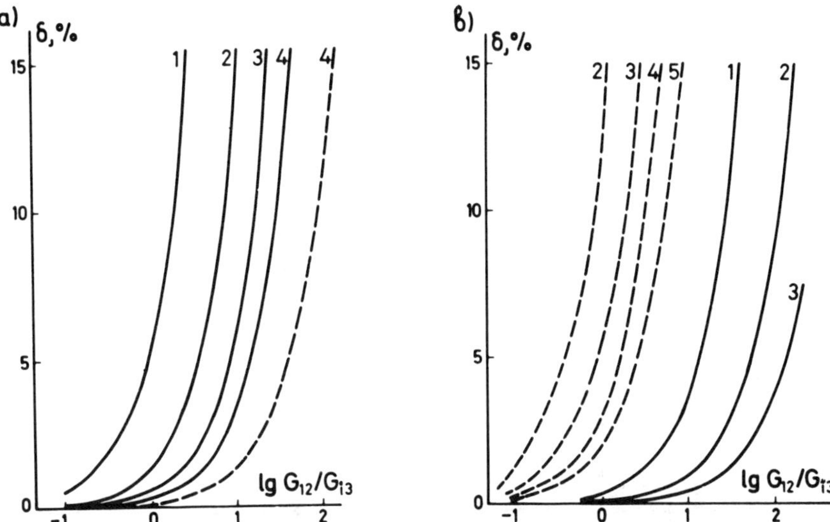

Fig. 3.9. The $\delta(\lg G_{12}/G_{i3})$ functions for shells with circumferential (a) and longitudinal (b) reinformcement at $n = 10$, $m = 1$ (———) and 20 (– – –). Numbers at the curves correspond to the magnitudes of R/h: 50 (1); 100 (2); 150 (3); 200 (4) and 250 (5).

3.4 EFFECT OF TANGENTIAL INERTIA AND INERTIA OF ROTATION ON NATURAL FREQUENCIES

Let us revert to matrix eqn (3.46). Let us multiply both its parts by matrix $\hat{M}^{-1}$ inverse to matrix $\hat{M}$ and represent the vector $\mathbf{f}_{mn}$ in the form

$$\mathbf{f}_{mn}(t) = \mathbf{F}_{mn} e^{i\omega_{mn} t}. \tag{3.54}$$

As a result of the insertion of (3.54) into (3.46) we obtain the equation

$$(\hat{A}_{mn} - \lambda_{mn} \hat{I}) \mathbf{F}_{mn} = 0, \tag{3.55}$$

where $\hat{A}_{mn} = \hat{M}^{-1}\hat{G}_{mn}$; $\hat{I}$ is a unit matrix; $\lambda_{mn} = \omega_{mn}^2$.

The problem of calculation of the eigenvalues λ_{mn} and eigenvectors $\mathbf{F}_{mn}$ is sufficiently well worked out. We have used for this sake the software for EC computers. To increase confidence of the results, some variants of standard programs, corresponding to various algorithms, described by J. Wilkinson,[246] were used.

Henceforth we shall concentrate on the analysis of five natural frequencies, obtainable from eqn (3.55) for a fixed pair of wave numbers $\{m, n\}$, without considering especially their identification (i.e., their correspondence to a definite kind of vibration). Speaking henceforth

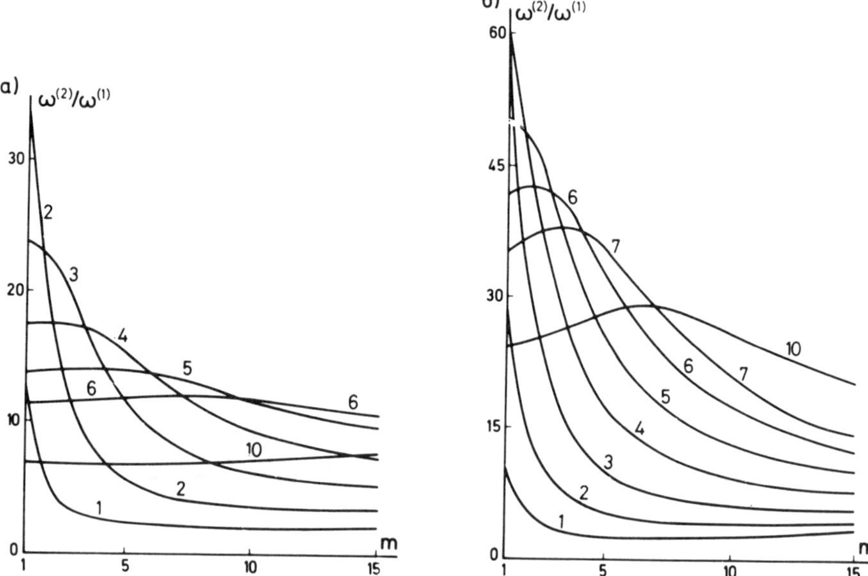

Fig. 3.10. The dependencies of $\omega^{(2)}/\omega^{(1)}$ on m for a shell with circumferential (a) and longitudinal (b) reinforcement at $L/R = 10$, $R/h = 100$. Numbers at the curves correspond to the magnitudes of n.

about the lowest natural frequency — $\omega_{mn}^{(1)}$ and highest frequencies $\omega_{mn}^{(2)}$, $\omega_{mn}^{(3)}$, $\omega_{mn}^{(4)}$, $\omega_{mn}^{(5)}$, we shall bear in mind that the latter two frequencies are connected with the inertia of rotation of the normal element, but the frequency $\omega_{mn}^{(3)}$ corresponds preferably to longitudinal vibrations. As to the frequencies $\omega_{mn}^{(1)}$ and $\omega_{mn}^{(2)}$, then within definite ranges of the parameters involved into calculation, it is possible by using special analysis to establish correlation of each of these frequencies preferably either to flexural or rotational vibrations. But there could also be such situations when it is impossible to make the identification of these two frequencies due to closeness of two corresponding amplitudes of vibrations. Let us add that such combinations of parameters of the shell are possible, at which preferably the torsional mode of vibrations corresponds to the lowest frequency $\omega_{mn}^{(1)}$. This is typical for axisymmetric ($n = 0$) and beam-type ($n = 1$) vibrations.[288,324] The problem of identification of the first three frequencies to definite modes of vibrations were first considered by M. L. Baron and H. H. Bleich,[288] where the frequencies $\omega_{mn}^{(1)}$, $\omega_{mn}^{(2)}$, $\omega_{mn}^{(3)}$ and the relations of torsional, flexural and longitudinal vibration amplitudes were tabulated for an isotropic cylindrical shell.

Let us dwell on the results of numerical calculations for carbon plastic

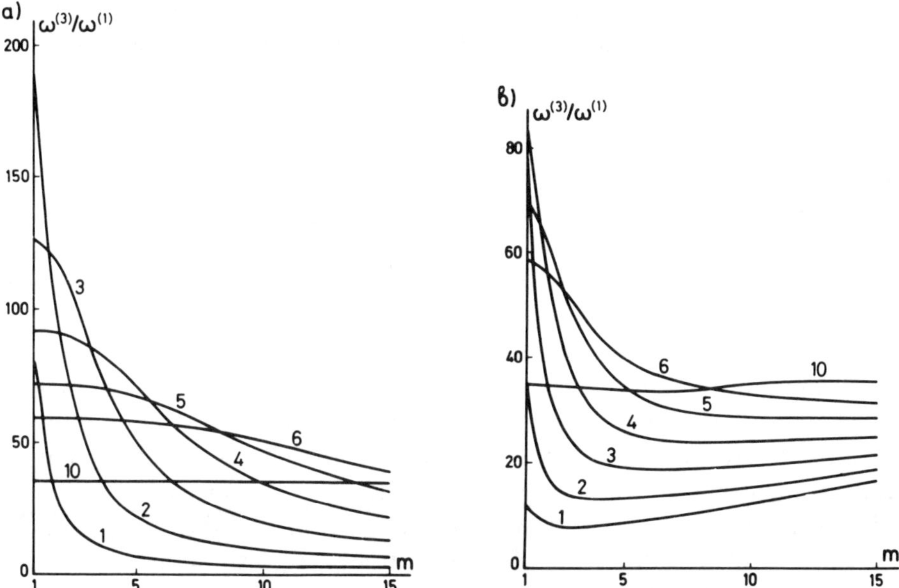

Fig. 3.11. The dependencies of $\omega^{(3)}/\omega^{(1)}$ on m for a shell with circumferential (a) and longitudinal (b) reinforcement at $L/R = 10$, $R/h = 100$. Numbers at the curves correspond to the magnitudes of n.

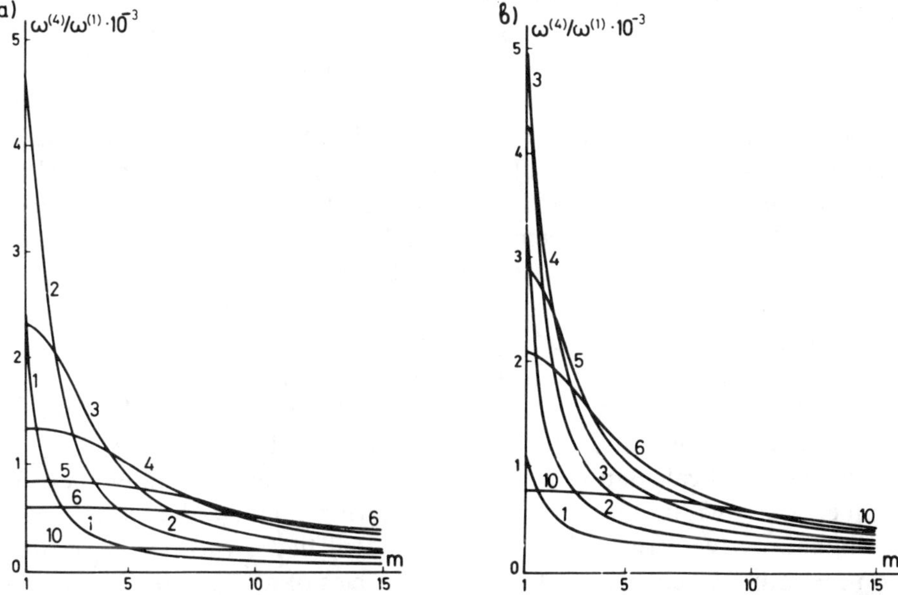

Fig. 3.12. The dependencies of $\omega^{(4)}/\omega^{(1)}$ on m for a shell with circumferential (a) and longitudinal (b) reinforcement at $L/R = 10$, $R/h = 100$. Numbers at the curves correspond to the magnitudes of n.

shells with the material characteristics (3.23). As before, let us treat the two extreme cases of unidirectional layer orientation: along the generatrix and along the circumference.

In Figs. 3.10–3.12 dependencies of the relations $\omega^{(2)}/\omega^{(1)}$, $\omega^{(3)}/\omega^{(1)}$, $\omega^{(4)}/\omega^{(1)}$ on the number of axial harmonic m, for various numbers of circumferential harmonic n at $L/R = 10$, $R/h = 100$ are presented. We shall not treat the relation $\omega^{(5)}/\omega^{(1)}$ specially, since the frequency $\omega^{(5)}$ is very close to $\omega^{(4)}$.

As is seen from Fig. 3.10, the character of relations $(\omega^{(2)}/\omega^{(1)})(m)$ differ qualitatively for $n = 1$–3 and for $n \geq 6$: in the first case, there is a minimum as to m, in the second one — the maximum. The positions of these minimum and maximum with increasing n are transferred to higher values of m. Let us note that for shells of medium length the minimum corresponds to $m = 2$ at $n = 1$, $m = 3$ at $n = 2$ for circumferential reinforcement and $m = 1$ at $n = 1$, $m = 2$ at $n = 2$ — for axial one.

Further we shall pay attention to the fact that at $n \geq 3$ the frequency $\omega^{(2)}$ exceeds $\omega^{(1)}$ in a wide range of m magnitudes no less than 5-fold. The flexural mode of vibrations corresponds preferentially to the frequency $\omega^{(1)}$. In accordance with the results presented in Ref. 288, in a series of other works and our calculations, the amplitude of the displacement component w in such cases exceeds four times, as a minimum, the amplitude of displacement component v.

The curves presented in Fig. 3.11(a), are quantitatively identical to the curves of Fig. 3.10(a). Let us only note that the relationship $\omega^{(3)}/\omega^{(1)}$ is greater than $\omega^{(2)}/\omega^{(1)}$ and $\omega^{(3)}/\omega^{(2)}$ is increasing with the growth of n. The curves indicated in Fig. 3.11(b) are rather different from the corresponding curves in Fig. 3.10(b) and 3.11(a). Namely, the minimum is reached at much lower values of m in this case. As a result, for the shells of medium length with an axial reinforcement the relationship $\omega^{(3)}/\omega^{(1)}$ is increasing with the increase in m for the majority of n-values.

It is obvious from comparison of Fig. 3.10 and 3.11 that the frequencies $\omega^{(3)}$ exceed considerably $\omega^{(2)}$ practically within the whole range of variation of parameters m and n. It follows directly from this fact that the amplitude value of displacement u is less than the amplitude value of displacement v. The ratio u/v is the smaller, the higher is $\omega^{(3)}/\omega^{(2)}$.

The results of calculating $\omega^{(4)}/\omega^{(1)}$, presented in Fig. 3.12, show that the frequency $\omega^{(4)}$ within the whole range of varying m and n exceeds $\omega^{(1)}$, as a minimum, by two orders. It can be also noted that a rather complicated non-monotonous character of the dependencies of $\omega^{(4)}/\omega^{(1)}$ on m and n is in general features analogous to the character revealed in Figs. 3.10 and 3.11. Let us note in addition that the frequencies $\omega^{(4)}$ and $\omega^{(1)}$ become closer with decreasing R/h and G_{i3}/G_{12}; the frequencies $\omega^{(5)}$, $\omega^{(4)}$ usually differ only in the third significant digit.

Further let us investigate the effect of tangential inertia and inertia of rotation on the value of the lowest natural frequency $\omega^{(1)}$. In the absence of all these inertia factors, $\omega^{(1)}$ is calculated by the formula (3.47). If we take into account the longitudinal and circumferential inertia but neglect inertia of rotation, then $\omega^{(1)}$ is obtained from the eqn (3.55); dimensionality of the problem is three. Further we shall designate the respective value of $\omega^{(1)}$ by $\omega_{\mathrm{TI}}^{(1)}$. If we take the inertia of rotation into account but disregard longitudinal and circumferential inertia, then $\omega^{(1)}$ is also determined from eqn (3.55) (with another matrix $\hat{A}$ and vector $\mathbf{F}$); dimensionality of the problem in this case is three also. The corresponding value of $\omega^{(1)}$ is designated as $\omega_{\mathrm{RI}}^{(1)}$. Let us introduce the relative corrections into the values of the lowest natural frequency by taking these two separate groups of inertia terms into account:

$$\Delta_{\mathrm{TI}} = \frac{\omega^{(1)} - \omega_{\mathrm{TI}}{}^{(1)}}{\omega^{(1)}} \cdot 100\%; \quad \Delta_{\mathrm{RI}} = \frac{\omega^{(1)} - \omega_{\mathrm{RI}}{}^{(1)}}{\omega^{(1)}} \cdot 100\% \tag{3.56}$$

The dependencies of value Δ_{TI} on n and m for carbon plastic shells with longitudinal and circumferential reinforcement are presented in Fig. 3.13. The main conclusion is that the maximum of Δ_{TI}, approximately 30%, is

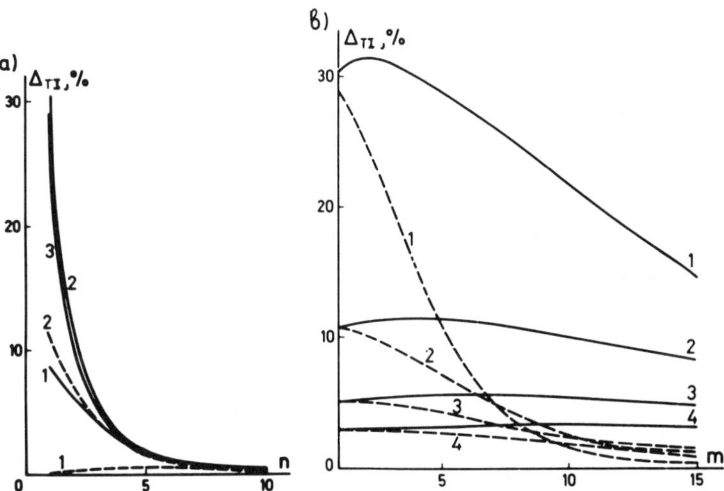

Fig. 3.13. The relative correction for the lowest natural frequency, introduced with tangential inertia taken into account, depending on n (a) and m (b) for a shell with circumferential (———) and longitudinal (– – –) reinforcement; $m = 1$, $R/h = 50$ (a) and $R/h = 50$, $L/R = 10$ (b). Numbers at the curves correspond to $L/R = 0.5$ (1); 2 (2); and 10 (3) for (a) and to magnitudes of n for (b).

reached at $n = 1$. The effect of tangential inertia decreases sharply with increasing n. In the case of circumferential reinforcement, the presence of the maximum on the curves $\Delta_{TI}(m)$ also attracts attention. Thereby, the effect of tangential inertia will reach a maximum for shells of fairly definite length (decreasing with growing n). In the case of axial reinforcement, the curves $\Delta_{TI}(m)$ decrease monotonously at all the treated values of n. Let us note that in the case of circumferential reinforcement Δ_{TI} is higher than in the case of axial reinforcement; the difference increases with increasing m and decreasing n.

The dependence of Δ_{RI} on n and m in the cases of circumferential and axial reinforcements are presented in Fig. 3.14. As is seen, Δ_{RI} is considerably lower than Δ_{TI} and does not exceed 0.5% at $R/h = 50$, $G_{i3}/G_{12} = 0.1$ within the investigated range of variation of n and m. Let us note that in the case of circumferential reinforcement Δ_{RI} depends slightly on n, but increases considerably with increasing m. In the case of axial reinforcement, on the contrary, the dependence $\Delta_{RI}(m)$ is insignificant, but the growth of Δ_{RI} with increasing n is evident. As is seen from the results, the value of Δ_{RI} increases with decreasing R/h. Let us add that the dependence $\Delta_{RI}(G_{12}/G_{i3})$ appears close to the linear one. Taking this result into account, one can compare the values of Δ_{RI} (Fig. 3.14) and δ (Figs 3.6 and 3.7) and conclude that, if other conditions are equal, Δ_{RI} is approximately two orders lower than δ. This quantitative conclusion

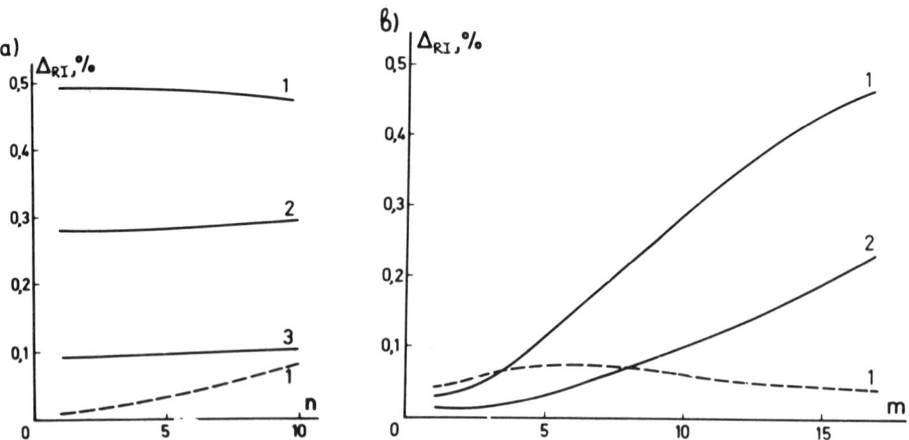

Fig. 3.14. The relative correction of the lowest natural frequency, introduced with inertia of rotation taken into account, depending on n (a) and m (b) for a shell with circumferential (———) and longitudinal (– – –) reinforcement at $L/R = 2$; $G_{i3}/G_{12} = 0.1$; $m = 20$ (a); $n = 5$ (b). Numbers at the curves correspond to $R/h = 50$ (1); 100 (2); and 200 (3).

confirms the results obtained earlier for isotropic[344] and orthotropic[45] cylindrical shells.

Let us summarize the analysis of the effects associated with the tangential inertia terms and inertia of rotation presented in this paragraph. The most important correction factor in solving the problems of vibrations, dynamic stability, non-axisymmetric dynamic deformation of orthotropic cylindrical shells could be, obviously, the circumferential inertia. The frequency of natural vibrations $\omega^{(2)}$ and amplitude of corresponding displacement v are the closest in magnitude to respective characteristics of flexural motion $\omega^{(1)}$ and w. The circumferential inertia affects also most strongly the frequency of natural flexural vibrations of a shell.

As it follows from the numerical results, the effect of circumferential inertia is most strongly pronounced for lowest circumferential harmonics. In solving the problems, in which the harmonics with $n \geq 3$ play the main role and the flexural deformations prevail, it is, evidently, possible to neglect the circumferential inertia.

The effects of taking the longitudinal inertia into account, from the qualitative viewpoint, are considerably weaker than those of taking the circumferential inertia into account: the frequency $\omega^{(3)}$ is usually considerably higher than $\omega^{(2)}$, but a contribution of the longitudinal inertia into the correction Δ_{TI}, as it was indicated by special calculations, is considerably smaller than the contribution of circumferential inertia.

Taking into account the inertia of rotation gives two additional frequencies for each pair of values $\{m, n\}$ which over the range of shell parameters and modes of vibration exceed the lowest frequency $\omega^{(1)}$, as a minimum, by two orders. As a result, the inertia of rotation affects very slightly the amplitude values of displacements of a shell. The correction introduced into the value of $\omega^{(1)}$, with the inertia of rotation taken into account, is also negligible. Let us add that the effects, associated with the inertia of rotation, gain force with the increase in relative thickness of a shell and decrease of transverse shear moduli.

The analysis performed has also shown that the quantitative effects associated with tangential inertia and inertia of rotation, depend greatly on the parameters of the material orthotropy. Therefore, the evaluation of the possibility of neglecting one or another inertia terms in particular problems, being obtained from the analysis of an isotropic shell, can be erroneous for shells of fibrous composites.

4

Parametric Vibrations of Orthotropic Cylindrical Shells

The case of a circular cylindrical shell loaded with uniformly distributed at the end faces axial vibrational tractions is under consideration. The tractions lead, at definite combinations of amplitude and frequency, to parametric resonance. The methods of calculating the dynamic instability regions (DIR) of an orthotropic elastic and viscoelastic cylindrical shells are treated. Thereby, the transverse shear deformations are taken into account under the assumption that the viscoelastic material properties are revealed only under the effect of shear stresses. The solutions, by the use of the Bubnov-Galerkin method, are elaborated for the problems of parametric vibrations of elastic and viscoelastic shells in geometrically non-linear statement. The characteristic features of the state of stress, having arisen in a shell at various regimes of parametric resonance are analyzed. The possibilities of a purposeful variation of the lowest frequency part of the DIR spectra by varying the mechanical characteristics of a composite material, the number of layers and layer layup angles in a package are illustrated. The statement of the problems of optimal design of the laminated composite cylindrical shells is given. Several versions of physical restrictions reflecting the peculiarities of parametric resonance are considered.

At the juncture of two problems — natural vibrations and static stability of thin-wall structural elements — there arises the problem of parametric vibrations. Starting with N. M. Belyaev's[41] work, the parametric vibrations became the subject of numerous investigations as applied to diverse mechanical systems with distributed parameters (in particular, to bars, plates and shells). The results of these investigations were summarized in a number of reviews, monographs and manuals.[39,65,72,79,268,317,377] In Ref. 268 there is practically a complete list of literature on this problem, having been published prior to 1972. A great number of papers of a mathematical character was devoted to the study of stability of ordinary differential equations with periodic coefficients and systems of such equations. This mathematical problem arises, particularly, from the mechanical problem of parametric vibrations. Let us mention here only the first fundamental researches,[11,183,235] a review[234] and a monograph[272] in which the mathematical results on this problem have been summarized.

The fact that in neglecting tangential inertia and inertia of rotation the boundaries of the DIR for a thin-wall structural element are fully determined by the frequency spectra and critical static loads were reported by N. M. Belyaev, B. A. Bodner and V. N. Tchelomei in the study of parametric vibrations of bars and plates. This permitted, already in the first work by A. N. Markov[186] to obtain a sufficiently complete solution of the linearized problem for an orthotropic cylindrical shell — to derive the formula for calculating the boundaries of the principal, the second, and the third DIR. From among subsequent works on the determination of DIR for cylindrical shells let us note the work of J. C. Yao[460] which is comprised of extensive numerical material; Ref. 449 in which the calculated results were compared with the experimental data from kinematic excitation of parametric vibrations; papers[214,257,258] which were aimed at studying the effect of damping on DIR boundaries; a theoretical treatise[188] showing that the DIR boundaries are greatly affected by eccentricity of load application; and Refs 116 and 168, where the effect of initial imperfections on DIR was treated. In Ref. 45 the effect of transverse shear deformations as well as the effects of tangential inertia and inertia of rotation on the location and width of DIR for orthotropic elastic cylindrical shells was studied. Analogous problems have been subsequently treated in Ref. 418.

Calculations and experiments on parametric vibrations of viscoelastic bars were presented in Refs 187 and 427–430. In Ref. 46, a problem of calculating DIR for a cylindrical shell of the viscoelastic material was first treated. In Ref. 50, the effects of the reinforcing fiber elastic characteristics, the reinforcement coefficient, the number of layers and their layup angles in a multilayered package in the low-frequency part of DIR spectra were investigated. Several glass fiber, carbon and boron plastic cylindrical shells were considered. On the basis of the results, obtained in Ref. 50, the problems of optimal design of reinforced multi-layer cylindrical shells with physical constraints, reflecting specific features of parametric resonance have been formulated in Ref. 51. Let us also mention the experimental researches on parametric vibrations of glass fiber reinforced plastic cylindrical shells.[77,169]

On the whole, by now the problem of defining the boundaries of DIR for cylindrical shells (elastic, viscoelastic, orthotropic, multilayer) has been thoroughly studied. The theoretical results are fairly well confirmed by experimental data.

Prior to directing our attention to the problem of non-linear parametric vibrations, let us briefly dwell upon important, from a methodological viewpoint, problems of calculating natural and forced non-linear vibrations of cylindrical shells. The first works of E. I. Grigolyuk[133] and I. I. Vorovich[106] should be mentioned. In Ref. 133 a problem of forced vibrations in freely supported circular cylindrical panels was solved on the basis of equations of medium bending of shallow shells. In Ref. 106 the problems of existence and uniqueness of solution for the equations describing non-linear vibrations of shallow shells were investigated. A series of works started by H. N. Chu,[303] J. L. Nowinski,[392] D. A. Evensen[318] were devoted to the problem of free and forced non-linear vibrations of a closed circular cylindrical shell. Their attention has been focussed on the calculation of an amplitude-frequency response and discussion of the question — which kind of non-linearity ('rigid' or 'soft') should it be referred to? In the first two studies[303,392] a shell was considered as a system having only one degree of freedom, and the obtained Duffing type equation assessed the non-linearity of a 'rigid' type. Thereby in Ref. 303 the condition of continuity of circumferential displacement was not satisfied while in Ref. 392 the equality to zero of shell deflection at the end faces was not fulfilled. On the basis of two-term approximation of deflection the solution has been obtained in Ref. 318 which satisfies both these requirements. In this paper a shell was also reduced to the system with one degree of freedom whereas the equation obtained for vibration amplitude was of essentially another type: it contained the non-linear inertia terms. As a result of the solution of this equation the amplitude-frequency characteristic of a 'soft' type was obtained. The existence of such a type of characteristic was confirmed in M. D. Olson's experiments[396] on very thin shells. A further evolution of Evensen's approach was given in Refs 319 and 320.

The work by E. H. Dowell and C. S. Ventres[313] was a principal step forward. A shell was considered as a system with three degrees of freedom; the conditions of continuity of circumferential displacement and the equality of axial displacement at the end faces to zero were satisfied in an integral sense. It has been noted in this work that Evensen's solution does not satisfy the condition of equality to zero of bending moment at the end faces and some definite kind of boundary condition imposed on longitudinal displacement. The approach presented in Ref. 313 was further developed by S. Atluri,[287] where on the basis of the same three-term approximation of deflection, the amplitude-frequency curves have been calculated. It was shown that they refer to a 'rigid' type (more pronounced than in solutions[303,392]). The theoretical and experimental study of the same problem was carried on in work by T. Matsuzaki and S. Kobayashi.[372] The solution was based on the approximation of deflection, permitting to satisfy the conditions of full end restraint. The calculated and experimentally obtained amplitude-frequency curves for very thin shells were of a 'soft' type. Let us also mention,[327,328] where the solution was obtained on the basis of a variational principle by the use of a disturbance method. In a first approximation, the condition of equality of axial force to zero at the end was satisfied. It has been marked that both 'soft' and 'rigid' types of non-linearities are possible. According to J. H. Ginsberg's[328] viewpoint, the difference between his results and those obtained by S. Atluri[287] was not surprising, since imposing of the constraint on the axial displacement mentioned above makes the shell more rigid. The non-linear vibrations of cylindrical shells were also considered in Refs 339 and 401.

As a result of investigations in the works mentioned, the authors have failed to find the answer on the most principal question: at what parameters of a shell and modes of vibration does either 'soft' or 'rigid' amplitude-frequency characteristics exist? The work by J. C. Chen and C. D. Babcock[301] has thrown some light on this problem. The solution was obtained by the use of the disturbance method with a small parameter, equal to the relation of deflection versus radius of the shell. The possibility of non-linearity of both 'rigid' and 'soft' types, depending on the circumferential wave number, has been theoretically established and experimentally confirmed.

It must be especially emphasized that one of the aspects of the problem — interaction of non-axisymmetric modes of vibrations corresponding to different circumferential spatial harmonics — was not investigated in any of the above mentioned works.

Among the first works on non-linear parametric vibrations in thin-wall structural elements the papers by V. V. Bolotin[64] and F. Weidenhammer[452] should be mentioned. In V. V. Bolotin's monograph[65] the principal statements of the theory of non-linear parametric vibrations of mechanical systems have been outlined. In particular, the admissibility of assessing the boundaries of DIR when using the linearized equations of motion was postulated.

Early in the 60s, V. Ts. Gnuni[119,121,122] and G. V. Mishenkov[190,191] obtained a number of solutions on the problem of geometrically non-linear parametric vibrations of shallow cylindrical shells. In Ref. 30, an analogous solution of the problem for a shell of revolution was performed. In subsequent numerous works the non-linear parametric vibrations of a closed cylindrical shell[32,103,120,402,461] were treated. In Refs 119, 121, 122, 190, 191, the solutions were obtained on the basis of the approximation of radial deflection by only one term taken of double Fourier series. The deflection approximation and the procedure of subsequent calculations, used in Refs 32, 103, 120 correspond to the method used in Ref. 4 for the problem of dynamic stability of a cylindrical shell loaded with axial compression and external pressure. In all the works mentioned above, the initial system of equations of motion was ultimately reduced to some modifications of the non-linear ordinary differential equation with a periodic coefficient. Certain qualitative conclusions as regards the character of the amplitude-frequency characteristic have been drawn from the analysis of such equations, performed by different methods.

Let us especially mark the solution,[461] in which the well-known approximation of deflection according to Karman-Chien was used with retaining three degrees of freedom. Through numerical integration of the system of non-linear ordinary differential equations obtained in Ref. 461 the dependence of deflection on time was calculated. It was of a typical beating character in the cases if the exciting force fell within DIR. It has been noted in Ref. 461 that the additional two terms in the approximation having been introduced with the aim to reflect the specific non-linear effects, do not practically affect the character of the solution. The latter result served as a starting point for formulation in our paper[53] of the new approach to the solution of non-linear parametric vibration problems. This approach principally differed from anything used previously. It was generalized later in Ref. 54 for viscoelastic shells.

In a series of works[78,110,114] the effect of simultaneous excitation of several spatial forms of vibrations, corresponding to different numbers of a circumferential harmonic was experimentally established. Earlier it was predicted theoretically in Ref. 53. It can be concluded, on the whole, that in the last years the problem of non-linear parametric vibrations has attracted great interest both in theoretical and experimental aspects.

4.1 DETERMINATION OF DYNAMIC INSTABILITY REGIONS FOR AN ELASTIC SHELL

Let us consider an orthotropic circular cylindrical shell loaded at both end faces by equal uniformly distributed longitudinal forces

$$T_{11}{}^0 \Big|_{x=0,L} = -(P_0 + P_t \cos \Theta t). \tag{4.1}$$

Under the assumption that the radial displacement of end faces is not constrained, the forced vibrations of a shell can be described by the system of momentless equations of motion:

$$C_{11} \frac{\partial^2 u_0}{\partial x^2} + \frac{C_{12}}{R} \frac{\partial w_0}{\partial x} = \mu \frac{\partial^2 u_0}{\partial t^2}; \tag{4.2}$$

$$\frac{C_{22}}{R^2} w_0 + \frac{C_{12}}{R} \frac{\partial u_0}{\partial x} = -\mu \frac{\partial^2 w_0}{\partial t^2}; \quad \mu = \rho h.$$

The solution of (4.2) with the boundary conditions (4.1) has the form

$$u_0 = \frac{P_0 C_{22}}{C_{11} C_{22} - C_{12}{}^2} \left(\frac{L}{2} - x \right) + \frac{P_t (C_{22} - R^2 \mu \Theta^2)}{\lambda (C_{11} C_{22} - C_{12}{}^2 - C_{11} R^2 \mu \Theta^2)} \times$$

$$\times \left(\frac{1 - \cos \lambda L}{\sin \lambda L} \cos \lambda x - \sin \lambda x \right) \cos \Theta t; \tag{4.3}$$

$$w_0 = \frac{P_0 C_{12} R}{C_{11} C_{22} - C_{12}{}^2} + \frac{P_t C_{12} R}{C_{11} C_{22} - C_{12}{}^2 - C_{11} R^2 \mu \Theta^2} \times$$

$$\times \left[\cos \lambda x + \frac{\sin \lambda x}{\sin \lambda L} (1 - \cos \lambda L) \right] \cos \Theta t.$$

The longitudinal and circumferential tractions arising in a shell can be written in the form

$$T_{11}{}^0 = -P_0 - P_t \left[\cos \lambda x + \frac{\sin \lambda x}{\sin \lambda L} (1 - \cos \lambda L) \right] \cos \Theta t; \tag{4.4}$$

$$T_{22}{}^0 = \frac{P_t C_{12} R^2 \mu \Theta^2}{C_{11} C_{22} - C_{12}{}^2 - C_{11} R^2 \mu \Theta^2} \left[\cos \lambda x + \frac{\sin \lambda x}{\sin \lambda L} (1 - \cos \lambda L) \right] \cos \Theta t,$$

where

$$\lambda = \Theta \left[\frac{\Theta^2 - \dfrac{C_{22}}{R^2 \mu}}{\Theta^2 \dfrac{C_{11}}{\mu} - \dfrac{C_{11} C_{22} - C_{12}{}^2}{R^2 \mu^2}} \right]^{1/2}. \tag{4.5}$$

The momentless stress–strain state of a shell, described by formulas (4.3) and (4.4), is nonuniform as to the x coordinate. However, this nonuniformity depends essentially on the frequency Θ of exciting force. Thus, within the range of frequency, bounded from above by the condition

$$\Theta \ll \frac{1}{R} \left(\frac{C_{22}}{\mu} \right)^{1/2} = \frac{1}{R} \left[\frac{E_2}{\rho (1 - \nu_{12} \nu_{21})} \right]^{1/2}, \tag{4.6}$$

the nonuniformity appears to be weak, since

$$\lambda L \approx \left(\frac{C_{22}\mu L^2\Theta^2}{C_{11}C_{22}-C_{12}^2} \right)^{1/2} \ll \frac{L}{R} \frac{C_{22}}{\sqrt{C_{11}C_{22}-C_{12}^2}}. \tag{4.7}$$

Really, if the right-hand part of (4.7) is of the order of unity, then $\lambda L \ll 1$ and in accordance with formulas (4.3) and (4.4)

$$\begin{aligned} u_0 &\approx \frac{C_{22}}{C_{11}C_{22}-C_{12}^2} \left(\frac{L}{2} - x \right) (P_0 + P_t \cos \Theta t); \\ w_0 &\approx \frac{C_{12}R}{C_{11}C_{22}-C_{12}^2} (P_0 + P_t \cos \Theta t); \\ T_{11}{}^0 &\approx -(P_0 + P_t \cos \Theta t); \\ T_{22}{}^0 &\approx \frac{C_{12}R^2\mu\Theta^2}{C_{11}C_{22}-C_{12}^2} P_t \cos \Theta t. \end{aligned} \tag{4.8}$$

Let us note that the ratio of the amplitude of T_{22}^0 to the amplitude of the vibrational component of T_{11}^0 is

$$\frac{C_{12}R^2\mu\Theta^2}{C_{11}C_{22}-C_{12}^2} \ll 1. \tag{4.9}$$

The forced axisymmetric vibrations of a shell, loaded with vibrational longitudinal force, at definite combinations of Θ, P_t, P_0 can be unstable even if very small 'disturbances' of various kinds exist, because these 'disturbances' initiate flexural deformations. The unstable vibrations of such a nature refer to the class of 'parametrically excited vibrations' (the terminology was introduced in Ref. 28), or are simply called 'parametric' vibrations.

The procedure for solving the geometrically linear problem of parametric vibrations in any thin-wall structural element is carried out as follows. At first, the linearized equations of motion are reduced to ordinary differential equations with periodic coefficients (the simplest particular cases being Mathieu and Hill's equations). Further through studying the stability of the solutions of these equations, the boundaries of DIR are calculated.

If the amplitude and frequency of the external load correspond to the point within DIR, the parametric vibrations of the structure are of resonance character — their amplitude increases unrestrictedly in time. If they correspond to the point outside DIR, the amplitude of the parametric vibrations will remain sufficiently small during any time interval. In a particular case, when the appearance of parametric vibrations is initiated by the initial deflection w^0, the amplitude of deflection for parametric vibrations outside DIR is usually of the order w.

In such a way, the study of parametric vibrations in the geometrically linear statement allows one to separate resonance and out-of resonance regimes. This study is a necessary preliminary stage in solving the geometrically non-linear problem of calculating the amplitudes of vibrations.

Let us consider the stability problem for axisymmetric forced vibrations of an orthotropic cylindrical shell, describing with the formulas (4.8). Let us suppose that non-axisymmetric 'disturbances' exist.

Further we shall use the linearized equations of Timoshenko type improved theory. Taking into account the results of paragraph 3.4, we shall henceforth disregard the tangential inertia terms and the inertia of rotation. This simplification is valid in the study of the low-frequency part of DIR spectrum corresponding mainly to the flexural form of vibrations. Let us also omit the tangential 'loading' terms X_1 and X_2. Taking into account the fact that in axisymmetric momentless state there exist tractions $T_{11}^0(t)$ and $T_{22}^0(t)$, determined by the formulas (4.8), we have for X_3 in accordance with (2.55) the following expression

$$X_3 = T_{11}{}^0 \frac{\partial^2 w}{\partial x^2} + T_{22}{}^0 \left[\frac{\partial^2 w}{\partial y^2} - \frac{1}{R}\left(\frac{\partial v}{\partial y} + \frac{\partial u}{\partial x} \right) \right] .$$

Bearing in mind that the parametric vibrations under investigation here are mainly of a flexural type, let us employ the simplifying expression for X_3 corresponding to the technical theory approach:

$$X_3 = T_{11}{}^0 \frac{\partial^2 w}{\partial x^2} + T_{22}{}^0 \frac{\partial^2 w}{\partial y^2} . \tag{4.10}$$

A further solution is based on the system (3.43) through addition to the left-hand part of the third equation, the expression (4.10). If we assume that at the end faces of a shell there exist the conditions of free support (3.4) and expand the unknown functions into series (3.5) and (3.44), then for every fixed pair of $\{m, n\}$ we obtain an ordinary differential equation

$$\frac{d^2 W_{mn}}{dt^2} + \Omega_{mn}{}^2 \left\{ 1 - 2\delta_{mn} \cos \Theta t \times \right.$$
$$\left. \times \left[1 - \left(\frac{\beta_n}{\alpha_m} \right)^2 \frac{C_{12} R^2 \mu \Theta^2}{C_{11} C_{22} - C_{12}{}^2} \right] \right\} W_{mn} = 0, \tag{4.11}$$

where

$$\Omega_{mn}{}^2 = \omega_{mn}{}^2 - \frac{\alpha_m{}^2}{\mu} P_0; \quad \delta_{mn} = \frac{P_t}{2(P^*{}_{mn} - P_0)}; \quad P^*{}_{mn} = \frac{\omega_{mn}{}^2 \mu}{\alpha_m{}^2} . \tag{4.12}$$

Here, Ω_{mn} is the frequency of vibrations of a shell, loaded with axial static compressive force P_0; P^*_{mn} is the critical axial compressive force. With (4.9) taken into account, one can simplify the equation (4.11) by introducing the condition

$$\left(\frac{\beta_n}{\alpha_m}\right)^2 \frac{C_{12}R^2\mu\Theta^2}{C_{11}C_{22}-C_{12}^2} \ll 1, \tag{4.13}$$

which imposes at $\beta_n > \alpha_m$ a more rigid restriction from above on the frequency Θ than the restriction (4.9). Let us remark that by omitting the second term in square brackets in the formula (4.11), we somehow overestimate the magnitude of P_t. Thus in fulfilling (4.13) we arrive at Mathieu's equation

$$\frac{d^2 W_{mn}}{dt^2} + \Omega_{mn}{}^2(1-2\delta_{mn}\cos\Theta t)\,W_{mn}=0. \tag{4.14}$$

The equations for lines, separating the regions of stable and unstable solutions of the eqn (4.14), are well-known (they were presented, in particular, in a monograph by V. V. Bolotin)[65]. The principal DIR is encompassed between the $\Theta^{(1)}(\delta)$ curves specified by the equations

$$\Theta_1{}^{(1)} = 2\Omega\sqrt{1+\delta}; \quad \Theta_2{}^{(1)} = 2\Omega\sqrt{1-\delta}. \tag{4.15}$$

The second DIR — between the curves

$$\Theta_1{}^{(2)} = \Omega\sqrt{1+\frac{\delta^2}{3}}; \quad \Theta_2{}^{(2)} = \Omega\sqrt{1-2\delta^2}, \tag{4.16}$$

the third — between the curves

$$\Theta_1{}^{(3)} = \frac{2}{3}\Omega\sqrt{1-\frac{9\delta^2}{8+9\delta}}; \quad \Theta_2{}^{(3)} = \frac{2}{3}\Omega\sqrt{1-\frac{9\delta^2}{8-9\delta}}. \tag{4.17}$$

As is seen from (4.15) to (4.17), in the given statement of the problem, the boundaries of DIR for a fixed pair of $\{m, n\}$ numbers are completely determined by only two parameters of a shell: frequency ω_{mn} and critical force P^*_{mn}, specified by the formulas (3.47) and (3.34). On employing the kinematic Kirchhoff-Love model the eqn (4.14) and formulas (4.15)–(4.17) remain without changes, while ω_{mn} is calculated according to (3.12). Respectively, in calculating the DIR boundaries on the basis of Flügge's theory the expressions (3.14) and (3.30) should be used. We can conclude then that the error in calculating the DIR boundaries according to one or another theory is completely determined by the error made in calculating the values ω_{mn} and P^*_{mn}. Both of these values were thoroughly investigated in Chapter 3.

4.2 DETERMINATION OF DYNAMIC INSTABILITY REGIONS FOR A VISCOELASTIC SHELL

The problem of parametric vibrations of a viscoelastic orthotropic cylindrical shell was first dealt with in Ref. 46 on the basis of linearized equations of Ambartsumian's improved theory. It was assumed that the viscoelastic properties of the material display themselves only at shear deformations (both in the plane of a layer and in transverse directions). Such a model was first employed by G. I. Brizgalin[86] in creep calculations for an orthotropic plate and G. A. Teters and B. L. Pelekh[244] in static stability calculations for a viscoelastic orthotropic cylindrical shell.

Let us assume the relation between shear stresses and deformations in a form

$$\sigma_{12}=\hat{G}_{12}\varepsilon_{12};\quad \sigma_{13}=k'\hat{G}_{13}\varepsilon_{13};\quad \sigma_{23}=k''\hat{G}_{23}\varepsilon_{23}, \tag{4.18}$$

where $\hat{G}_{12}$, $\hat{G}_{13}$, $\hat{G}_{23}$ are integral operators:

$$\begin{aligned} \hat{G}_{12}&=G_{12}{}^0-(G_{12}{}^0-G_{12}{}^\infty)\int_0^t R_6(t-\tau)\,d\tau;\\ \hat{G}_{13}&=G_{13}{}^0-(G_{13}{}^0-G_{13}{}^\infty)\int_0^t R_5(t-\tau)\,d\tau;\\ \hat{G}_{23}&=G_{23}{}^0-(G_{23}{}^0-G_{23}{}^\infty)\int_0^t R_4(t-\tau)\,d\tau; \end{aligned} \tag{4.19}$$

$R_4(t)$, $R_5(t)$, $R_6(t)$ are the kernels of relaxation, while G^0_{12}, G^0_{13}, G^0_{23} and G^∞_{12}, G^∞_{13}, G^∞_{23} are instantaneous and long-term values of shear moduli, respectively.

Assuming the improved Timoshenko type theory as a basis, let us present the linearized equations of motion for a viscoelastic shell in the same form as (3.43), only replacing the constants of the material C_{66}, C_{13}, C_{23}, D_{66}, by integral operators $\hat{C}_{66}$, $\hat{C}_{13}$, $\hat{C}_{23}$, $\hat{D}_{66}$. These operators are determined through (4.19) in an obvious way. Confining in this system to the inertia term, entering the third equation, adding the load $X_3=-(P_0+P_t\cos\Theta t)(\partial^2 w/\partial x^2)$ to the left-hand part of this equation and substituting into the system obtained, the expansions of unknown functions (3.5) and (3.44) we get the system of equations:

$$-(\alpha_m^2 C_{11}+\beta_n^2 \hat{C}_{66})\,U_{mn}+\alpha_m\beta_n\,(C_{12}+\hat{C}_{66})\,V_{mn}+\frac{C_{12}}{R}\,\alpha_m W_{mn}=0;$$

$$\alpha_m\beta_n\,(C_{12}+\hat{C}_{66})\,U_{mn}-(\alpha_m^2\hat{C}_{66}+\beta_n^2 C_{22})\,V_{mn}-\frac{C_{22}}{R}\,\beta_n W_{mn}=0;$$

$$-\left(\frac{C_{22}}{R^2}+\hat{C}_{13}\alpha_m^2+\hat{C}_{23}\beta_n^2\right)W_{mn}+\frac{C_{12}}{R}\,\alpha_m U_{mn}-\frac{C_{22}}{R}\,\beta_n V_{mn}- \tag{4.20}$$

$$-\hat{C}_{13}\alpha_m X_{mn}+\hat{C}_{23}\beta_n Y_{mn}+(P_0+P_t\cos\Theta t)\,\alpha_m^2 W_{mn}=\mu\frac{d^2W_{mn}}{dt^2};$$

$$-(\alpha_m^2 D_{11}+\beta_n^2\hat{D}_{66}+\hat{C}_{13})\,X_{mn}+\alpha_m\beta_n\,(D_{12}+\hat{D}_{66})\,Y_{mn}-\hat{C}_{13}\alpha_m W_{mn}=0;$$

$$\alpha_m\beta_n\,(D_{12}+\hat{D}_{66})\,X_{mn}-(\alpha_m^2\hat{D}_{66}+\beta_n^2 D_{22}+\hat{C}_{23})\,Y_{mn}+\hat{C}_{23}\beta_n W_{mn}=0.$$

In further solution, we shall employ the same approach as in Ref. 46. We shall apply Laplace transformation to the first, second, fourth and fifth equations of the system (4.20) and express the Laplace transform of functions $\bar{U}_{mn}(p)$, $\bar{V}_{mn}(p)$, $\bar{X}_{mn}(p)$, $\bar{Y}_{mn}(p)$ in terms of $\bar{W}_{mn}(p)$ (p is the parameter of Laplace transformation):

$$\bar{U}_{mn}=\bar{\Gamma}_{mn}{}^{(1)}\bar{W}_{mn};\qquad \bar{V}_{mn}=\bar{\Gamma}_{mn}{}^{(2)}\bar{W}_{mn};$$
$$\bar{X}_{mn}=\bar{\Gamma}_{mn}{}^{(5)}\bar{W}_{mn};\qquad \bar{Y}_{mn}=\bar{\Gamma}_{mn}{}^{(4)}\bar{W}_{mn}, \tag{4.21}$$

where

$$\bar{\Gamma}_{mn}{}^{(1)}(p)=\frac{R\alpha_m\bar{C}_{66}(\alpha_m^2 C_{12}-\beta_n^2 C_{22})}{\bar{\Delta}_{mn}};$$

$$\bar{\Gamma}_{mn}{}^{(2)}(p)=-R\beta_n\frac{\alpha_m^2(C_{11}C_{22}-C_{12}^2)+\beta_n^2 C_{22}\bar{C}_{66}-\alpha_m^2 C_{12}\bar{C}_{66}}{\bar{\Delta}_{mn}};$$

$$\bar{\Gamma}_{mn}{}^{(4)}(p)=\frac{\beta_n}{\bar{B}_{mn}}\{\alpha_m^2[\bar{C}_{23}D_{11}-\bar{C}_{13}(D_{12}+\bar{D}_{66})]+\beta_n^2\bar{C}_{23}\bar{D}_{66}+$$
$$+\bar{C}_{23}\bar{C}_{13}\}; \tag{4.22}$$

$$\bar{\Gamma}_{mn}{}^{(5)}(p)=-\frac{\alpha_m}{\bar{B}_{mn}}\{\alpha_m^2\bar{C}_{13}\bar{D}_{66}+\beta_n^2[\bar{C}_{13}D_{22}-\bar{C}_{23}(D_{12}+\bar{D}_{66})]+$$
$$+\bar{C}_{23}\bar{C}_{13}\}.$$

The values $\bar{\Delta}_{mn}$ and $\bar{B}_{mn}$ are determined by the formulas (3.13) and (3.48) upon substitution of C_{66}, C_{13}, C_{23}, D_{66} by $\bar{C}_{66}$, $\bar{C}_{13}$, $\bar{C}_{23}$, $\bar{D}_{66}$. The latter values have the form

$$\bar{C}_{66}=C_{66}{}^0-(C_{66}{}^0-C_{66}{}^\infty)\bar{R}_6;\qquad \bar{C}_{13}=C_{13}{}^0-(C_{13}{}^0-C_{13}{}^\infty)\bar{R}_5;$$
$$\bar{C}_{23}=C_{23}{}^0-(C_{23}{}^0-C_{23}{}^\infty)\bar{R}_4;\qquad \bar{D}_{66}=D_{66}{}^0-(D_{66}{}^0-D_{66}{}^\infty)\bar{R}_6. \tag{4.23}$$

And thus, in accordance with (4.21), the original functions U_{mn}, V_{mn}, X_{mn}, Y_{mn} can be represented in the form of convolution integrals:

$$\begin{aligned} U_{mn}(t) &= \int_0^t \Gamma_{mn}^{(1)}(t-\tau)\, W_{mn}(\tau)\, d\tau; \\ V_{mn}(t) &= \int_0^t \Gamma_{mn}^{(2)}(t-\tau)\, W_{mn}(\tau)\, d\tau; \\ X_{mn}(t) &= \int_0^t \Gamma_{mn}^{(5)}(t-\tau)\, W_{mn}(\tau)\, d\tau; \\ Y_{mn}(t) &= \int_0^t \Gamma_{mn}^{(4)}(t-\tau)\, W_{mn}(\tau)\, d\tau. \end{aligned} \tag{4.24}$$

Substitution of (4.24) into the third equation of the system (4.20) leads to the following integro-differential equation:

$$\frac{d^2 W_{mn}}{dt^2} + Q_{mn} W_{mn} - \frac{1}{\mu}(P_0 + P_t \cos \Theta t)\alpha_m^2 W_{mn} - \\ - \int_0^t S_{mn}(t-\tau) W_{mn}(\tau)\, d\tau - \int_0^t \int_0^\tau T_{mn}(t-\tau, \tau - s) W_{mn}(s)\, d\tau ds = 0, \tag{4.25}$$

where

$$Q_{mn} = \frac{1}{\mu}\left(\frac{C_{22}}{R^2} + C_{13}{}^0 \alpha_m{}^2 + C_{23}{}^0 \beta_n{}^2 \right);$$

$$S_{mn}(t) = \frac{1}{\mu}\Big[(C_{13}{}^0 - C_{13}{}^\infty)\alpha_m{}^2 R_5(t) + (C_{23}{}^0 - C_{23}{}^\infty)\beta_n{}^2 R_4(t) + \\ + \frac{C_{12}}{R}\alpha_m \Gamma_{mn}{}^{(1)}(t) - \frac{C_{22}}{R}\beta_n \Gamma_{mn}{}^{(2)}(t) - C_{13}{}^0 \alpha_m \Gamma_{mn}{}^{(5)}(t) + \\ + C_{23}{}^0 \beta_n \Gamma_{mn}{}^{(4)}(t) \Big]; \tag{4.26}$$

$$T_{mn}(t,\tau) = \frac{1}{\mu}[(C_{13}{}^0 - C_{13}{}^\infty)\alpha_m R_5(t) \Gamma_{mn}{}^{(5)}(\tau) - \\ - (C_{23}{}^0 - C_{23}{}^\infty)\beta_n R_4(t) \Gamma_{mn}{}^{(4)}(\tau)].$$

If we confine ourselves to the viscoelastic behavior of the material under shear in $\{x,y\}$ plane only, then we have to insert $C_{13}^0 = C_{13}^\infty$, $C_{23}^0 = C_{23}^\infty$ into expressions (4.26). Then we have

$$S_{mn}(t)=\frac{1}{\mu}\left[\frac{C_{12}}{R}\alpha_m\Gamma_{mn}{}^{(1)}(t)-\frac{C_{22}}{R}\beta_n\Gamma_{mn}{}^{(2)}(t)-\right.$$

$$\left.-C_{13}{}^0\alpha_m\Gamma_{mn}{}^{(5)}(t)+C_{23}{}^0\beta_n\Gamma_{mn}{}^{(4)}(t)\right]; \qquad (4.27)$$

$$T_{mn}(t,\tau)\equiv 0,$$

and the eqn (4.25) will take the form

$$\frac{d^2W_{mn}}{dt^2}+Q_{mn}W_{mn}-\frac{1}{\mu}(P_0+P_t\cos\Theta t)\alpha_m{}^2W_{mn}-$$

$$-\int_0^t S_{mn}(t-\tau)\,W_{mn}(\tau)\,d\tau=0. \qquad (4.28)$$

The expressions for $\bar{\Gamma}^{(1)}_{mn}$, $\bar{\Gamma}^{(2)}_{mn}$ (4.22) are preserved, while $\bar{\Gamma}^{(4)}_{mn}$, $\bar{\Gamma}^{(5)}_{mn}$ are written down in the form:

$$\bar{\Gamma}_{mn}{}^{(4)}(p)=\frac{\beta_n}{\bar{B}_{mn}}\{\alpha_m{}^2[C_{23}D_{11}-C_{13}(D_{12}+\bar{D}_{66})]+\beta_n{}^2C_{23}\bar{D}_{66}+C_{23}C_{13}\};$$

$$\bar{\Gamma}_{mn}{}^{(5)}(p)=-\frac{\alpha_m}{\bar{B}_{mn}}\{\alpha_m{}^2C_{13}\bar{D}_{66}+\beta_n{}^2[C_{13}D_{22}-C_{23}(D_{12}+\bar{D}_{66})]+C_{23}C_{13}\}$$

$$(C_{23}=C_{23}{}^0;\ C_{13}=C_{13}{}^0). \qquad (4.29)$$

Let us present now the calculation of a viscoelastic shell made on the basis of the classical Kirchhoff-Love theory. In this case, the viscoelastic behavior of the material can be taken into account only in respect of shear deformation in $\{x,y\}$ plane. It is necessary then to substitute the modulus G_{12} by integral operator $\hat{G}_{12}$ in the linearized equations of motion (4.19). Let us supplement the left-hand part of the third equation in the system (3.10) by the term $(P_0+P_t\cos\Theta t)\alpha_m^2W_{mn}$. The expressions for $U_{mn}(t)$ and $V_{mn}(t)$, having been obtained from the first two equations, take the form (4.24). On inserting them into the third equation and after a series of manipulations, we obtain the following integro-differential equation

$$\frac{d^2W_{mn}}{dt^2}+\Omega_{mn}{}^2(1-2\delta_{mn}\cos\Theta t)\,W_{mn}-g_{mn}{}^{(1)}\int_0^t R(t-\tau)\,W_{mn}(\tau)\,d\tau-$$

$$-g_{mn}{}^{(2)}\int_0^t \chi_{mn}(t-\tau)\,W_{mn}(\tau)\,d\tau=0, \qquad (4.30)$$

in which the notations have been introduced:

$$g_{mn}{}^{(1)}=\frac{4\alpha_m{}^2\beta_n{}^2}{\mu}\,(D_{66}{}^0-D_{66}{}^\infty);$$

$$g_{mn}{}^{(2)}=\frac{R^2(C_{66}{}^0-C_{66}{}^\infty)\,\alpha_m{}^6\beta_n{}^2(C_{11}C_{22}-C_{12}{}^2)^2}{\mu\Delta_{mn}{}^2}\;. \tag{4.31}$$

The values Ω_{mn}, δ_{mn} and P^*_{mn} are determined by the formulas (4.12); ω_{mn} and Δ_{mn} by (3.12) and (3.13). The transform of kernel $\chi_{mn}(t)$ is equal to

$$\bar{\chi}_{mn}(p)=\frac{\bar{R}(p)}{1-\bar{R}(p)\,\dfrac{(C_{66}{}^0-C_{66}{}^\infty)\,(\alpha_m{}^4C_{11}+\beta_n{}^4C_{22}-2\alpha_m{}^2\beta_n{}^2C_{12})\,R^2}{\Delta_{mn}}}\;. \tag{4.32}$$

In the particular case of the exponential kernel of relaxation

$$R(t)=\frac{1}{\eta}\,e^{-\frac{t}{\eta}} \tag{4.33}$$

according to formula (4.32) we obtain

$$\bar{\chi}_{mn}(p)=\frac{1}{\eta\left(p+\dfrac{1}{\beta_{mn}}\right)}\,, \tag{4.34}$$

where

$$\beta_{mn}=\frac{\eta}{1-\dfrac{(C_{66}{}^0-C_{66}{}^\infty)\,(\alpha_m{}^4C_{11}+\beta_n{}^4C_{22}-2\alpha_m{}^2\beta_n{}^2C_{12})\,R^2}{\Delta_{mn}}}\;. \tag{4.35}$$

In such a way,

$$\chi_{mn}(t)=\frac{1}{\eta}e^{-(t/\beta_{mn})}$$

and eqn (4.30) takes the form

$$\frac{d^2W_{mn}}{dt^2}+\Omega_{mn}{}^2(1-2\delta_{mn}\cos\Theta t)\,W_{mn}-\frac{g_{mn}{}^{(1)}}{\eta}\int_0^t e^{-\frac{t-\tau}{\eta}}\,W_{mn}(\tau)\,d\tau-$$

$$-\frac{g_{mn}{}^{(2)}}{\eta}\int_0^t e^{-\frac{t-\tau}{\beta_{mn}}}\,W_{mn}(\tau)\,d\tau=0. \tag{4.36}$$

In the case of a purely elastic behavior of the material we have $G^0_{12}=G^\infty_{12}$, $g^{(1)}_{mn}=g^{(2)}_{mn}=0$ and eqn (4.36) turns into (4.14).

If we introduce the notation

$$\Pi_{mn}(t) = g_{mn}^{(1)} R(t) + g_{mn}^{(2)} \chi_{mn}(t), \tag{4.37}$$

then (4.30) can be written as follows

$$\frac{d^2 W_{mn}}{dt^2} + \Omega_{mn}^2 (1 - 2\delta_{mn} \cos \Theta t) W_{mn} - \int_0^t \Pi_{mn}(t-\tau) W_{mn}(\tau) d\tau = 0. \tag{4.38}$$

The eqn (4.24) is close in its form to the one obtained by V. I. Matyash[187] where the problem of parametric vibrations of a thin viscoelastic bar was solved. The stability of solution of eqn (4.38) has been thoroughly studied and it has been shown that the regions of unboundedly increasing and attenuating solutions are separated by the lines, on which the solutions are close to the periodic ones with the periods $4\pi/\Theta$ and $2\pi/\Theta$. The analogous result was established earlier by V. V. Bolotin[65] in studying the differential equation

$$f'' + 2\varepsilon f' + \Omega^2 (1 - 2\delta \cos \Theta t) f = 0, \tag{4.39}$$

describing the parametric vibrations of a single-mass system with the viscous friction forces. Under the condition that (4.39) has a periodic solution with a period $4\pi/\Theta$, this solution can be written down as

$$f(t) = \sum_{k=1,3,5}^{\infty} \left(a_k \sin \frac{k\Theta t}{2} + b_k \cos \frac{k\Theta t}{2} \right). \tag{4.40}$$

In an analogous form the solution of integro-differential equation (4.38) was undertaken by V. I. Matyash.[187] In this paper, the system of uniform algebraic equations in respect to a_k and b_k from which the equations for the boundaries of DIR follow, was presented.

The boundaries of DIR for viscoelastic structural elements are determined, in the general case, in terms of rather complicated transcendental equations and we failed in our attempt to express the critical frequency Θ in terms of disturbance coefficient δ in an obvious form. Thus, for example, restricting ourselves to the first approximation $k = 1$ in (4.40), for the case of exponential kernel (4.33) we obtained the following equation for the boundaries of the main DIR:

$$\delta_{mn}^2 = \left\{ 1 - \left(\frac{\Theta}{2\Omega_{mn}} \right)^2 - \frac{g_{mn}^{(1)}}{\Omega_{mn}^2 \left[1 + \left(\frac{\Theta\eta}{2} \right)^2 \right]} - \right.$$

$$
-\frac{g_{mn}^{(2)}\beta_{mn}}{\Omega_{mn}^{2}\left[1+\left(\frac{\Theta\beta_{mn}}{2}\right)^{2}\right]\eta}\Bigg\}^{2}+\left\{\frac{g_{mn}^{(1)}\eta\Theta}{2\Omega_{mn}^{2}\left[1+\left(\frac{\Theta\eta}{2}\right)^{2}\right]}+\right.
$$

$$
\left.+\frac{g_{mn}^{(2)}\beta_{mn}^{2}\Theta}{2\Omega_{mn}^{2}\left[1+\left(\frac{\Theta\beta_{mn}}{2}\right)^{2}\right]\eta}\right\}^{2}. \tag{4.41}
$$

In the given case, the determination of the boundaries of the main DIR is reduced to calculating the roots of a sixth-degree polynomial as regards to Θ^2.

For an isotropic viscoelastic shell, the eqn (4.41) takes the form

$$
\delta_{mn}^{2}=\left\{1-\left(\frac{\Theta}{2\Omega_{mn}}\right)^{2}-\frac{g_{mn}}{\Omega_{mn}^{2}\left[1+\left(\frac{\Theta\eta}{2}\right)^{2}\right]}\right\}^{2}+
$$

$$
+\left\{\frac{g_{mn}\Theta\eta}{2\Omega_{mn}^{2}\left[1+\left(\frac{\Theta\eta}{2}\right)^{2}\right]}\right\}^{2},
$$

where $g_{mn}=[(E_0-E_\infty)/(E_0)]\omega_{mn}^2$; E_0 and E_∞ are instantaneous and long-term magnitudes of elastic modulus of the material.

Let us remark that in the ultimate cases $\eta\to 0$ and $\eta\to\infty$ the formula $\Theta_{1,2}=2\Omega_{mn}\sqrt{1\pm\delta_{mn}}$ follows from (4.41) but Ω_{mn} and δ_{mn} are determined through long-term (at $\eta\to 0$) or instantaneous (at $\eta\to\infty$) modulus of the material.

The numerical study presented in Ref. 46 showed that the existence of viscous properties of the material lead to transformation of the lower sections of DIR, corresponding to low magnitudes of the amplitude of the periodical force, into the regions of stable solutions. This effect is analogous to that obtained in studying eqn (4.39) in Ref. 65. For a viscoelastic material, this effect is more strongly manifested, the smaller is the ratio of the long-term shear moduli magnitudes to the instantaneous ones.

The relaxation time η affects the DIR in a more complicated manner. In Fig. 4.1, the results illustrating the change in the location and the shape of the principal DIR with increasing η from zero to infinity are presented (the values Ω_{mn} and δ_{mn} are determined through the instantaneous magnitude of elastic modulus). The given example refers to an isotropic shell with $E_0/E_\infty=5$. The right-hand side ultimate DIR, marked by a dash-dot line, corresponds to the solution of the elastic problem with the instantaneous modulus, the left-hand side ultimate DIR (a dashed line) — to the

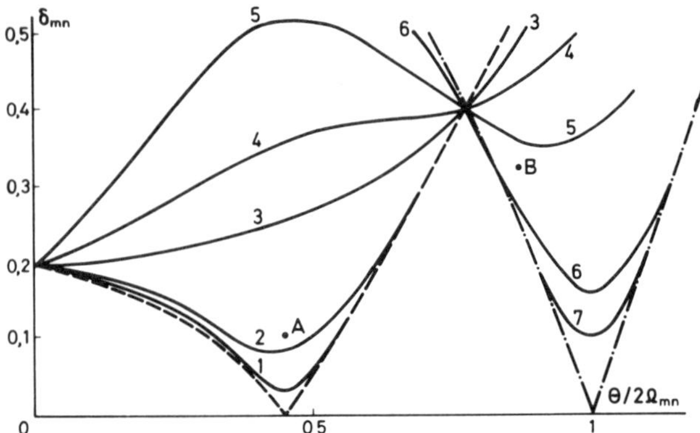

Fig. 4.1. DIR diagram for a viscoelastic shell. Numbers at curves correspond to values η/t_0: 0.1 (1); 0.3 (2); 1 (3); 1.5 (4); 2.5 (5); 5 (6); 10 (7); $t_0 = 2L/[E_0/\rho(1-\nu^2)]^{1/2}$.

solution of the elastic problem with the long-term modulus. As is shown, the decrease in relaxation time of the material results in the shift of DIR toward lower frequencies and the increase of its width.

From the viewpoint of general interpretation of the dynamic processes in viscoelastic media the indicated results are comprehensible. The response of a viscoelastic material to high-frequency (with a period $T \ll \eta$) and low-frequency ($T \gg \eta$) vibrations coincides with the response of elastic material (with the instantaneous and long-term moduli, respectively). At the period $T \approx \eta$, the viscoelastic behavior exerts the strongest effect on the amplitude of vibrations and leads to the largest phase shift between stress and strain and to maximum dissipation energy.

4.3 CALCULATION OF NON-LINEAR PARAMETRIC VIBRATIONS OF AN ELASTIC SHELL

Let us pass over to the investigation of the problem of parametric vibrations for an elastic orthotropic cylindrical shell in a geometrically non-linear statement. Let us make use of the equations in mixed form (2.51) and (2.52), neglecting the rotational inertia and taking into account the fact that the radial component of the mass force $P_3^* = -\mu(\partial^2 w/\partial t^2)$. According to paragraph 4.1, let us assume that in the case of axisymmetric forced vibrations of a shell, the momentless uniform stress state arises and this stress state is presented by the traction $T_{11}^0 = -(P_0 + P_t \cos \Theta t)$, to which corresponds the function of tractions $F^0(t) = -(P_0 + P_t \cos \Theta t)(y^2/2)$. Then the 'additional' function of tractions $\Phi(x, y, t)$, corresponding to the stress state, having arisen in the case of non-axisymmetric parametric vibrations is equal to

$$\Phi(x, y, t) = F(x, y, t) + (P_0 + P_t \cos \Theta t) \frac{y^2}{2}. \qquad (4.42)$$

As a result of inserting (4.42) the system of eqns (2.51) and (2.52) takes the form

$$D_{11} \frac{\partial^4 (w - w^0)}{\partial x^4} + 2(D_{12} + 2D_{66}) \frac{\partial^4 (w - w^0)}{\partial x^2 \partial y^2} + D_{22} \frac{\partial^4 (w - w^0)}{\partial y^4} + $$
$$+ \frac{1}{R} \cdot \frac{\partial^2 \Phi}{\partial x^2} - \frac{\partial^2 w}{\partial x^2} \frac{\partial^2 \Phi}{\partial y^2} - \frac{\partial^2 w}{\partial y^2} \frac{\partial^2 \Phi}{\partial x^2} + 2 \frac{\partial^2 w}{\partial x \partial y} \frac{\partial^2 \Phi}{\partial x \partial y} +$$
$$+ (P_0 + P_t \cos \Theta t) \frac{\partial^2 w}{\partial x^2} + \mu \frac{\partial^2 w}{\partial t^2} = 0; \qquad (4.43)$$

$$A_{22} \frac{\partial^4 \Phi}{\partial x^4} + (2A_{12} + A_{66}) \frac{\partial^4 \Phi}{\partial x^2 \partial y^2} + A_{11} \frac{\partial^4 \Phi}{\partial y^4} = \frac{1}{R} \frac{\partial^2 (w - w^0)}{\partial x^2} +$$
$$+ \left(\frac{\partial^2 w}{\partial x \partial y} \right)^2 - \left(\frac{\partial^2 w^0}{\partial x \partial y} \right)^2 - \frac{\partial^2 w}{\partial x^2} \frac{\partial^2 w}{\partial y^2} + \frac{\partial^2 w^0}{\partial x^2} \frac{\partial^2 w^0}{\partial y^2}. \qquad (4.44)$$

Since the tractions acting at the end faces of the shell are axisymmetric, the boundary conditions for the Φ function are uniform accordingly to (4.42).

Practically all the known solutions of the problem of non-linear parametric vibrations of cylindrical shells[32,103,120,402,461] have been obtained by the Bubnov-Galerkin method. As is known, the correctness and efficiency of solutions, obtained by this method is, firstly, determined by successive selection of the expressions, approximating the unknown functions. In the linear problem of parametric vibrations, the equations of motion and boundary simple support conditions are satisfied term-by-term by a double trigonometric series:

$$w(x, y, t) = \sum_{m=1}^{\infty} \sum_{n=1}^{\infty} W_{mn}(t) \sin \alpha_m x \cos \beta_n y; \qquad (4.45)$$

$$\Phi(x, y, t) = \sum_{m=1}^{\infty} \sum_{n=1}^{\infty} f_{mn}(t) \sin \alpha_m x \cos \beta_n y. \qquad (4.46)$$

In the solution of the non-linear system (4.43) and (4.44) by direct substituting of (4.45) and (4.46) into them, all the longitudinal and circumferential harmonics turn out to be interrelated in an infinite system of non-linear ordinary differential equations. In order to put down this system in the form acceptable, from the viewpoint of a subsequent numerical solution, it is necessary to perform extremely awkward analytical calculations. The subsequent numerical integration, even when a comparatively small number of harmonics is taken into account, is not a simple problem either.

Therefore, an attempt to lead the solution to numerical results compelled practically all the authors to restrict themselves to one-to-four terms in a series (4.45). Thereby the approximations used were built according to the same scheme as in the earliest solutions of the problems of static stability 'at large'.[97] One of the terms of a series (4.45), corresponding to a fixed pair of numbers $\{m_0, n_0\}$ was selected as the first. The additional terms were selected proportional either to squares of trigonometric functions with the same arguments as in the first, basic term, or to trigonometric functions with double arguments. Their mission was to reflect the specifically non-linear properties of the solution, in particular, the so-called preferential buckling inwards.

From the viewpoint of physics of parametric vibrations, such a direct transfer of the methodology used for static problems seems groundless. And it is directly confirmed by the numerical results presented in Ref. 461, where a typical four-term approximation of deflection was assumed

$$\begin{aligned} w(x, y, t) = {} & f_1(t)\cos\alpha x\cos\beta y + \\ & + f_2(t)\cos 2\alpha x + f_3(t)\cos 2\beta y + f_0(t); \\ & \alpha = \frac{\pi m}{L}; \quad \beta = \frac{n}{R}. \end{aligned} \tag{4.47}$$

Finally, the primary system of equations in partial derivatives was reduced in Ref. 461 to the system of four non-linear ordinary differential equations. Through their numerical integration the functions $f_1(t)$, $f_2(t)$, $f_3(t)$ were calculated. The magnitudes of loading parameters in the calculations presented were inside the DIR for a selected pair of wave numbers $\{m, n\}$. The numerical results have shown that within the time interval that was under consideration the function $f_1(t)$ exceeds greatly $f_2(t)$ and $f_3(t)$. Besides, the amplitudes of the latter two functions did not show the tendencies to increase in time at all, but this is the main sign of resonance parametric vibrations. The author of Ref. 461 has left these important results without further discussion. Besides, it is impossible now to re-establish the numerical magnitudes of $f_2(t)$ and $f_3(t)$ from the illustrative material of Ref. 461. Nevertheless, the character of dependencies of these functions on time evokes earnest doubt about the reasonableness of taking the second and third terms of the approximation (4.47) into account.

In order to reveal the essence of the problem, let us consider the solution for the problem of non-linear parametric vibrations of an isotropic cylindrical shell on the basis of the approximation which is practically equivalent to (4.47)

$$w(x, y, t) = a(t) \sin \alpha x \cos \beta y + \\ + b(t) \cos 2\alpha x + c(t) \cos 2\beta y + d(t). \tag{4.48}$$

Substituting (4.48) into eqn (4.44), we obtain the expression for the function of tractions

$$\Phi(x, y, t) = A_1(t) \sin \alpha x \cos \beta y + A_2(t) \cos 2\alpha x + \\ + A_3(t) \cos 2\beta y + A_4(t) \sin \alpha x \cos 3\beta y + A_5(t) \sin 3\alpha x \cos \beta y + \\ + A_6(t) \cos 2\alpha x \cos 2\beta y, \tag{4.49}$$

where $A_i(t)$ is found through comparison of coefficients at respective trigonometric functions:

$$A_1(t) = -Eh\alpha^2 a \frac{1 - 2b^2 R(b-c)}{R(\alpha^2+\beta^2)^2}; \quad A_2(t) = -Eh \frac{8b - \beta^2 R a^2}{32 R \alpha^2};$$

$$A_3(t) = -Eh \frac{\alpha^2 a^2}{32\beta^2}; \quad A_4(t) = -Eh \frac{2\alpha^2\beta^2 ac}{(\alpha^2+9\beta^2)^2};$$

$$A_5(t) = -Eh \frac{2\alpha^2\beta^2 ab}{(9\alpha^2+\beta^2)^2}; \quad A_6(t) = -Eh \frac{\alpha^2\beta^2 bc}{(\alpha^2+\beta^2)^2}.$$

Further, substitution of (4.48) and (4.49) into the equation of motion (4.43) and application of the orthogonalization procedure result in the system of ordinary differential equations:

$$\frac{d^2 a}{dt^2} + \Omega_{mn}^2 (1 - 2\delta_{mn} \cos \Theta t) a + F_1(a, b, c) = 0;$$
$$\frac{d^2 b}{dt^2} + \Omega_{2m,0}^2 (1 - 2\delta_{2m,0} \cos \Theta t) b + F_2(a, b, c) = 0; \tag{4.50}$$
$$\frac{d^2 c}{dt^2} + \Omega_{0,2n}^2 c + F_3(a, b, c) = 0,$$

where

$$F_1 = -\frac{2\alpha^4\beta^2 E a(b-c)}{\rho R(\alpha^2+\beta^2)^2} - \frac{2\alpha^2\beta^2}{\rho h}(aA_3 - cA_1 - bA_1 - aA_2 + bA_5 + cA_4);$$

$$F_3 = -\frac{\alpha^2\beta^2}{\rho h}(aA_1 + aA_4 + 8bA_6);$$

$$F_2 = -\frac{E}{8R\rho}\beta^2 a^2 - \frac{\alpha^2\beta^2}{\rho h}(aA_5 - aA_1 + 8cA_6).$$

Let us analyze the possibility of appearance of parametric resonance in the system described by eqns (4.50). Thereby the principal consideration of Ref. 65 (this is confirmed by the results of direct numerical integration of the system (4.50)), consisting in that the boundaries of DIR are determined by the linear part of (4.50) and do not depend on the functions F_1, F_2 and F_3, should be taken into account.

For the first equation of (4.50) the main DIR corresponds to a certain pair of $\{m, n\}$ numbers characterizing the spatial form of vibration, for the second one — to axisymmetric $\{2m, 0\}$ form. The calculations done for a number of different cylindrical shells with various magnitudes of geometric and mechanical characteristics show that these two DIRs never intersect. Consequently the parametric resonance cannot simultaneously arise for two spatial forms corresponding to $\{m, n\}$ and $\{2m, 0\}$. If the loading parameters Θ, P_t, P_0 are selected such that the corresponding point in the $\{\Theta, \delta\}$ plane lies within DIR for $\{m, n\}$ form, then the amplitude of $a(t)$ function increases exponentially during the initial stage of parametric vibrations and reaches the magnitudes of the same order as the thickness of a shell. Thereby the maximum magnitudes of $b(t)$ and $c(t)$ remain of the order of their initial magnitudes. If Θ, P_t, P_0 are selected such that they lie within DIR for $\{2m, 0\}$ form, then the function $b(t)$ exceeds considerably $a(t)$ and $c(t)$. The result, obtained in Ref. 461, from this point of view, seems to be absolutely reasonable: the functions $f_2(t)$ and $f_3(t)$ in approximation (4.47) are small in comparison to $f_1(t)$ because the calculations were performed for those loading parameters which fall within DIR for $\{m, n\}$ form.

The example under consideration shows that the main principle for selecting the approximation of deflection in solving the problem of non-linear parametric vibrations should completely differ from that used in the non-linear problems of static stability. The number of the retained terms of the series (4.45) in the approximation and the respective numbers of spatial harmonics should be determined from the previously calculated DIR spectrum for the given structure.

Let us formulate the basic principle. If the specified characteristics of the external load correspond to the point in the $\{\Theta, \delta\}$ plane which lies within only one DIR (say, for $\{m_1, n_1\}$ form), then it is fully sufficient to approximate deflection by the respective one term of Fourier series (4.45). If in this point the two DIRs are intersecting, corresponding to $\{m_1, n_1\}$ and $\{m_2, n_2\}$ forms of vibrations, then the sum of two fairly definite terms of Fourier series should be employed for approximation of deflection:

$$w(x, y, t) = f_1(t) \sin \alpha_1 x \cos \beta_1 y + f_2(t) \sin \alpha_2 x \cos \beta_2 y, \qquad (4.51)$$

where

$$\alpha_1=\frac{\pi m_1}{L};\quad \alpha_2=\frac{\pi m_2}{L};\quad \beta_1=\frac{n_1}{R};\quad \beta_2=\frac{n_2}{R}.$$

Let us emphasize that the second term in (4.51) is not an additional one, correcting the solution, but is of equal rights with the first one. The maximum value of $f_2(t)$ can be both smaller and greater than the maximum value of $f_1(t)$ depending on the magnitudes of amplitude and frequency of the load. In the case when three, four, etc. DIR are intersecting in the given point within the plane of parameters, the sum of respective three, four, etc. terms of Fourier series should be used for approximation of deflection.

In such a way, in solving the problem of non-linear parametric vibrations, the complication of deflection approximation brought about by adding the terms with double harmonics in the cases when DIR, corresponding to these additional terms, do not intersect with the basic one, will result neither in qualitative, nor any notable quantitative change in the calculated deflection.

To illustrate the formulated general principle, let us consider the results of solving the problem for non-linear parametric vibrations of a cylindrical shell on the basis of two-term approximation (4.51). Such an approximation was first employed in the paper.[53]

Substitution of (4.51) into eqn (4.44) yields the following expression for the function of tractions:

$$\begin{aligned}\Phi(x,y,t) &= (C_1\cos 2\alpha_1 x + C_2\cos 2\beta_1 y)(f_1^2 - f_1^{0^2}) +\\ &+ (C_3\cos 2\alpha_2 x + C_4\cos 2\beta_2 y)(f_2^2 - f_2^{0^2}) +\\ &+ [C_5\cos(\alpha_1+\alpha_2)x\cos(\beta_1+\beta_2)y +\\ &+ C_6\cos(\alpha_1-\alpha_2)x\cos(\beta_1+\beta_2)y +\\ &+ C_7\cos(\alpha_1+\alpha_2)x\cos(\beta_1-\beta_2)y +\\ &+ C_8\cos(\alpha_1-\alpha_2)x\cos(\beta_1-\beta_2)y](f_1f_2 - f_1^0f_2^0) +\\ &+ C_9(f_1-f_1^0)\sin\alpha_1 x\cos\beta_1 y + C_{10}(f_2-f_2^0)\sin\alpha_2 x\cos\beta_2 y,\end{aligned}\tag{4.52}$$

where f_1^0 and f_2^0 are the amplitudes of initial imperfections, corresponding to the $\{m_1, n_1\}$ and $\{m_2, n_2\}$ spatial forms;

$$C_1=\frac{\beta_1^2}{32A_{22}\alpha_1^2};\quad C_2=-\frac{\alpha_1^2}{32A_{11}\beta_1^2};$$

$$C_3=\frac{\beta_2^2}{32A_{22}\alpha_2^2};\quad C_4=-\frac{\alpha_2^2}{32A_{11}\beta_2^2};$$

$$C_5=\frac{(\alpha_1\beta_2-\alpha_2\beta_1)^2}{A_{22}(\alpha_1+\alpha_2)^4+(A_{66}+2A_{12})(\alpha_1+\alpha_2)^2(\beta_1+\beta_2)^2+A_{11}(\beta_1+\beta_2)^4};$$

$$C_6=\frac{-(\alpha_1\beta_2+\alpha_2\beta_1)^2}{A_{22}(\alpha_1-\alpha_2)^4+(A_{66}+2A_{12})(\alpha_1-\alpha_2)^2(\beta_1+\beta_2)^2+A_{11}(\beta_1+\beta_2)^4};$$

$$C_7=\frac{(\alpha_1\beta_2+\alpha_2\beta_1)^2}{A_{22}(\alpha_1+\alpha_2)^4+(A_{66}+2A_{12})(\alpha_1+\alpha_2)^2(\beta_1-\beta_2)^2+A_{11}(\beta_1-\beta_2)^4}; \qquad (4.53)$$

$$C_8=\frac{-(\alpha_1\beta_2-\alpha_2\beta_1)^2}{A_{22}(\alpha_1-\alpha_2)^4+(A_{66}+2A_{12})(\alpha_1-\alpha_2)^2(\beta_1-\beta_2)^2+A_{11}(\beta_1-\beta_2)^4};$$

$$C_9=-\frac{\alpha_1^2}{R[A_{22}\alpha_1^4+(A_{66}+2A_{12})\alpha_1^2\beta_1^2+A_{11}\beta_1^4]};$$

$$C_{10}=-\frac{\alpha_2^2}{R[A_{22}\alpha_2^4+(A_{66}+2A_{12})\alpha_2^2\beta_2^2+A_{11}\beta_2^4]}.$$

Further, on substituting (4.51) and (4.52) into (4.43) and employing the procedure of orthogonalization, we obtain the following system of two non-linear ordinary differential equations with periodic coefficients

$$\frac{d^2f_i}{dt^2}+\Omega_i^2(1-2\delta_i\cos\Theta t)f_i-\omega_i^2f_i^0+d_if_i(f_i^2-f_i^{0^2})+$$
$$+\frac{q_{ij}}{4\mu}f_j(f_if_j-f_i^0f_j^0)=0. \qquad (4.54)$$

Here, ω_i, P_i^* are the frequency of natural vibrations and critical static axial traction for the $\{m_i, n_i\}$ spatial form

$$d_i=\frac{1}{16\mu}\left(\frac{\alpha_i^4}{A_{11}}+\frac{\beta_i^4}{A_{22}}\right); \qquad (4.55)$$

$$\begin{aligned}q_{ij}={}&\alpha_j^2[(\beta_i+\beta_j)^2(C_5-C_6)+(\beta_i-\beta_j)^2(C_7-C_8)]+\\&+\beta_j^2[(\alpha_i+\alpha_j)^2(C_5+C_7)-(\alpha_i-\alpha_j)^2(C_6+C_8)]+\\&+2\alpha_j\beta_j[-(\alpha_i+\alpha_j)(\beta_i+\beta_j)C_5-(\alpha_i-\alpha_j)(\beta_i+\beta_j)C_6+\\&+(\alpha_i+\alpha_j)(\beta_i-\beta_j)C_7+(\alpha_i-\alpha_j)(\beta_i-\beta_j)C_8].\end{aligned} \qquad (4.56)$$

In (4.54)–(4.56), $i=1,2$; $j=2$ at $i=1$ and $j=1$ at $i=2$. The system (4.54) is supplemented with initial conditions

$$f_i|_{t=0}=f_i^0; \qquad \left.\frac{df_i}{dt}\right|_{t=0}=0. \qquad (4.57)$$

For numerical integration of Cauchy problem (4.54) and (4.57), the Runge-Kutta method of the fourth order of accuracy was used.

Let us consider as an example the shell with the following characteristics: $R=1$ m, $R/h=100$, $L/R=2$, $E=4\cdot10^{10}$ N/m^2; $\nu=0.3$; $\rho=2.5\cdot10^3$ kg/m^3. Let us designate the minimum in respect to $\{m,n\}$ magnitude of axial critical traction by P^* and introduce the designation

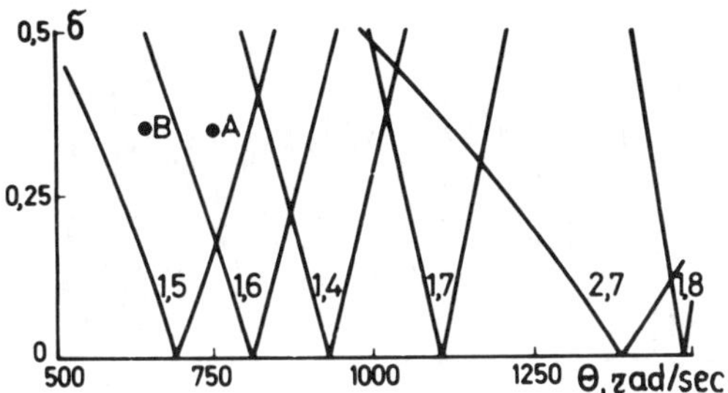

Fig. 4.2. Initial part of the main DIR spectrum for an elastic shell at $P_0 = 0.5P^*$. Numbers correspond to the values of axial and circumferential modes of vibrations.

$\delta = (P_t)/(2(P^* - P_0))$. In the calculations it was assumed that $P_0 = 0.5P^*$ and $f_1^0 = f_2^0 = 0.01\,\mathrm{h}$.

In Fig. 4.2, the low-frequency part of the spectrum of main DIR is shown. It is calculated according to the formulas (4.15). Let us mark that a considerable part of the plane of parameters is occupied by the intersections of two and more DIR. In Fig. 4.3, the results of numerical integration of Cauchy problem (4.54) and (4.57) are presented for $\delta = 0.35$ and two magnitudes of Θ: 750 rad/s (Fig. 4.3a) and 650 rad/s (Fig. 4.3b). In the first case, the load parameters correspond to the point A (Fig. 4.2) lying inside the intersection of two DIRs for the forms of vibrations $\{1,5\}$ and $\{1,6\}$. In the second — to the point B, lying within one DIR for the form $\{1,5\}$. As is shown, in the first case the parametric resonance is observed for both the forms, while in the second — only for the form

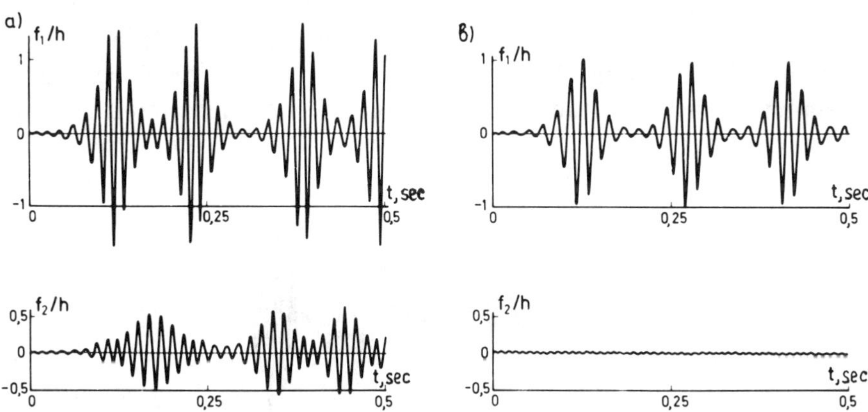

Fig. 4.3. Dependencies of $f_1(t)$ and $f_2(t)$ at the load parameters corresponding to: a: intersection of two DIR ($\Theta = 750$ rad/s); b: one DIR ($\Theta = 650$ rad/s).

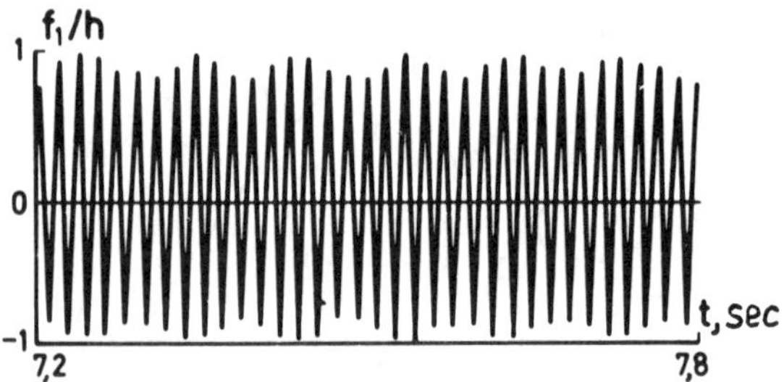

Fig. 4.4. A fragment of $f_1(t)$ function at $\Theta = 650$ rad/s and $\delta = 0.35$ after several seconds from the moment of load application.

$\{1, 5\}$. Thereby the amplitude of vibrations for the form $\{1, 6\}$ in the whole time interval is of the order of initial magnitude f_2^0.

The numerical results presented confirm the validity of the above formulated principle for selecting the approximation of deflection. Namely, at specified load parameters Θ and δ, only those spatial harmonics should be taken into account for which with respect to these harmonics DIR include the specified parameters Θ and δ. In accordance with the solution obtained, in the case when the load parameters lie at the intersection of several DIR, the shell deflection is determined by summation of several terms of Fourier series and represents itself the sum of several beats, occurring simultaneously in several corresponding spatial forms.

The non-linear parametric vibrations (Fig. 4.4) retain their character of beats in time, but these beats, however, are becoming less pronounced upon approaching to stationary vibrations with the frequency $\Theta/2$. The analogous result was first observed in experiments performed on bars.[446]

The study of the functions $f_i(t)$, calculated at various magnitudes of initial imperfections f_i^0 has shown that the decrease in f_i^0 leads to slower build-up of vibrations. The time interval, prior to reaching the first maximum by the amplitude f_i, is prolonged, but the maximum itself does not practically depend on the value f_i^0 (cf. Figs 4.3(b) and 4.5).

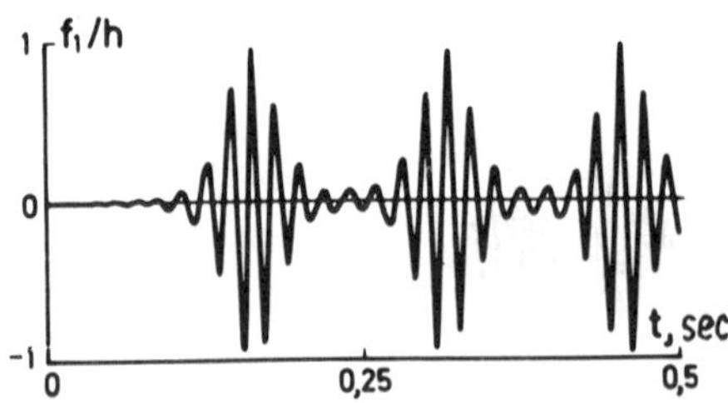

Fig. 4.5. Dependency $f_1(t)$ at $\Theta = 650$ rad/s; $\delta = 0.35$; $f_1^0 = 10^{-3}$ h.

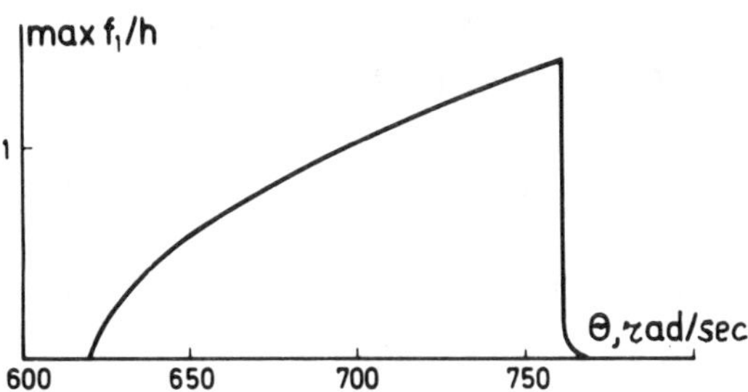

Fig. 4.6. Amplitude-frequency response for parametric vibrations of a shell in $\{1,5\}$ spatial mode; $\delta = 0.2$.

The amplitude-frequency response in the regime of parametric vibrations corresponding to the form $\{1,6\}$ is presented in Fig. 4.6. On the ordinate axis the absolute maximum of $f_1(t)$ found for each fixed frequency is plotted over the time interval $t \in [0, 10\,\mathrm{s}]$. As shown, the obtained response refers to a 'rigid' type, which is explained by the absence of quadratic terms in the system (4.54). It can also be noted that if frequency Θ goes out of the DIR on both sides, the amplitude of vibrations sharply falls (to the value f_1^0). This result is a direct confirmation of the conclusion made above that the DIR boundaries are being determined by the linear part of the system (4.54).

In completion of discussing the specific features of non-linear parametric vibrations of an elastic cylindrical shell let us emphasize that the magnitudes of amplitudes $f_i(t)$ increase with increasing δ. Besides, in the examples under consideration the beats occur around the 'undisturbed' momentless state. This is observed right up to $\delta = 0.5$. At $\delta > 0.5$, the non-linear parametric vibrations are accompanied by loss of stability: there occurs swift buckling to the new state of equilibrium and the non-linear vibrations proceed further around this state. Methodology of solving this problem, in a general way, is sufficiently clear while its direct solution needs a special approach which would combine the ideas of the solutions obtained on the basis of multi-term approximations for the problems of non-linear parametric vibrations and static loss of stability. We shall not discuss here this problem further.

The calculation procedure presented above and based on the two-term approximation of deflection (4.51) can be generalized for the case of an arbitrary number of interacting spatial harmonics. Thereby it is sufficient to obtain the system, generalizing (4.54) for the case of three harmonics. A further generalization can be easily carried out by the induction method.

4.4 CALCULATION OF NON-LINEAR PARAMETRIC VIBRATIONS OF A VISCOELASTIC SHELL

The methodology for solving the problem about parametric vibrations of a viscoelastic orthotropic cylindrical shell in linear formulation has been described in paragraph 4.2. Let us consider now the solution of this problem with the geometric non-linearity taken into account.

At first, it should be emphasized that a particular form of the functional relation between stress and strain does not principally affect the selection of deflection approximation in solving the problem of non-linear parametric vibrations. Therefore, the approach outlined in paragraph 4.3 for an elastic shell, remains valid also for the viscoelastic one. Namely, the amount and the numbers of harmonics, which must be taken into account in approximation, should be evaluated from the character of DIR spectrum for the shell (in this case, a viscoelastic one) under consideration.

The problem about the non-linear parametric vibrations of a viscoelastic cylindrical shell was first reported by A. E. Bogdanovich and E. G. Feldmane.[54] Following the approach described in this paper, let us assume that the orthotropic material reveals its viscoelastic properties only in shear deformation. The following equations written in the mixed form are assumed as primary ones:

$$D_{11}\frac{\partial^4(w-w^0)}{\partial x^4}+2(D_{12}+2\hat{D}_{66})\frac{\partial^4(w-w^0)}{\partial x^2\partial y^2}+D_{22}\frac{\partial^4(w-w^0)}{\partial y^4}+$$

$$+\frac{1}{R}\frac{\partial^2\Phi}{\partial x^2}-\frac{\partial^2 w}{\partial x^2}\frac{\partial^2\Phi}{\partial y^2}-\frac{\partial^2 w}{\partial y^2}\frac{\partial^2\Phi}{\partial x^2}+2\frac{\partial^2 w}{\partial x\partial y}\frac{\partial^2\Phi}{\partial x\partial y}+$$

$$+(P_0+P_t\cos\Theta t)\frac{\partial^2 w}{\partial x^2}+\mu\frac{\partial^2 w}{\partial t^2}=0; \tag{4.58}$$

$$A_{22}\frac{\partial^4\Phi}{\partial x^4}+(2A_{12}+\hat{A}_{66})\frac{\partial^4\Phi}{\partial x^2\partial y^2}+A_{11}\frac{\partial^4\Phi}{\partial y^4}=\frac{1}{R}\frac{\partial^2(w-w^0)}{\partial x^2}+$$

$$+\left(\frac{\partial^2 w}{\partial x\partial y}\right)^2-\left(\frac{\partial^2 w^0}{\partial x\partial y}\right)^2-\frac{\partial^2 w}{\partial x^2}\frac{\partial^2 w}{\partial y^2}+\frac{\partial^2 w^0}{\partial x^2}\frac{\partial^2 w^0}{\partial y^2}, \tag{4.59}$$

where the following integral operators are introduced

$$\hat{D}_{66}=\frac{h^3}{12}\hat{G}_{12};\quad \hat{A}_{66}=\frac{1}{h}\hat{G}_{12}^{-1}. \tag{4.60}$$

Here, $\hat{G}_{12}$ is determined by the formula (4.19), while the inverse operator is written as

$$\hat{G}_{12}^{-1}=\frac{1}{G_{12}^0}+\frac{G_{12}^0-G_{12}^\infty}{G_{12}^0G_{12}^\infty}\int_0^t K(t-\tau)\,d\tau; \tag{4.61}$$

$K(t)$ is the creep kernel.

Let us treat the simplest solution based on a single term approximation of deflection:

$$w(x,y,t)=W_{mn}(t)\sin\alpha_m x\cos\beta_n y. \tag{4.62}$$

Substituting (4.62) into (4.59) we find the function of tractions

$$\Phi(x,y,t)=F_{mn}^{(1)}(t)\sin\alpha_m x\cos\beta_n y+F_{mn}^{(2)}(t)\cos 2\alpha_m x+ \\ +F_{mn}^{(3)}(t)\cos\beta_n y, \tag{4.63}$$

where

$$[\alpha_m^4A_{22}+\alpha_m^2\beta_n^2(2A_{12}+\hat{A}_{66})+\beta_n^4A_{11}]F_{mn}^{(1)}(t)= \\ =-\frac{\alpha_m^2}{R}[W_{mn}(t)-W_{mn}^0]; \tag{4.64}$$

$$F_{mn}^{(2)}(t)=\frac{\beta_n^2}{32A_{22}\alpha_m^2}[W_{mn}^2(t)-W_{mn}^{0^2}]; \tag{4.65}$$

$$F_{mn}^{(3)}(t)=-\frac{\alpha_m^2}{32\beta_n^2A_{11}}[W_{mn}^2(t)-W_{mn}^{0^2}].$$

In order to express $F_{mn}^{(1)}(t)$ in terms of $W_{mn}(t)$, let us solve the integral eqn (4.64) by using Laplace transform. Taking into account (4.61), let us rewrite (4.64) in the form

$$A_{mn}F_{mn}^{(1)}(t)+\alpha_m^2\beta_n^2\gamma\int_0^t K(t-\tau)F_{mn}^{(1)}(\tau)\,d\tau= \\ =-\frac{\alpha_m^2}{R}\cdot[W_{mn}(t)-W_{mn}^0]. \tag{4.66}$$

Here,

$$\gamma=\frac{1}{h}\frac{G_{12}^0-G_{12}^\infty}{G_{12}^0G_{12}^\infty};\quad A_{mn}=\alpha_m^4A_{22}+\alpha_m^2\beta_n^2(A_{66}+2A_{12})+\beta_n^4A_{11}. \tag{4.67}$$

On applying Laplace transform to (4.66) we obtain the expression for $\bar{F}_{mn}^{(1)}(p)$:

$$\bar{F}_{mn}^{(1)}(p)=-\frac{\alpha_m^2}{A_{mn}R}\left[\bar{W}_{mn}(p)-\frac{W_{mn}^0}{p}\right]\times \\ \times\left\{1-\frac{\alpha_m^2\beta_n^2\gamma}{A_{mn}\left[\frac{1}{\bar{K}(p)}+\frac{\alpha_m^2\beta_n^2\gamma}{A_{mn}}\right]}\right\}. \tag{4.68}$$

Inverting this expression, we obtain

$$F_{mn}^{(1)}(t) = -\frac{\alpha_m^2}{A_{mn}R}[W_{mn}(t) - W_{mn}^0] + \frac{\alpha_m^4\beta_n^2\gamma}{A_{mn}^2R}\int_0^t \Psi_{mn}(t-\tau)\times$$
$$\times[W_{mn}(\tau) - W_{mn}^0]\,d\tau, \qquad (4.69)$$

where $\Psi_{mn}(t)$ is the inverse Laplace transform of the function

$$\overline{\Psi}_{mn}(p) = \frac{1}{\dfrac{1}{\bar{K}(p)} + \dfrac{\alpha_m^2\beta_n^2\gamma}{A_{mn}}}.$$

After insertion of (4.62) and (4.63) into (4.58) and employing the procedure of orthogonalization we obtain the following integro-differential equation:

$$\frac{d^2W_{mn}}{dt^2} + \Omega_{mn}^2(1-2\delta_{mn}\cos\Theta t)\,W_{mn} - \omega_{mn}^2W_{mn}^0 + d_{mn}W_{mn}\times$$
$$\times(W_{mn}^2 - W_{mn}^{0^2}) - g_{mn}^{(1)}\int_0^t R(t-\tau)[W_{mn}(\tau) - W_{mn}^0]\,d\tau -$$
$$-g_{mn}^{(2)}\int_0^t \chi_{mn}(t-\tau)[W_{mn}(\tau) - W_{mn}^0]\,d\tau = 0, \qquad (4.70)$$

where

$$d_{mn} = \frac{1}{16\mu}\left(\frac{\alpha_m^4}{A_{11}} + \frac{\beta_n^4}{A_{22}}\right), \qquad (4.71)$$

$\chi_{mn}(t)$ is inverse Laplace transform of the function (4.32), and the coefficients $g_{mn}^{(1)}$ and $g_{mn}^{(2)}$ are determined according to (4.31).

In the particular case of exponential relaxation kernel (4.33) we have

$$\chi_{mn}(t) = \frac{1}{\eta}e^{-\frac{t}{\beta_{mn}}}, \qquad (4.72)$$

where the value β_{mn} is specified by the formula (4.35). Then the eqn (4.70) takes the form

$$\frac{d^2W_{mn}}{dt^2} + \Omega_{mn}^2(1-2\delta_{mn}\cos\Theta t)\,W_{mn} - \omega_{mn}^2W_{mn}^0 + d_{mn}W_{mn}\times$$
$$\times(W_{mn}^2 - W_{mn}^{0^2}) - \int_0^t\left[\frac{g_{mn}^{(1)}}{\eta}e^{-\frac{t-\tau}{\eta}} + \frac{g_{mn}^{(2)}}{\eta}e^{-\frac{t-\tau}{\beta_{mn}}}\right]\times$$
$$\times[W_{mn}(\tau) - W_{mn}^0]\,d\tau = 0. \qquad (4.73)$$

With the expression (4.37) taken into account, this equation can be rewritten in the following form:

$$\frac{d^2W_{mn}}{dt^2}+\Omega_{mn}{}^2(1-2\delta_{mn}\cos\Theta t)\,W_{mn}-\omega_{mn}{}^2W_{mn}{}^0+d_{mn}W_{mn}\times$$
$$\times\,(W_{mn}{}^2-W_{mn}{}^{02})-\int_0^t\Pi_{mn}(t-\tau)\,[W_{mn}(\tau)-W_{mn}{}^0]\,d\tau=0. \tag{4.74}$$

Let us note that at $d_{mn}=g_{mn}^{(1)}=g_{mn}^{(2)}=W_{mn}^0=0$, eqn (4.74) turns into the Mathieu eqn (4.14) obtained in solving the linear elastic problem; at $W_{mn}^0=d_{mn}=0$, it turns into eqn (4.38), corresponding to the linear viscoelastic problem, while at $g_{mn}^{(1)}=g_{mn}^{(2)}=0$, into equation corresponding to the non-linear elastic problem in a single term approximation approach (this equation follows from the system (4.54) at $f_1\equiv W_{mn}$, $f_2\equiv 0$). Let us attach our attention to the fact that the terms additional to eqn (4.14), which reflect viscosity of the material and geometric non-linearity, enter (4.74) as a sum of two separate terms.

The solution obtained can be generalized without difficulty for the case of employing the multi-term approximations of deflection. For example, in the case of two-term approximation (4.51) the value A_{66} should be substituted by $\hat{A}_{66}$ in the formulas (4.53) for constants C_5, C_6, C_7, C_8, C_9 and C_{10}, which become the time functions and are found from integral equations of the type (4.64). At the final point, the system (4.54) is supplemented by a number of integral terms.

The calculation of non-linear parametric vibrations for a viscoelastic shell is accomplished, therefore, in three stages:

(1) assessment of DIR boundaries in accordance with the methodology presented in 4.2;
(2) selection of the deflection approximation for specified parameters of loading Θ, δ;
(3) numerical solution of integro-differential equations of the type (4.70) or, in a general case, systems of such equations.

In solving the integro-differential eqn (4.70), the Runge-Kutta method can be used in a similar way as it was used in solving the non-linear ordinary differential equations in the case of the elastic problem. Let us rewrite (4.70) as follows

$$\frac{d^2W_{mn}}{dt^2}=F_{mn}(W_{mn},t). \tag{4.75}$$

A specific feature of the viscoelastic problem consists in the fact that the expression for F_{mn} contains integrals of $W_{mn}(t)$ functions with varying upper boundary. Consequently, F_{mn} depends not only on the current time moment t and the value of W_{mn} at this time moment, but also on the previously calculated W_{mn} values in all the preceding time moments.

The numerical integration of (4.74), corresponding to the case of exponential relaxation kernel, was performed according to the scheme of Runge-Kutta method of the fourth-order accuracy; the integrals entering the right-hand part were calculated according to the trapezium method. Let us assume the characteristics of the shell equal to those in the numerical example considered in paragraph 4.2. The low-frequency part of the main DIR spectrum for this shell is shown in Fig. 4.7. Below the results, obtained for a fixed frequency $\Theta = 730\,\text{rad/s}$ at $\delta = 0.2$ and 0.03, are presented. The corresponding points at $\{\Theta, \delta\}$ plane are lying within the DIR for $\{1, 5\}$ mode of vibration in the case of elastic shell. The point corresponding to $\delta = 0.2$ and the same frequency, lies within DIR also for a viscoelastic shell, whereas the point, corresponding to $\delta = 0.03$, is not inside the region of unstable solutions in the case of a viscoelastic shell. In these calculations, $W_{15}^0 = 0.01\,\text{h}$ was assumed.

Figure 4.8 allows to compare the results of solving the non-linear elastic and viscoelastic problems. As is seen, in the case when the load parameters are inside the DIR for a viscoelastic shell, both the solutions are qualitatively similar. It should only be noted that in the viscoelastic case the beats become less pronounced already in the initial stage (cf. Fig. 4.8(a) and Fig. 4.4), while the maximum value of $W_{15}(t)$ diminishes slowly with time.

Comparison of results, presented in Figs 4.8(c) and 4.8(d), show that the solutions of elastic and viscoelastic problems at $\delta = 0.03$ differ qualitatively. In the elastic case, $W_{15}(t)$ increases in the initial time

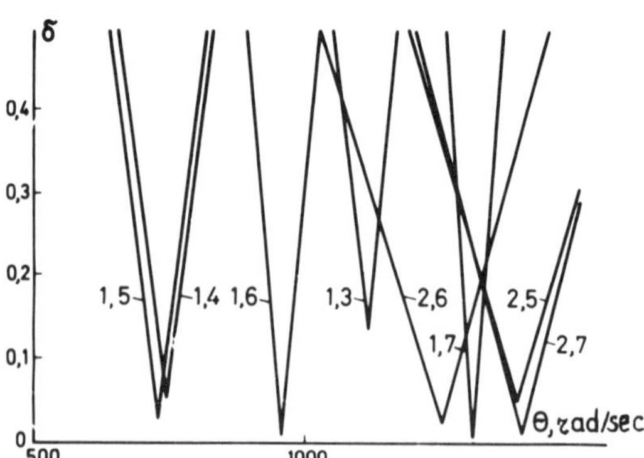

Fig. 4.7. The initial part of the main DIR spectrum of a viscoelastic shell. Numbers at curves are the numbers of axial and circumferential spatial modes of vibrations.

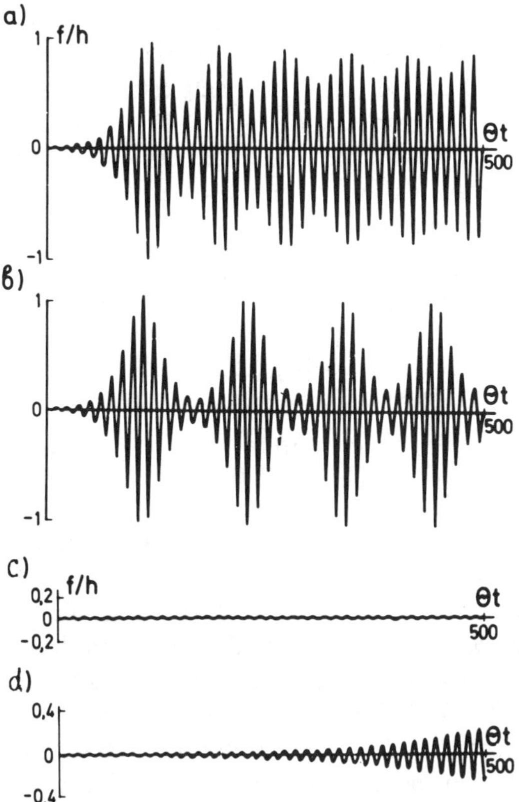

Fig. 4.8. Dependencies $W_{15}(t)$ for a viscoelastic (a, c) and elastic (b, d) shells at $\Theta = 730\,\text{rad/s}$; $\delta = 0.2$ (a, b) and 0.03 (c, d).

interval, though more slowly than at $\delta = 0.2$. Further the vibrations acquire the usual character of beats. In the viscoelastic case, the $W_{15}(t)$ function retains its amplitude value of the same order as W_{15}^0. This result serves as an evident confirmation of the accuracy of calculation of DIR boundaries for a viscoelastic shell according to eqn (4.41).

4.5 ANALYSIS OF A STATE OF STRESS IN A SHELL UNDER CONDITIONS OF PARAMETRIC RESONANCE

The methodologies outlined in paragraph 4.3 and 4.4, allow one to determine deflection w of the shell and function of tractions Φ in an arbitrary point $\{x, y\}$ at any time moment t. This is realized by summing the corresponding number of terms in the Fourier series (4.45) and (4.46).

Thus in employing two-term approximation of deflection, the functions w and Φ are calculated according to the formulas (4.51) and (4.52). Further, we can calculate all the necessary characteristics of stress–strain state. In accordance with the formulas (2.16), (2.41), (2.46) and (2.50) in the case of a homogeneous through the thickness orthotropic shell we obtain the following expressions for normal stresses: σ_{xx} along x-axis, σ_{yy} along y-axis and shear stress σ_{xy}:

$$\sigma_{xx}=\frac{T_{11}{}^0}{h}+\frac{1}{h}\frac{\partial^2\Phi}{\partial y^2}-z\left[B_{11}\frac{\partial^2(w-w^0)}{\partial x^2}+B_{12}\frac{\partial^2(w-w^0)}{\partial y^2}\right];$$
$$\sigma_{yy}=\frac{T_{22}{}^0}{h}+\frac{1}{h}\frac{\partial^2\Phi}{\partial x^2}-z\left[B_{12}\frac{\partial^2(w-w^0)}{\partial x^2}+B_{22}\frac{\partial^2(w-w^0)}{\partial y^2}\right]; \tag{4.76}$$
$$\sigma_{xy}=-\frac{1}{h}\frac{\partial^2\Phi}{\partial x\partial y}-2zB_{66}\frac{\partial^2(w-w^0)}{\partial x\partial y},$$

Here the momentless components of tractions T_{11}^0 and T_{22}^0 are calculated by the formulas (4.8).

Let us consider some results by calculating the stress state in an elastic shell with parameters presented in paragraph 4.3. Let us assume $P_0=0.5P^*, \delta=0.35, \Theta=650\,\mathrm{rad/s}$, corresponding to the point in the plane of parameters within DIR for a mode of vibrations $\{1,5\}$ (see Fig. 4.2).

Figure 4.9 reflects the dependence of axial and circumferential stresses on the y coordinate at the time moment $t=0.125$ s. As is clearly seen from Figs 4.9(c, d), there are ten nodal points along the circumference. Additional calculations revealed that in other time moments the location of nodal points does not change with time. Therefore, if the load parameters are inside only one DIR, the resonance parametric vibrations proceed according to certain stationary spatial form. Let us also note that, in accordance with Figs 4.9(a, b), the momentless component presents the main contribution to stress σ_{xx}, while in σ_{yy} (Figs 4.9(c, d)) the flexural component prevails. The stresses σ_{xx} on both the surfaces are compressive, while σ_{yy} in each fixed y-section is of opposite signs on the inner and outer surfaces.

In Fig. 4.10 the stress distribution along the y coordinate in two time moments are presented. The load frequency is assumed as $\Theta=750\,\mathrm{rad/s}$. In this case, in accordance with Fig. 4.2, the two modes of vibrations should be of resonance-type, namely, $\{1,5\}$ and $\{1,6\}$. And really, at $t=0.13$ s (see Fig. 4.10(a, c)) both the stresses have five distinctly expressed maxima and minimums, whereas at $t=0.17$ s (see Figs 4.10(b, d)) there are six of them. In such a way, under the given conditions of parametric resonance, the location of nodal points along the circumference changes continuously

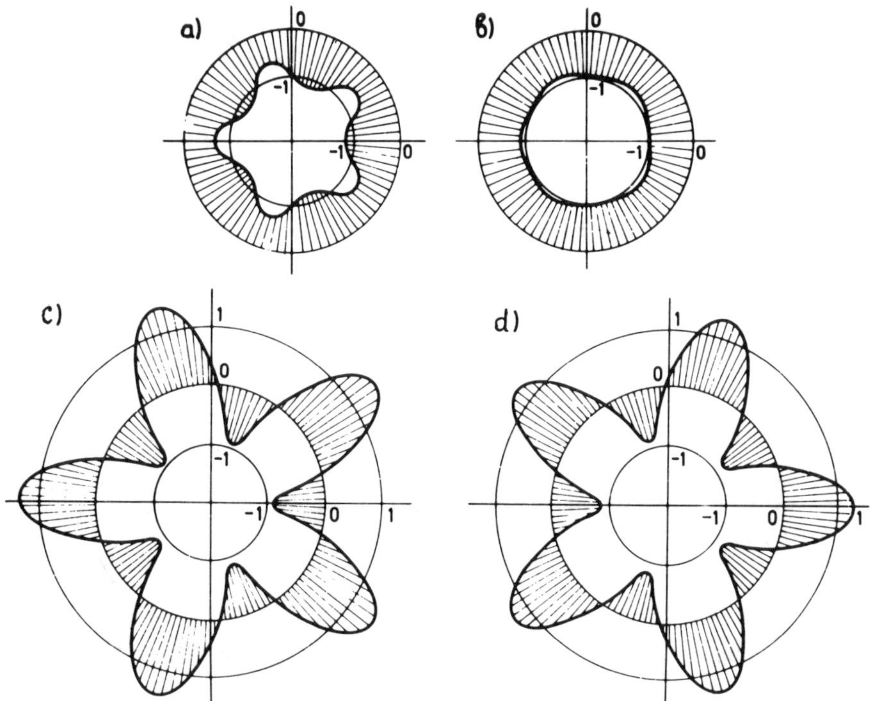

Fig. 4.9. Diagrams of axial $\sigma_{xx}/(2 \cdot 10^8\ \mathrm{N/m^2})$ (a, b) and circumferential $\sigma_{yy}/(5 \cdot 10^7\ \mathrm{N/m^2})$ (c, d) stresses in $x = L/2$ section (at the load parameters corresponding to the case of falling within one DIR) on the inner (a, c) and outer (b, d) surfaces of a shell.

with time. At certain moments, it is possible to define the presence of five or six nodal lines along the circumference of a shell, whereas in the intervals between these moments the spatial distribution of characteristics of stress–strain state is determined as a combination of two spatial modes of vibration with amplitudes being comparable in magnitude. It is natural to assume that such spatially-non-stationary vibrations are more dangerous for a structure than the vibrations in one fixed spatial mode.

As a confirmation of this fact we shall retrace the change occurring in stresses at fixed magnitude $\delta = 0.35$ when the frequency Θ is increasing from $\Theta_1 = 560\,\mathrm{rad/s}$ (the left boundary of DIR for $\{1, 5\}$ mode) up to $\Theta_4 = 905\,\mathrm{rad/s}$ (the right boundary of DIR for $\{1, 6\}$ mode). The frequencies $\Theta_2 = 700\,\mathrm{rad/s}$ and $\Theta_3 = 805\,\mathrm{rad/s}$ correspond to the left and right boundaries of the intersection of these two DIR (see Fig. 4.2).

In Fig. 4.11 the maximum magnitudes of stresses σ_{xx} and σ_{yy} calculated for each fixed frequency value are presented. As is seen, within the range of

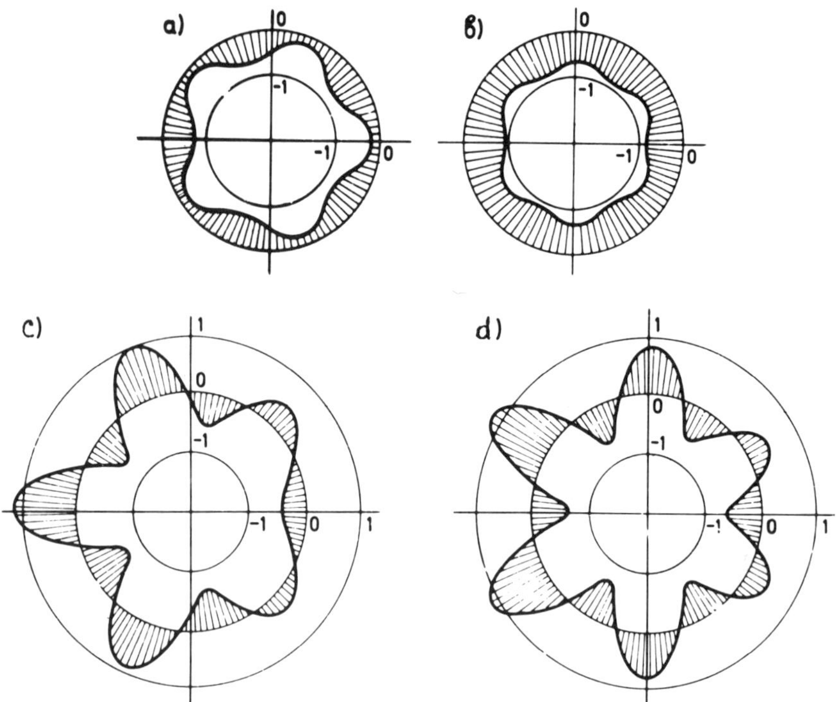

Fig. 4.10. Diagrams of axial $\sigma_{xx}/(2 \cdot 10^8\ \text{N/m}^2)$ (a, b) and circumferential $\sigma_{yy}/(5 \cdot 10^7\ \text{N/m}^2)$ (c, d) stresses in $x = L/2$ section (at the load parameters corresponding to the case of intersection of two DIR) at the time instants $t = 0.13$ s (a, c) and 0.17 (b, d).

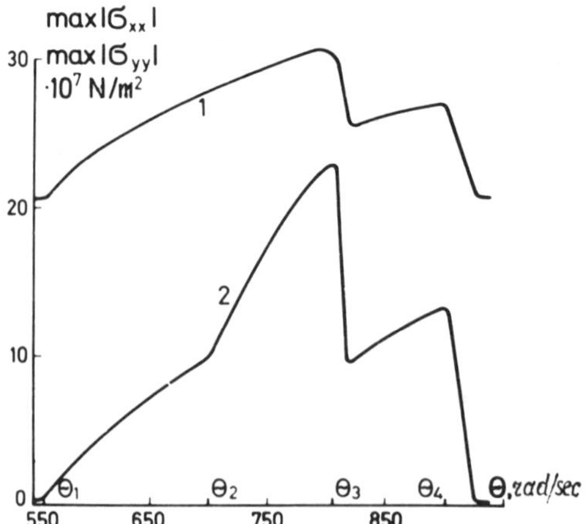

Fig. 4.11. The maximum stresses dependence on the frequency of external load.

frequencies $[\Theta_2, \Theta_3]$ corresponding to the DIR intersection, a considerable increase in stresses is observed, it being more strongly pronounced for σ_{yy} (the less pronounced increase in σ_{xx} is connected, obviously, with the great contribution of momentless component to this stress). Thus the maximum magnitude of circumferential stress (corresponding to the frequency somewhat lower than Θ_3) exceeds nearly twice the maximum of σ_{yy} being for vibrations out of the area of DIR intersection. The example under consideration shows that the conditions of parametric resonance, under which the amplitude and frequency of external load fall within the area of intersection of several DIR, should be the most dangerous from the viewpoint of fatigue failure of the shell.

In conclusion we shall note that the theoretically established and investigated effect of interaction of spatial modes of vibrations for cylindrical shells during parametric resonance (the respective results were first outlined in Ref. 53) was observed experimentally in Refs 110 and 114 later. In particular, the resonance conditions were found, under which three DIR, corresponding to two, three and four circumferential waves, intersected. It was noted in Ref. 144 that the motion of a shell involved simultaneously three modes of flexural vibrations. At separate time moments, sufficiently 'pure' flexural modes with $n = 3$, 4 and 2 could be identified on the background of otherwise non-stationary character of shell motion.

As is seen, qualitatively these experimental results correlate fully with the theoretical results revealed in Figs 4.9 and 4.10.

4.6 ON GOVERNING A SPECTRUM OF DYNAMIC INSTABILITY REGIONS OF COMPOSITE MULTI-LAYERED CYLINDRICAL SHELLS

The appearance of parametric resonance phenomenon in thin-wall structural elements is, naturally, undesirable. If the operating frequency range $[\Theta_1, \Theta_2]$ for a shell is known, but the expected amplitudes of axial vibrational force are limited from above by the excitation coefficient δ^*, it is possible to avoid parametric resonance by 'purifying' of DIR the rectangle $[\Theta_1, \Theta_2] \times [0, \delta^*]$ within the plane $\{\Theta, \delta\}$ parameters. For this sake, in accordance with the results of paragraphs 4.1 and 4.2, there appear two ways:

(1) variation of frequencies of natural vibrations ω_{mn} and critical tractions P^*_{mn};

TABLE 4.1

Mechanical Characteristics of Reinforcing Fibers (E_A, ν_A, ρ_A) and Orthotropic Laminae ($E_1, E_2, G_{12}, \nu_{12}, \nu_{21}, M$) with Reinforcement Coefficients μ', μ''

Fiber characteristics		*Boron*			*Graphite*			*Glass-994*			*E-glass*	
$E_A \cdot 10^{-10}$, N/m^2		37.8			16.6			8.52			7.2	
ν_A		0.2			0.2			0.33			0.22	
$\rho_A \cdot 10^{-3}$, kg/m^3		2.6			1.55			2.49			2.55	
Lamina characteristics	*Variant of lamina reinforcement*											
	1	2	3	*1*	2	3	*1*	2	3	*1*	2	3
μ'	0	0.3	0.6	0	0.3	0.6	0	0.3	0.6	0	0.3	0.6
μ''	0.6	0.3	0	0.6	0.3	0	0.6	0.3	0	0.6	0.3	0
$E_1 \cdot 10^{-10}$, N/m^2	0.97	11.7	22.8	0.95	5.37	10.1	0.92	2.94	5.25	0.90	2.54	4.46
$E_2 \cdot 10^{-10}$, N/m^2	22.8	11.7	0.97	10.1	5.37	0.95	5.25	2.94	0.92	4.46	2.54	0.90
$G_{12} \cdot 10^{-10}$, N/m^2	0.50	0.43	0.50	0.48	0.42	0.48	0.45	0.39	0.45	0.44	0.38	0.44
ν_{12}	0.26	0.07	0.01	0.26	0.08	0.02	0.34	0.15	0.06	0.27	0.11	0.05
ν_{21}	0.01	0.07	0.26	0.02	0.08	0.26	0.06	0.15	0.34	0.05	0.11	0.27
M, kg/m^2	9.4	9.4	9.4	6.5	6.5	6.5	9.1	9.1	9.1	9.2	9.2	9.2

(2) the use of the material with sufficiently high viscosity, affecting efficiently the DIR within the frequency range of interest.

In certain situations, it is enough to ensure durability of the shell, subjected to vibrational loading, over a fairly definite time interval. Such a statement of the problem permits the occurrence of parametric resonance but imposes restrictions on amplitudes of the stressses initiated in the shell. As is demonstrated by the study presented in paragraph 4.5, the regimes of parametric vibrations at which both frequency and amplitude of load fall within the zone of intersection of several DIR are the most dangerous. Obviously, the greater the number of DIR intersect in the given point of plane of parameters, the higher will be the level of stresses. This circumstance allows one to consider the problem of governing the DIR spectra with a milder restriction: it is expedient that within the range of specified rectangle $[\Theta_1, \Theta_2] \times [0, \delta^*]$ there should be no DIR intersections.

Let us consider several examples illustrating the ample possibilities of composite materials in respect to solving the problems indicated. In Table 4.1, the mechanical characteristics of four types or reinforcing fibers as well as those of unidirectionally and orthogonally reinforced composites are presented. The composites are based on an epoxy matrix with $E_m = 3.48 \cdot 10^9\,\mathrm{N/m^2}$, $\nu_m = 0.35$, $\rho_m = 1.3 \cdot 10^3\,\mathrm{kg/m^3}$. The values μ' and μ'' are reinforcement coefficients of an individual layer in a shell along x- and y-axis, respectively. The overall coefficient of reinforcement for all layers is maintained constant: $\mu_A = \mu' + \mu'' = 0.6$. The characteristics E_1, E_2, G_{12}, ν_{12}, ν_{21} presented in Table 4.1 were calculated according to the formulas in Refs 1 and 185.

The calculations were made for all variants at geometric parameters: $R = 1\,\mathrm{m}$, $L/R = 2.4$, $R/h = 144$ and one and the same constant component of load $P_0 = 1.1 \cdot 10^6\,\mathrm{N/m}$. In subsequent figures, the value $\tilde{P}_t = (P_t)/(P_t + P_0)$ is plotted on ordinate axis.

In Fig. 4.12, the DIR spectra within the range of frequencies right up to 400 Hz are presented. The shells with longitudinal-transversal reinforcement ($\mu' = \mu'' = 0.3$) with graphite and two types of glass fibers are considered. It is seen from these parts of DIR spectra that the increase in stiffness of the reinforcing fibers shifts the first (left) DIR toward higher frequencies and enlarges greatly the distance between adjacent regions. The width of each DIR reduces at the same time. In such a way, the increase in stiffness of the material allows one to considerably extend the frequency range, free of DIR, and avoid the DIR intersections within the rectangle $0 \leq (\Theta/2\pi) \leq 280\,\mathrm{Hz}$; $0 \leq \tilde{P}_t \leq 0.3$, in the example treated.

The effect of fiber direction on the low-frequency part of DIR spectrum is illustrated in Fig. 4.13. As is seen, from three variants under

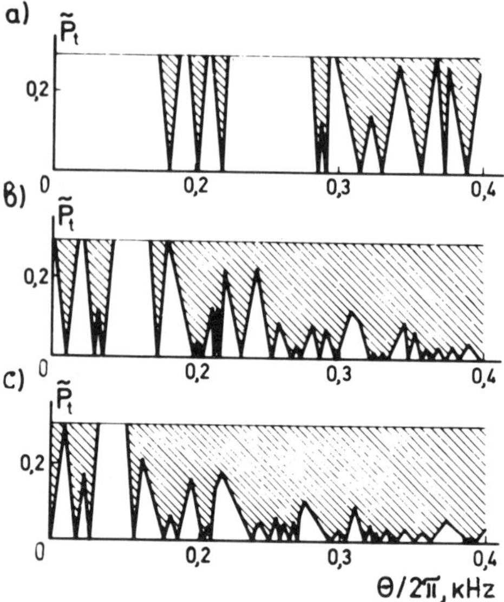

Fig. 4.12. Initial sections of DIR spectra for single-ply orthotropic shells with various types of reinforcing fibers: (a) graphite; (b) glass-994; (c) E-glass.

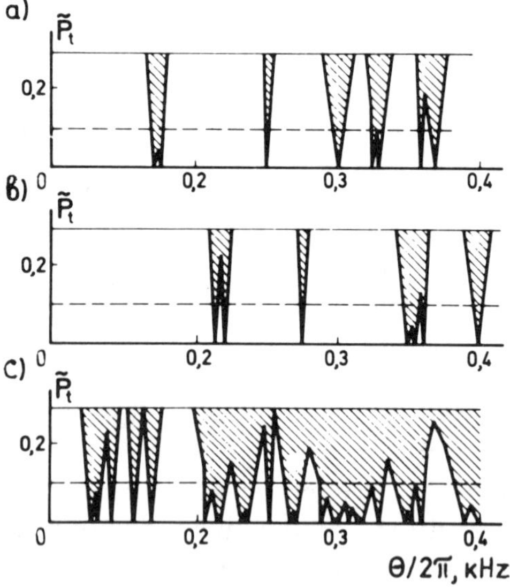

Fig. 4.13. Initial part of DIR spectra for boron plastic single-ply orthotropic shells reinforced in various directions: (a) $\mu' = 0$, $\mu'' = 0.6$; (b) $\mu' = \mu'' = 0.3$; (c) $\mu' = 0.6$, $\mu'' = 0$.

consideration, the variant of longitudinal-transversal reinforcement is preferential. For an axial reinforcement, the DIR spectrum is displaced toward lower frequencies; besides, within the frequency range considered, the spectrum is considerably denser than that for circumferential and longitudinal-transversal reinforcements. This example shows that unsuccessful selection of fiber layup can reduce the advantages of high-modulus fibers to zero. The axially reinforced boron plastic shell proved to be practically equivalent as regards its low-frequency DIR to a glass fiber shell with a longitudinal-transversal reinforcement (cf. Fig. 4.13(c) and 4.12(b)), although the elastic modulus of fibers is more than four times lower.

The use of a multi-layer structure gives extra possibilities for varying the DIR spectra as it is necessary for practical purposes. In particular, it is possible to obtain the following results:

(a) to shift the very left boundary of the DIR spectrum toward higher frequencies;
(b) to enlarge the distance between the adjacent DIR;
(c) to reduce the width of each DIR.

To illustrate this, in Fig. 4.14 the low-frequency part of DIR spectrum for four multilayer carbon plastic shells with symmetrical and self-balanced laminate structures are presented. The membrane C_{ij} and flexural D_{ij} stiffness of the laminates were calculated by the use of characteristics of carbon plastic laminae (see Table 4.1) according to the well-known formulas.[21,24] In the left side of Fig. 4.13 a composition of the upper halves of laminates is presented. Here, φ_i is the angle between x-axis and the fiber direction in the i-th layer.

A visual comparison of various composite shells can be based on DIR diagrams. However, in the cases when an exact assessment of the quality of one or another shell is required (for example, in solving the problems of optimal design), it is necessary to have a quantitative characteristic of DIR spectra. Such a characteristic should respect the location of the first DIR (the beginning of the spectrum), spectrum density, as well as the change in width of each DIR with varying load amplitude $\tilde{P}_t$. This can be done by introducing such a function which characterizes the degree of filling-in of $\{\Theta, \tilde{P}_t\}$ plane with DIR.

Let us consider a straight line, parallel to Θ axis at a fixed magnitude of $\tilde{P}_t$. It is possible to calculate the total length of intercepts, enclosed within DIR, for any frequency range $[\Theta_1, \Theta_2]$. Let us evaluate the function of filling degree $\xi(\Theta_0, \Theta)$ as a relation of this sum calculated within the $[\Theta_0, \Theta]$ range over the value of $\Theta - \Theta_0$ (in particular, we can set $\Theta_0 = 0$). In such a

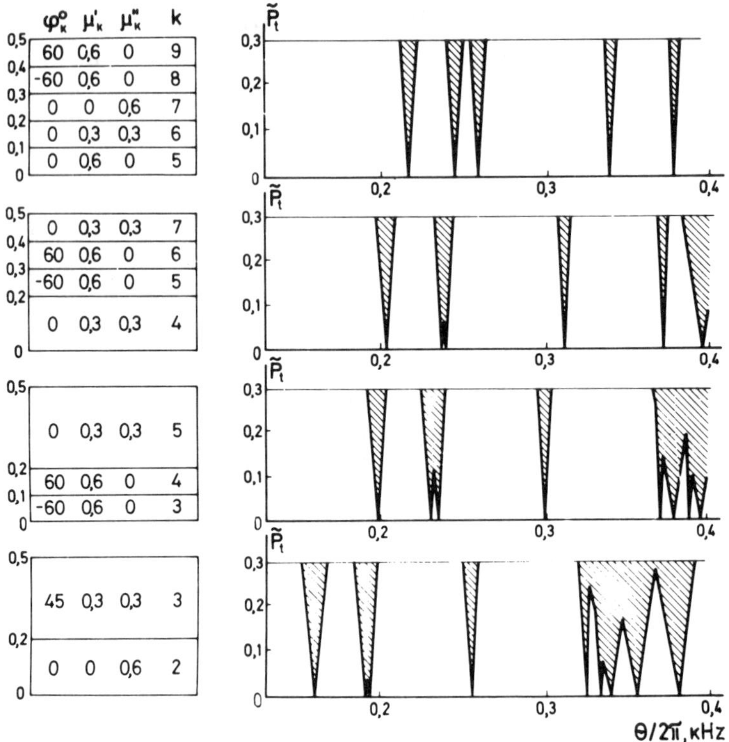

Fig. 4.14. Stacking sequence and initial parts of DIR spectra for carbon plastic shells with different number of layers.

way, if at the given level of $\tilde{P}_t$ inside the $[\Theta_0, \Theta]$ range there is no DIR, then $\xi(\Theta_0, \Theta) = 0$ and, on the contrary, if all the region of parameters at this level of $\tilde{P}_t$ is occupied by DIR, $\xi(\Theta_0, \Theta) = 1$. Henceforth, we shall employ more often the $\eta(\Theta_0, \Theta)$ function related with $\xi(\Theta_0, \Theta)$ by the formula $\eta = 1 - \xi$. In such a way, if $\xi = 0$, then $\eta = 1$ and, on the contrary, at $\xi = 1$, the function $\eta = 0$. Further, let us present some examples of calculating the function $\eta(0, \Theta) = \eta(\Theta)$.

In Fig. 4.15, the $\eta(\Theta)$ functions at $\tilde{P}_t = 0{\cdot}3$ are presented for various single-layer shells, the low-frequency parts of DIR spectra for which are shown in Figs 4.12 and 4.13. The beginning of $\eta(\Theta)$ curve corresponds to the left boundary of the first DIR. The magnitude of $\eta(\Theta)$ function at fixed Θ depends both on the number of regions within $[0, \Theta]$ range and on their width. The uppermost curve in Fig. 4.15 corresponds to the spectrum of Fig. 4.13(b), the lowest — to the spectrum of Fig. 4.12(c).

In Fig. 4.16, the $\eta(\Theta)$ functions for multi-layer carbon plastic shells are plotted. The low-frequency parts of their DIR are shown in Fig. 4.14. The

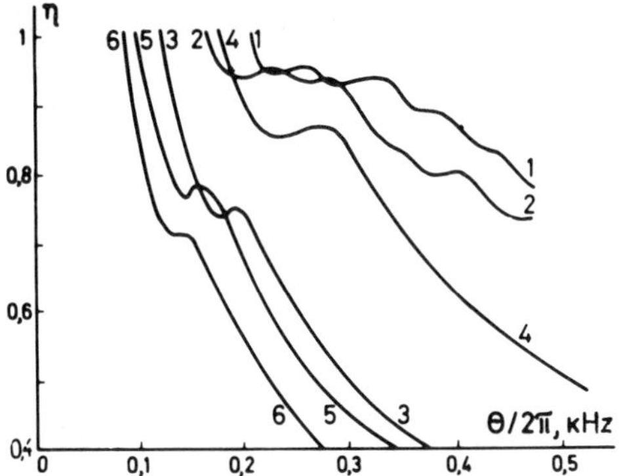

Fig. 4.15. Dependencies $\eta(\Theta)$ corresponding to DIR spectra of Figs. 4.12 and 4.13: 1 — 4.13, b; 2 — 4.13, a; 3 — 4.13, c; 4 — 4.12, a; 5 — 4.12, b; 6 — 4.12, c.

numbers at the curves correspond to the number of layers in the shell laminate. The uppermost curve corresponds to nine-ply shell, the lowest to a single-ply shell. Let us note that the $\eta(\Theta)$ function depends on the magnitudes of P_0, P_t as on parameters. The greater P_t, the lower is the $\eta(\Theta)$ curve. Upon increasing P_0, the curve is shifted toward lower magnitudes of Θ.

As a supplement to $\xi(\Theta_0, \Theta)$ and $\eta(\Theta_0, \Theta)$ functions yielding a sufficiently full quantitative information not only about the location and width of the first DIR, but also about the character of the entire DIR spectrum, we shall introduce the integral characteristic

$$I(\Theta_0, \Theta) = \int_{\Theta_0}^{\Theta} [1 - \eta(\Theta)]\, d\Theta. \tag{4.77}$$

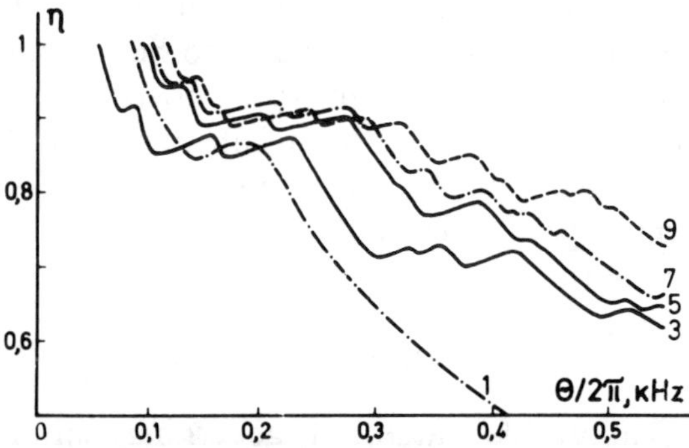

Fig. 4.16. Dependencies $\eta(\Theta)$ corresponding to DIR spectra of Fig. 4.14. Numbers at curves are equal to the number of layers.

This is convenient as a physical constraint in solving the problems of optimal design.

Let us emphasize in the conclusion that, although only the main DIR have been investigated above, the ways of generalizing the methodologies for the case of consideration the second, third and other accessory DIR and for the combined resonances, stand out fairly obviously.

4.7 EXAMPLES OF SOLVING THE PROBLEMS OF OPTIMAL DESIGN FOR SHELLS OPERATING IN THE REGIME OF PARAMETRIC RESONANCE

On the basis of methods presented in paragraphs 4.1–4.5 for the calculation of elastic and viscoelastic shells under axial vibrational load several problems of optimal design of a multi-layer cylindrical shell with orthotropic reinforced layers can be formulated. In a general statement the problem consists in finding out such a vector of design parameters

$$\mathbf{x}^* = (x^*_1, \ \ldots, \ x^*_n), \tag{4.78}$$

at which the selected objective function $C(\mathbf{x})$ reaches its optimal magnitude within the range D of allowable magnitudes of governing parameters $\mathbf{x}$, i.e.

$$C(\mathbf{x}^*) = \underset{\mathbf{x} \in D}{\mathrm{opt}}\, C(\mathbf{x}), \tag{4.79}$$

where

$$D = \{:\mathbf{x}: g_j(\mathbf{x}) \leqslant 0, \ j = 1, 2, \ \ldots, \ N\}; \tag{4.80}$$

$g_j(\mathbf{x}) = (g_1, g_2, \ldots, g_N)$ is the vector of constraint functions; N is the total number of constraints.

Assuming the length and the reference surface to be fixed, let us postulate the mass M of a shell surface unit to be the objective function. Then at the fixed characteristics of reinforcing fibers and matrix, the design parameters will be the volume fractions of fibers in both the directions μ_i', μ_i'' and thickness of layers Δ_i. At a fixed number of layers n in a shell the number of governing parameters therefore is $3n$, and the corresponding vector assumes the form

$$\mathbf{x} = (\mu'_1, \mu''_1, \Delta_1, \ \ldots, \ \mu'_n, \mu''_n, \Delta_n). \tag{4.81}$$

The region of allowable values of governing parameters (4.80) is restricted from above and below by technologically accessible reinforcement coefficients and reasonable values of layer thickness. Besides, this region is restricted according to the constraints for static stability and other physical constraints, taking into account the specific features of mechanical behavior of a shell under axial vibrational loading.

Let us write down the constraints which are to be fulfilled for each individual layer:

1) $$g_j(\Delta_i) = h_0 - \Delta_i \leqslant 0, \tag{4.82}$$

where $j = 1, 2, \ldots, n$, while h_0 is the allowable minimum layer thickness,

2) $$g_j(\mu'_i) = -\mu'_i \leqslant 0, \quad j = n+1, \ldots, 2n, \tag{4.83}$$

3) $$g_j(\mu''_i) = -\mu''_i \leqslant 0, \quad j = 2n+1, \ldots, 3n, \tag{4.84}$$

4) $$g_j(\mu'_i, \mu''_i) = \mu_0 - \mu'_i - \mu''_i \leqslant 0, \quad j = 3n+1, \ldots, 4n \tag{4.85}$$

(here, μ_0 is the minimum total volume fraction of fibers),

5) $$g_j(\mu'_i, \mu''_i) = \mu'_i + \mu''_i - \mu^0 \leqslant 0, \quad j = 4n+1, \ldots, 5n, \tag{4.86}$$

where μ^0 is the maximum total volume fraction of fibers.

For the shell as a whole we suggest that its thickness should not exceed some given value h^0:

6) $$g_{5n+1}(\Delta_i) = \sum_{i=1}^{n} \Delta_i - h^0 \leqslant 0. \tag{4.87}$$

The constraints (4.82)–(4.87) are not in need of any further comments, as well as the constraint in stability

7) $$g_{5n+2}(\mathbf{x}) = P_0 + P_t - P^*(\mathbf{x}) \leqslant 0, \tag{4.88}$$

where $P^*(\mathbf{x})$ is the critical static force for a shell under axial compression.

Let us dwell in detail on the selection of physical constraints in terms of which the specific mechanical behavior of the shell operating under axial vibrational load is reflected.

The condition that within the shell operating range of frequencies $[0, \Theta_0]$ there should be no DIR, serves as the first variant of physical constraints. It is expressed in the following form

$$g_{5n+3}(\mathbf{x}, \Theta_0) = \Theta_0 - \Theta_{\text{H}}(\mathbf{x}) \leqslant 0, \tag{4.89}$$

where Θ_L are the frequencies corresponding to the left DIR boundaries. The constraint (4.89) is essentially the generalization of the widely used constraint imposed on the lowest natural frequency.

The second, weakest variant of the physical constraint can be formulated by admitting the existence of one or several DIR within the frequency range $[0, \Theta_0]$, but with the limited total width of these DIR. It can be mathematically formulated by employing the $\eta(\Theta)$ function introduced in paragraph 4.6 or the respective integral $I(\Theta) = I(0, \Theta)$. If, for example, I_0 is the upper allowable value of $I(\Theta)$ at $\Theta = \Theta_0$, then the constraint is written down as

$$g_{5n+3}(\mathbf{x}, \Theta_0) = I(\mathbf{x}, \Theta_0) - I_0 \leqslant 0. \qquad (4.90)$$

The third type of the physical constraints can be formulated according to the requirement that within the specified frequency range the adjacent DIR should not intersect. In such a way, the condition is laid down that the lowest frequency Θ_{I}, at which two adjacent DIRs intersect, exceeds Θ_0:

$$g_{5n+3}(\mathbf{x}, \Theta_0) = \Theta_0 - \Theta_{\text{II}}(\mathbf{x}) \leqslant 0. \qquad (4.91)$$

Thus let us formulate the problem of optimal design: such a vector of design parameters should be found

$$\mathbf{x}^* = (\mu'^*_1, \mu''^*_1, \Delta^*_1, \ldots, \mu'^*_n, \mu''^*_n, \Delta^*_n), \qquad (4.92)$$

at which the objective function — mass of the shell M — reaches its minimum value

$$M(\mathbf{x}^*) = \min_{\mathbf{x} \in D} M(\mathbf{x}) \qquad (4.93)$$

within the region of allowable magnitudes of governing parameters (4.80). This region is defined by the constraints $g_j(\mathbf{x})$ according to (4.82)–(4.88) and one of the three physical constraints $g_j(\mathbf{x}, \Theta_0)$ (4.89)–(4.91).

In order to choose an efficient method of solving the problem formulated it is necessary to reveal the basic features of an extremum problem. The problem under consideration has the following features: the objective function — mass $M(\mathbf{x})$ is strictly monotonous on each of the design parameters, and the extremum of the objective function belongs to the boundary Γ_D of the region D of allowable magnitudes of governing parameters. Let us remark also that because of the properties of the extremal problem, the constraints (4.86) and (4.87) are fulfilled automatically through selection of the initial vector of design parameters $\mathbf{x}_0$. The lowest magnitudes of these parameters, defined by the functions of constraints (4.82)–(4.85), are practically unattainable because of the functions of constraints (4.88)–(4.91). A simple expression of the objective function makes it easier to use the well developed gradient methods evolved for the search of optimum by means of reducing the design parameters.

Another alternative statement of the problem of optimal design is also possible. According to this statement, at fixed mass of the shell, such a structure of its laminate is being searched for, at which certain characteristics of the shell operating under axial vibrational load reach their extremal magnitudes within the region D' of allowable magnitudes of governing parameters. Then one of the following functions can be conceived as the objective function:

(a) the extent Θ_L of the frequency range $[0, \Theta_L]$, free of DIR; this value is reasonable to maximize:

$$\Theta_L(\mathbf{x}^*) = \max_{x \in D'} \Theta_L(\mathbf{x}); \tag{4.94}$$

(b) the integral $I(0, \Theta_0) = I(\Theta_0)$, which it is reasonable to minimize:

$$I(\mathbf{x}^*, \Theta_0) = \min_{\mathbf{x} \in D'} I(\mathbf{x}, \Theta_0); \tag{4.95}$$

(c) the extent Θ_I of the frequency range $[0, \Theta_I]$ which does not comprise intersections of DIR; it is reasonable to maximize this value:

$$\Theta_I(\mathbf{x}^*) = \max_{x \in D'} \Theta_I(\mathbf{x}). \tag{4.96}$$

The vector of design parameters maintains the form (4.81), while the region of allowable magnitudes of governing parameters

$$D' = \{:\mathbf{x} : g'_j(\mathbf{x}) \leqslant 0, \ j = 1, 2, \ \ldots, \ N'\} \tag{4.97}$$

will be defined by the new vector of constraint functions $g'_j(\mathbf{x})$; the number of these functions N' cannot coincide with N. The functions (4.82)–(4.89) and the stability constraint (4.88) are preserved also for the region of allowable magnitudes D', while instead of constraints (4.85)–(4.87) it is necessary to put the conditions of constancy of the shell mass. These conditions can be written, for example, as follows

$$g'_j(\mu'_i, \mu''_i) = \mu'_i + \mu''_i - \mu = 0, \ j = 3n+1, \ \ldots, \ 4n, \tag{4.98}$$

where μ is the total volume fraction of fibers, which is the same for all layers, and

$$g'_{4n+1}(\Delta_i) = \sum_{i=1}^{n} \Delta_i - h = 0, \tag{4.99}$$

where h is the shell thickness, which is maintained constant.

TABLE 4.2

Values of Optimal Vector of Design Parameters and Objective Function under Various Physical Constraints

Constraint	$\Theta_L(\mathbf{x})/2\pi \geq 200\ Hz$		$\Theta_I(\mathbf{x})/2\pi \geq 200\ Hz$		$I_0(\mathbf{x}, \Theta_0) \leq 5\ Hz$	
	$\mu_0 = 0$	$\mu_0 = 0.5$	$\mu_0 = 0$	$\mu_0 = 0.5$	$\mu_0 = 0$	$\mu_0 = 0.5$
Δ_1, mm	1.78	1.72	1.73	1.62	1.67	1.44
μ_1'	0.20	0.25	0.11	0.25	0.02	0.25
μ_1''	0.22	0.25	0.17	0.25	0.11	0.25
Δ_2, mm	0.97	0.92	0.91	0.82	0.85	0.63
μ_2'	0.22	0.18	0.18	0.17	0.13	0.16
μ_2''	0.39	0.37	0.37	0.36	0.34	0.35
Δ_3, mm	1.38	1.33	1.32	1.22	1.26	1.03
μ_3'	0.06	0.09	0	0.08	0	0.05
μ_3''	0.46	0.48	0.42	0.46	0.38	0.45
Δ_4, mm	0.57	0.51	0.50	0.40	0.43	0.21
μ_4'	0.49	0.46	0.46	0.42	0.43	0.37
μ_4''	0.23	0.22	0.21	0.19	0.20	0.16
M, kg/m^2	9.32	9.03	8.22	8.10	7.27	6.49

The complexity of the objective functions (4.94)–(4.96) excludes practically the use of gradient methods for solving the problems of optimal design in this statement. The use of methods of the theory of experiment planning can be conceived as one of the possible ways in such a situation.

Henceforth we shall restrict ourselves to some illustrative examples of solving the problem for finding out the minimum of the shell mass at physical constraints (4.89)–(4.91). In Table 4.2, the results of solving the problem of mass minimization for a four-ply shell with $R = 1$ m, $L/R = 2.4$ and orthogonally reinforced boron/epoxy layers at three types of physical constraints (4.89), (4.91) and (4.90) are presented. Two magnitudes of minimum allowable total volume fraction of fibers for every layer (4.85) have been considered: $\mu_0 = 0$ and $\mu_0 = 0.5$. At the same time, twelve control parameters underwent variation: μ_i', μ_i'', Δ_i, where $i = 1, 2, 3, 4$. The problem was solved by the method of gradient descent in combination with the method of exhaustive search for selecting the initial vector of design parameters. An insignificant deviation (approx. 3–4%) of the mass of optimal shell was revealed as the result of varying the initial point $\mathbf{x}_0$. By comparing the results of solving the problem of optimal design on the basis of twenty initial points, the vector $\mathbf{x}_0 = (0.5; 0.4; 2.0; 0.4; 0.5; 1.2; 0.3; 0.6; 1.6; 0.6; 0.3; 0.8)$ was selected as the initial vector of design parameters; the values of Δ_i are given in mm.

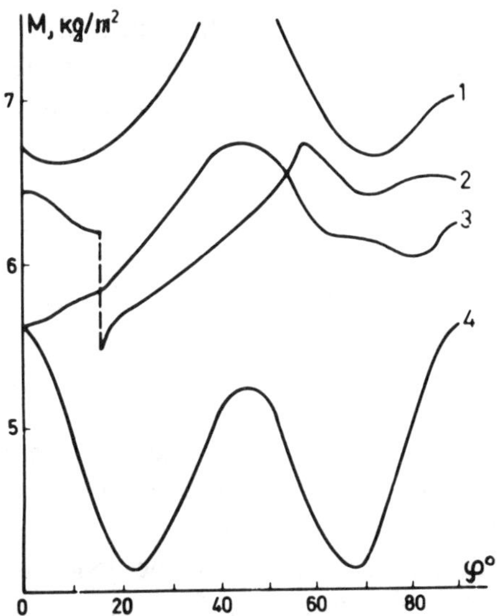

Fig. 4.17. The surface unit mass dependence on the reinforcement angle φ for optimal angle-ply carbon shells with various physical constraints: (1) $\Theta_L/2\pi \geq 200\,\text{Hz}$; (2) $\Theta_I/2\pi \geq 200\,\text{Hz}$; (3) $I_0 \leq 5\,\text{Hz}$; (4) $P^* \geq P_0 + P_t$.

The essential decrease in the optimal mass of the shell with weakening of the physical constraint is seen from the comparison of the obtained magnitudes of the objective function: under the constraint (4.91) the mass decreased by 10–12% and under the constraint (4.90) by 22–28% in comparison to the case of constraint (4.89). This result shows that the decrease in the shell mass can be attained at the expense of a reasonable weakening of physical constraints in accordance with the actual requirements imposed on the structure.

TABLE 4.3

Values of Optimal Parameters of Laminated Structure and Surface Unit Mass of Two-Ply Carbon Shell

Laminate characteristics	$\Theta_L/2\pi \geq 200\ Hz$	$I_0 \leq 5\ Hz$	$\Theta_I/2\pi \geq 200\ Hz$	$P^* \geq P_0 + P_t$
φ^0	10	0	16	68
$\mu'_{1,2}$	0.176	0.112	0.102	0.108
$\mu''_{1,2}$	0.442	0.403	0.398	0.401
h, mm	4.55	3.94	3.83	2.89
M, kg/m^2	6.62	5.63	5.46	4.12

In considering the structure of the optimal packets with $\mu_0 = 0$, described in Table 4.2, let us pay attention to the fact that the decrease in the shell mass is due, to a greater extent, to the reduction in the volume fraction of fibers than to the decrease in layer thickness. Imposing an additional constraint ($\mu_0 = 0.5$) might have led to the increase in mass. However, the result is directly opposite: for all three types of constraints with $\mu_0 = 0.5$, the mass of optimal shells turned out to be smaller than that for the case of $\mu_0 = 0$. On the one hand, this result indicates that by employing the gradient method only the local optimum can be found, and, on the other hand, it emphasizes the necessity of introducing constraints of the type (4.85) with the aim of regulating the optimal search in the direction of thinner and tougher (with higher volume fiber fraction) layers.

Further let us consider the results of solving the optimization problem for a two-ply graphite/epoxy balanced laminate shell with orthogonally reinforced layers, laid at an angle $\pm\varphi$ to the generatrix. These results are presented in Fig. 4.17 and Table 4.3. As is seen, under all four types of physical constraints (4.88)–(4.91) the effect of φ angle on the optimal mass is considerable. Thereby we shall remark that the $M(\varphi)$ functions for different physical constraints differ notably. Besides, at the given magnitudes of Θ_0 and I_0, the constraints (4.90) and (4.91) turn out to be competitive: which of them becomes active, it depends on the φ value. This is connected with the definition of the functions, according to which the intersection of adjacent DIR being in contrast to the condition (4.91) is favorable for fulfillment of (4.90), since such an intersection decreases the total DIR width within the frequency range under consideration.

In completing the discussion of the results of solving the optimal design problems for cylindrical shells made of composite laminates, operating under axial vibrational loading, let us emphasize once more the significance of imposing reasonable physical constraints. The effect of the decrease in mass at the expense of the allowable (from the exploitation conditions viewpoint) weakening of these constraints can be comparable to the effect of optimal design at fixed constraints of the structure itself.

5

Calculation of Deformation Processes for Orthotropic Cylindrical Shells under Dynamic Compressive Loads

The problem of deformation of orthotropic cylindrical shells loaded with uniformly distributed over the end faces longitudinal dynamic compressive forces and uniformly distributed over the outer surface dynamic lateral pressure is under consideration in geometrically non-linear statement. By employing the finite difference method the axisymmetric problem is solved for perfect and imperfect orthotropic shells at seven versions of end conditions. One of them is the version simulating the end conditions produced during the experiment of longitudinal rigid mass impact. Two calculation procedures are proposed for solving non-axisymmetric problems. The first procedure allows to take into account the process of stress wave propagation from the end faces and the non-linear axisymmetric edge effect. The Bubnov-Galerkin method along the circumferential coordinate and the finite difference method along the longitudinal coordinate are used in this procedure. The second procedure ignores the longitudinal inertia of a shell; the problem of non-axisymmetric dynamic buckling is solved on the basis of multi-term deflection approximations. The results obtained by these two procedures are compared. A series of novel effects associated with non-linear interaction of axisymmetric and non-axisymmetric deformation processes in a shell are established. Selection of appropriate dynamic instability criterion for imperfect shells is discussed. The solutions to geometrically non-linear problems, proposed in this chapter, are based on the assumption of linearly elastic behavior of material. One of the main objectives in analyzing the results concerns the validation of applicability of the linearly elastic model to the dynamic buckling analysis for cylindrical shells under axial compression and external pressure.

The problem of dynamic instability of thin-walled cylindrical shells has been attracting great interest since the late 50s. The relevant problem statement and the numerical results obtained were discussed in the monographs of A. S. Vol'mir,[97,100] R. Gryboš,[339] reports and review papers of V. V. Bolotin,[66,68] G. Yu. Dzhanelidze,[152] A. S. Vol'mir,[96,98,99] B. Budiansky,[295,296,350] N. J. Hoff,[346] V. I. Feodos'ev,[251] and V. L. Agamirov.[3] The works on the dynamic instability of cylindrical shells under longitudinal end face impact was comprehensively analyzed in a review paper by A. E. Bogdanovich.[47] Besides, practically a full list of publications on this problem (prior to 1977) is given in monograph by R. Gryboś.[339] Because of what was said above we shall restrict ourselves to a brief outline of the basic researches and principal results directly pertaining to the subject matter of this chapter and affecting author's philosophy.

At first, it is necessary to mention M. A. Lavrent'ev and A. Yu. Ishlinskii's work,[178] where it was first shown, by the example of the longitudinal dynamic bending problem for a thin bar, that comparatively high harmonics prevail in the process of dynamic buckling. This principal result holds good for any problem of dynamic buckling of thin-walled structural elements.

The theory of dynamic shell instability was developed in several directions. One of the directions was originated by A. S. Vol'mir,[95] devoted to the geometrically non-linear problem of axial dynamic compression of imperfect circular cylindrical panel. The motion of the panel was described by equations of medium bending written in the mixed form. The solution was obtained by the use of the Bubnov-Galerkin method with the approximation of both the overall and the initial deflections by only one, namely, the first term of Fourier series. An analogous problem for a closed circular cylindrical shell loaded by the combination of dynamic axial compression and dynamic external pressure was solved by V. L. Agamirov and A. S. Vol'mir.[4] They used the same method but with the two-term deflection approximation. V. V. Bolotin and G. A. Boichenko[74] treated the cylindrical panel by linearly increasing in time the longitudinal edge tractions. They obtained numerical results which revealed a series of important specific features in the non-linear behavior of the panel. The solutions of the problem analogous to Ref. 4 were obtained by using Bubnov-Galerkin method in Refs 44, 217, 218, 243 and 410. These solutions differ only by the basic deflection approximations and the procedure used for the unknown functions determination. In Refs 5 and 6, the problem of longitudinal rigid mass impact was treated in the statement analogous to Ref. 4. A further extension of these researches to orthotropic shells was reported by A. S. Vol'mir and L. N. Smetanina.[104] The problem of dynamic buckling of an orthotropic shell with stepwise increasing load in time applied to the end faces was treated in Ref. 270.

Among the most important works in this direction, Refs 434, 437 and 466 should be mentioned. Let us add that in Ref. 101, an attempt was made to take the finiteness of the velocity of perturbation propagation along the shell into account.

Within the framework of the direction in dynamic buckling theory, to which the above mentioned works belong, many important features of the non-linear dynamic buckling of imperfect cylindrical shells have been revealed: a specific deflection dependence on time, the effect of loading rate, amplitude of initial imperfections and relations between geometric parameters of a shell on this dependence. The load having caused dynamic instability was usually assessed from the criterion under which a sharp increase of the time-dependent coefficient at certain buckling mode starts. This buckling mode is indicated from the calculations and it is characterized as having the highest growth rate of all the modes. Another criterion supposed the condition that the above mentioned time-dependent coefficient would reach the specified ultimate value.

The method of solving the problem of non-axisymmetric dynamic buckling of imperfect cylindrical shells suggested in our papers[55–57] was, as a matter of fact, a further extension of this direction. The solution obtained by the Bubnov-Galerkin method was also based on equations of medium bending written in the mixed form. But two aspects were essentially novel. Firstly, an arbitrary field of initial imperfections were taken into account (specified in functional form or by means of Fourier series). Correspondingly, also the result of solving this problem (namely, the calculated displacements, deformations, and stresses in a shell), calculated by summing up the double Fourier series, involved in total all those spatial harmonics which should be taken into account in the summing up procedure. Secondly, the obtained time dependencies for any characteristic of stress–strain state in a shell allows to apply a number of different dynamic instability criteria and, in particular, any criterion used in the above mentioned works. A concrete form of the criterion can be selected, proceeding from the service requirements for the structure.

The theoretical works of the second direction in the theory of dynamic instability of cylindrical shells involve, in fact, the solution to the problems of dynamic instability on the basis of linearized equations of motion. For this sake, the methods of the classical theory of motion stability and various approximate analytical approaches have been used. They allow to analyze the stability of a shell in relation to minute 'disturbances'.

L. A. Movsisyan's[192] and V. I. Borisenko's works[80–82] were among the earliest ones belonging to this direction. In Ref. 82, in particular, the experimental and theoretical results for the case of longitudinal rigid mass impact on a cylindrical shell were compared. Obviously, in these works the problem of dynamic instability with the process of load wave propagation taken into account was first considered. Moreover, the effect of end restraint conditions on the dynamic instability process was investigated. From among the results obtained in subsequent years the ones by V. M. Darevskii,[149–151] I. A. Kiiko,[165,166] V. M. Kornev,[171,172] and I. Ya. Amiro[26] should be mentioned.

The works of the third direction in the theory of dynamic instability of cylindrical shells were aimed at analyzing the process of deformation of a shell (an ideal or imperfect one) during a certain time interval, following the moment of load application. This aim was realized through direct numerical integration of the equations of motion. The geometrically non-linear axisymmetric problems for isotropic and orthotropic shells were treated. The first work in this direction was performed by V. I. Borisenko and A. I. Klokova.[83] In a series of B. A. Gordienko's papers,[129–131] as well as in Ref. 83, the solution was obtained by the finite difference method. The problems of the same type were treated in Ref. 58, both for orthotropic and more general cross-ply laminated cylindrical shells; the specifics of non-linear axisymmetric edge effect at various combinations of boundary conditions has been comprehensively studied.

The algorithm for solving the problem of non-axisymmetric non-linear dynamic deformation of an isotropic imperfect cylindrical shell loaded by longitudinal impact, including the process of disturbances propagation, momentness and inhomogeneity of axisymmetric stress–strain state was evolved by V. G. Bazhenov and E. V. Igonicheva.[37] In our paper,[59] this algorithm was modified and generalized for the orthotropic material and different types of boundary conditions, specified on the end faces of a shell of finite length. A generalization of the algorithm,[37] with interaction of a finite number of circumferential harmonics taken into account, was carried out by the authors as described in Ref. 38. Let us note that the solution to the problem of non-linear non-axisymmetric deformation of a cylindrical shell loaded by axial dynamic compression, with such an interaction taken into account, was first obtained in our papers.[55,57]

Great attention has also been paid to experimental study on the behavior of cylindrical shells under axial dynamic loads. In this connection, let us mention the first paper of A. P. Coppa,[170] V. L. Agamirov and A. S. Vol'mir,[6] H. E. Lindberg and R. E. Herbert[368] and subsequent contributions.[82,105,128,131,201,248,249,349,385,386,438] On the basis of experimental and theoretical results, obtained on the problem of axial impact, A. P. Coppa in Ref. 170 and later, a number of other authors[128,156,157,368] put forward different variants of phenomenological analysis of the dynamic buckling process in a thin-walled cylindrical shell. Some new features of this process were revealed in Refs 57–59. The problem of selecting a proper system of equations of motion for a shell, applicable to the problems of impact buckling, was discussed by G. S. Nechiporuk and Ten En So.[202]

V. L. Agamirov and A. S. Vol'mir,[4] V. V. Bolotin *et al.*,[75] and Yu. P. Kadashevich and A. K. Pertsev[162] undertook the development of the geometrically non-linear dynamic problem for imperfect cylindrical shells loaded with impulsive lateral pressure. In Refs 4 and 75 the solutions were obtained by the use of approaches analogous to those employed in Refs 4, 74 and 95 for the problems of longitudinal dynamic compression. In Ref. 162, the effect of inertness of an axisymmetric pre-critical stress state in a shell on the process of dynamic buckling was discussed. It was suggested in this paper to assess the critical pulse according to the yield condition for the material. Let us also mention B. I. Slepov's work,[230] where the linear problem of dynamic stability of a perfect shell was studied. The most interesting researches in subsequent years were presented in the works of J. D. Wood and L. R. Koval,[458] G. R. Abrahamson and H. E. Lindberg,[273] D. L. Anderson and H. E. Lindberg,[277] Lindberg,[367] I. K. McIvor,[374] I. K. McIvor and E. G. Lovell,[375] E. I. Grigolyuk and A. I. Srebovskii,[139] A. S. Vol'mir and L. N. Smetanina,[104] V. E. Mineev,[189] and L. J. Mente.[376] Particularly, in Ref. 139 the solution, based on the approximation of deflection containing an arbitrary number of circumferential coordinate terms in Fourier series, was first treated. In Ref. 104, the calculation procedure proposed by V. L. Agamirov and A. S. Vol'mir[4] was extended for the case of an orthotropic shell. The first experiments on cylindrical shells loaded with dynamic external pressure were performed by A. F. Schmitt[415] and A. S. Vol'mir and V. E. Mineev.[102] From among the subsequent experimental works Refs 43, 87, 189 and 198 should be mentioned.

It follows from the results obtained in the above described theoretical and experimental works that three problems can be formulated depending on the rate of application of external pressure. For each of these problems we can define the loading as 'impulsive', 'dynamic' or 'quasi-static' ('slow dynamic') according to the classification suggested by D. L. Anderson and H. E. Lindberg.[277] The process of shell deformation under 'impulsive', short-term loads of high intensity is accompanied by considerable plastic deformations with very high circumferential modes of buckling being developed. Over the medium range of loading rate, plastic deformations in the material are initiated prior to the onset of intensive non-axisymmetric buckling, too.

But in this case the numbers of prevailing circumferential modes of buckling are considerably lower. Besides, at such loading rates the effect of inertness of an axisymmetric stress state in a shell is likely to be essential. Upon 'slow' dynamic load application, the numbers of prevailing circumferential modes of buckling are slightly greater than in the case of static instability: the inertness of axisymmetric stress state in a shell can be ignored. The critical value of pressure usually exceeds the static value several times. Within such a range of loading rate, the overall process of non-axisymmetric dynamic buckling can occur within the framework of elastic deformations. As a result, the solutions obtained in geometrically non-linear, but physically linear statement[4,55,56,60,75,104,139,162,189,376] prove to be correct.

The monograph by A. V. Karmishin *et al.*[164] was devoted to the methods for solving the problems of cylindrical shells' deformation under intensive short-term external pressure. The treated finite difference algorithms permit both the geometric and physical non-linearity to be taken into account. Let us also mention reports[34–36,348] devoted to the evolvement of the numerical algorithms for the non-linear shell dynamics problems. Variegated tasks about transient deformation of thin-walled structural elements were treated in Ref. 253.

The problem of initial imperfections is also in need of being briefly described here, because in this chapter, considerable attention will be paid to initial imperfections. Already in the late 1950s, V. V. Bolotin and I. I. Vorovich pointed out the actuality of studying various random factors (loads, shell surface geometric parameters, edge restraint conditions, material characteristics), defined by the specifics of manufacturing, transportation, assembling and exploitation of structures. Starting with A. P. Coppa's paper,[305] an extensive program for measuring, classification and theoretical describing of initial imperfections in cylindrical shells was underway. However, right up to the late 1960s, stability solutions (both static and dynamic) were performed according to a unique scheme: the initial deflection was assigned in advance for one fixed harmonic and for the same harmonic it was the previously preset loss of stability. Then by solving the geometrically non-linear problem, the effect of initial deflection amplitude on the critical load was studied. Of course, the assumption of peculiar 'resonance' between the initial deflection mode and critical instability mode permitted fairly vague information to be gained about the instability of imperfect shells. Practical needs required that the stability problems should be solved on the basis of real information about initial imperfections of each single unique shell or a group of uniform shells. Depending on the specific aim, such an investigation can be either deterministic or probabilistic.

A principally new, complex experimentally-theoretical approach to the static stability problem of imperfect cylindrical shells was formulated and illustrated in detail.[282] The deflection from the ideal cylindrical shell shape was measured and then used for studying the transformation of shell surface shape with increasing load. At each load magnitude the measured deflection field was analyzed through Fourier series expansion. Then the increase of each Fourier coefficient with increasing load was investigated. It was established for each concrete example that about forty of these coefficients tended to grow abruptly. Finally, a principally significant conclusion was made about the existence of several critical sets of instability modes instead of only one isolated 'critical mode' usually assumed. Besides, the 'critical', i.e. the most load-sensitive harmonics did not necessarily prevail in the initial deflection, measured prior to testing. In such a way, the analysis of Ref. 282 allowed one to trace all the process of quasi-static buckling of a shell — from the fixation of the distribution of initial imperfections till the moment of abrupt increase in amplitude of the most dangerous harmonics. This time moment, consequently, defines the load at which the loss of stability occurs.

Starting with late 1960s, scientific literature abounds with many works on measurement, classification and probabilistic analysis of initial imperfections in cylindrical shells and the use of the information obtained in solving the stability problems, particularly for shells made of polymeric and composite materials.[84,85,92,223,224,293,435,436] As was pointed out by G. A. Vanin *et al.*,[92] 'the main factor reducing the critical forces for GFRP shells, just as in the case of isotropic shells, are the initial imperfections of all kinds, in particular, initial deflection'. The developments in static stability analysis of imperfect shells realized in the late 1970s as well as shell imperfection data banks set up in a number of scientific centres[279–281,283] have permitted to draw the stability theory closer to the actual practical needs. The basic methodological aspects in studying static stability of imperfect cylindrical shells used in Ref. 282, and later in Refs 84 and 223 and many other works, were extended to the problems of dynamic buckling of cylindrical shells having a deterministic[55–57,59,60] or stochastic[61,62,373] field of initial imperfections.

5.1 AXISYMMETRIC DEFORMATION UNDER LONGITUDINAL COMPRESSION

It was established in a number of experimental works[128,131,201,249] that the buckling process of cylindrical shells subjected to axisymmetric impact loads, starts immediately after the impact and develops in the region close to the end face, at which the impact has been applied. In the first stage, buckling is of axisymmetric character. Thereby deflection in the vicinity of the end face usually develops in the direction of the outer normal. This is due to the restraint imposed at the end face of a shell in the radial direction, causing, due to Poisson's effect, well pronounced edge buckling. In later time instants, the deflection at a distance from the loaded end face turns to zero and the first circular fold is formed. Further the process of axisymmetric buckling can either propagate over the entire surface of a shell (this was observed in a number of experiments)[201,249] or transform into the process of non-axisymmetric buckling, which is possible due to the presence of non-axisymmetric initial imperfections. This process results in the generation of several bands of rhombic dents and bulges.

The process of non-axisymmetric deformation, in contrast to the axisymmetric one, starts its intensive development only after a considerable time interval after the external impact has been applied. Thus the stress state in a shell, initiated at the stage of axisymmetric dynamic buckling, serves as a background on which the non-axisymmetric deformations develop. Therefore, the study of axisymmetric edge effect and calculation of axisymmetric stress–strain state in a shell are necessary for further investigation of the process of non-axisymmetric deformation. Besides, the loss of the load carrying capacity of a shell, subjected to axial impact, can occur, under certain conditions, prior to a notable development of non-axisymmetric deformations. This attaches great importance to axisymmetric dynamic buckling analysis as a separate scientific problem.

Let us consider a closed orthotropic circular cylindrical shell of length L, thickness h and radius of the middle surface R. We shall describe the stress–strain state by using the Kirchhoff-Love model and a generalized Hooke's law for an orthotropic material.

Let us present several variants of equations for axisymmetric deformation in the order of their getting more complicated. The derivative with respect to the x coordinate is designated by a prime. The x coordinate origin is assumed to be at the left end of a shell.

1. Momentless equations:

$$N'_x = \mu \frac{\partial^2 u_0}{\partial t^2}; \qquad -\frac{N_y}{R} = \mu \frac{\partial^2 w_0}{\partial t^2} \tag{5.1}$$

and traction-displacement relations

$$N_x = C_{11} u'_0 + C_{12} \frac{w_0 - w_0^0}{R}; \quad N_y = C_{12} u'_0 + C_{22} \frac{w_0 - w_0^0}{R}; \tag{5.2}$$

$$w_0 = w_0^0 - \frac{R}{C_{11}C_{22}-C_{12}^2}(C_{12}N_x - C_{11}N_y); \quad u'_0 = \frac{C_{22}N_x - C_{12}N_y}{C_{11}C_{22}-C_{12}^2}, \tag{5.3}$$

where u_0 is the axial displacement; w_0 is the deflection, w_0^0 is the initial deflection; N_x and N_y are the axial and circumferential tractions; μ is a unit mass of shell surface; C_{ij} is the matrix of membrane stiffness of an orthotropic shell laminate.

Using (5.1) and (5.3) we can write the equations of motion relative to tractions as:

$$\mu\frac{\partial^2 N_x}{\partial t^2} = C_{11}N''_x - \frac{C_{12}}{R^2}N_y; \quad \mu\frac{\partial^2 N_y}{\partial t^2} = C_{12}N''_x - \frac{C_{22}}{R^2}N_y. \tag{5.4}$$

2. Linear moment equations:

$$N'_x = \mu\frac{\partial^2 u_0}{\partial t^2}; \quad -\frac{N_y}{R} - D_{11}(w_0 - w_0^0)^{\mathrm{IV}} = \mu\frac{\partial^2 w_0}{\partial t^2}, \tag{5.5}$$

where D_{11} is the flexural stiffness of the package. Relationships between tractions and displacements are determined by the formulas (5.2). After certain easy manipulations the system (5.5) can be written relative to the deflection and axial traction as:

$$\mu\frac{\partial^2 w_0}{\partial t^2} = -D_{11}(w_0 - w_0^0)^{IV} - \frac{C_{11}C_{22}-C_{12}^2}{C_{11}R^2}(w_0 - w_0^0) - \frac{C_{12}}{C_{11}R}N_x;$$

$$\mu\frac{\partial^2 N_x}{\partial t^2} = -\frac{D_{11}C_{12}}{R}(w_0 - w_0^0)^{IV} - \frac{C_{12}(C_{11}C_{22}-C_{12}^2)}{C_{11}R^3}(w_0 - w_0^0) - $$
$$- \frac{C_{12}^2}{C_{11}R^2}N_x + C_{11}N''_x. \tag{5.6}$$

3. The simplest variant of equations with geometric non-linearity taken into account can be obtained by incorporating the term $(N_x w'_0)'$ into the second eqns (5.5) and (5.6). As a result we obtain

$$N'_x = \mu\frac{\partial^2 u_0}{\partial t^2}; \quad -\frac{N_y}{R} - D_{11}(w_0 - w_0^0)^{IV} + (N_x w'_0)' = \mu\frac{\partial^2 w_0}{\partial t^2}; \tag{5.7}$$

$$\mu\frac{\partial^2 w_0}{\partial t^2} = -D_{11}(w_0 - w_0^0)^{IV} - \frac{C_{11}C_{22}-C_{12}^2}{C_{11}R^2}(w_0 - w_0^0) - $$
$$- \frac{C_{12}}{C_{11}R}N_x + (N_x w'_0)';$$

$$\mu\frac{\partial^2 N_x}{\partial t^2} = -\frac{D_{11}C_{12}}{R}(w_0 - w_0^0)^{IV} - \frac{C_{12}(C_{11}C_{22}-C_{12}^2)}{C_{11}R^3} \times$$
$$\times (w_0 - w_0^0) - \frac{C_{12}^2}{C_{11}R^2}N_x + C_{11}N''_x + \frac{C_{12}}{R}(N_x w'_0)'. \tag{5.8}$$

The traction-displacement relations remain linear (5.2).

4. Supplementing the system (5.7) with non-linear traction-displacement relations

$$N_x=C_{11}\left(u'_0+\frac{1}{2}w'^2_0\right)+C_{12}\frac{w_0-w_0{}^0}{R};$$
$$N_y=C_{12}\left(u'_0+\frac{1}{2}w'^2_0\right)+C_{22}\frac{w_0-w_0{}^0}{R}, \tag{5.9}$$

we arrive at equations, which are referred to as the equations of non-linear dynamic edge effect in analogy with the static case.[135]

Inserting (5.9) into (5.7) we obtain

$$C_{11}\left(u'_0+\frac{1}{2}w'^2_0\right)'+\frac{C_{12}}{R}(w_0-w_0{}^0)'=\mu\frac{\partial^2 u_0}{\partial t^2};$$
$$-\frac{C_{12}}{R}\left(u'_0+\frac{1}{2}w'^2_0\right)-\frac{C_{22}}{R^2}(w_0-w_0{}^0)-D_{11}(w_0-w_0{}^0)^{IV}+ \tag{5.10}$$
$$+\left[C_{11}\left(u'_0+\frac{1}{2}w'^2_0\right)w'_0+C_{12}\frac{w_0-w_0{}^0}{R}w'_0\right]'=\mu\frac{\partial^2 w_0}{\partial t^2}.$$

The equations of momentless theory will be integrated to variables N_x and N_y; the linear moment eqns (5.6) as well as eqns (5.8) — to variables N_x and w_0; the equations of non-linear dynamic edge effect (5.10) — to variables u_0 and w_0. The remaining unknown functions can be evaluated for each case according to the formulas (5.3), (5.2), (5.3) and (5.9), respectively.

Henceforth, the following types of boundary conditions at the end faces of a shell will be considered:

1) $w_0=w''_0=0,\quad N_x=-P(t)\quad$ at $\quad x=0,L;$ (5.11)

2) $w_0=w'_0=0,\quad N_x=-P(t)\quad$ at $\quad x=0,L;$ (5.12)

3) $w'_0=w'''_0=0,\quad N_x=-P(t)\quad$ at $\quad x=0,L;$ (5.13)

4) $w'_0=w'''_0=0,\quad N_x=-P(t)\quad$ at $\quad x=0,$
$w_0=w'_0=0,\quad N_x=-P(t)\quad$ at $\quad x=L;$ (5.14)

5) $w'_0=w'''_0=0,\quad N_x=-P(t)\quad$ at $\quad x=0,$
$w_0=w'_0=0,\quad u_0=0\quad$ at $\quad x=L;$ (5.15)

6) $w'_0=w'''_0=0,\quad N_x=-P(t)\quad$ at $\quad x=0,$
$w_0=w''_0=0,\quad u_0=0\quad$ at $\quad x=L.$ (5.16)

The initial conditions are specified as

$$w_0\Big|_{t=0}=w_0{}^0;\quad \frac{\partial w_0}{\partial t}\Big|_{t=0}=0;\quad u_0\Big|_{t=0}=\frac{\partial u_0}{\partial t}\Big|_{t=0}=0;$$
$$N_x\Big|_{t=0}=N_y\Big|_{t=0}=\frac{\partial N_y}{\partial t}\Big|_{t=0}=0; \tag{5.17}$$

$$\frac{\partial N_x}{\partial t}\Big|_{t=0}=0 \quad \text{at} \quad 0<x<L;$$
$$\frac{\partial N_x}{\partial t}\Big|_{t=0}=-\frac{dP}{dt}\Big|_{t=0} \quad \text{at} \quad x=0,L. \tag{5.18}$$

The numerical solution to the mixed boundary problem is performed according to the longitudinal scheme of straight lines method. The differential operators along the spatial coordinate, entering the equations and boundary conditions, are approximated by central differences of second-order accuracy on a uniform grid. For instance, the values of the derivatives of w_0 function in the i-th node of the grid is calculated by the formulas

$$\frac{\partial w_0{}^{(i)}}{\partial x}=\frac{w_0{}^{(i+1)}-w_0{}^{(i-1)}}{2\Delta};\quad \frac{\partial^2 w_0{}^{(i)}}{\partial x^2}=\frac{w_0{}^{(i-1)}-2w_0{}^{(i)}+w_0{}^{(i+1)}}{\Delta^2};$$
$$\frac{\partial^4 w_0{}^{(i)}}{\partial x^4}=\frac{w_0{}^{(i-2)}-4w_0{}^{(i-1)}+6w_0{}^{(i)}-4w_0{}^{(i+1)}+w_0{}^{(i+2)}}{\Delta^4},\quad i=1,\,\ldots,\,N.$$

Here, Δ is the value of a step along the x coordinate; N is the number of inner nodes of difference grid.

The values of unknown functions in contour $i=0$, $i=N+1$ and off-contour $i=-1$, $i=N+2$ points are determined from the boundary conditions. Thus the homogeneous conditions on the deflection (5.11)–(5.13) are reduced to the form

$$w_k{}^{(0)}=w_k{}^{(N+1)}=0;\quad w_k{}^{(-1)}=-w_k{}^{(1)};\quad w_k{}^{(N+2)}=-w_k{}^{(N)};$$
$$w_k{}^{(0)}=w_k{}^{(N+1)}=0;\quad w_k{}^{(-1)}=w_k{}^{(1)};\quad w_k{}^{(N+2)}=w_k{}^{(N)};$$
$$w_k{}^{(0)}=\frac{4w_k{}^{(1)}-w_k{}^{(2)}}{3};\quad w_k{}^{(N+1)}=\frac{4w_k{}^{(N)}-w_k{}^{(N-1)}}{3};$$
$$w_k{}^{(-1)}=w_k{}^{(1)};\quad w_k{}^{(N+2)}=w_k{}^{(N)},\; k=0,\,1.$$

The inhomogeneous boundary condition imposed on the longitudinal traction N_x is written as:

$$\left\{C_{11}\left[\frac{\partial u_0}{\partial x}+\frac{1}{2}\left(\frac{\partial w_0}{\partial x}\right)^2-\frac{1}{2}\left(\frac{\partial w_0{}^0}{\partial x}\right)^2\right]+\right.$$
$$\left.+C_{12}\frac{w_0-w_0{}^0}{R}\right\}_{x=0,\,L}=-P(t).$$

As is seen, this condition is non-linear and comprises the derivatives of the unknown functions. At the same time, the high accuracy in satisfying this condition, plays an especially important role in the problem under consideration. For this reason, the differential operators, entering this condition, are approximated to the fourth-order accuracy. For instance,

$$\left.\frac{\partial u_0}{\partial x}\right|_{x=0}=\frac{1}{\Delta}\left[-\frac{25}{12}u_0^{(0)}+4u_0^{(1)}-3u_0^{(2)}+\frac{4}{3}u_0^{(3)}-\frac{3}{12}u_0^{(4)}\right];$$

$$\left.\frac{\partial u_0}{\partial x}\right|_{x=L}=\frac{1}{\Delta}\left[\frac{25}{12}u_0^{(N+1)}-4u_0^{(N)}+3u_0^{(N-1)}+\frac{4}{3}u_0^{(N-2)}-\right.$$
$$\left.-\frac{3}{12}u_0^{(N-3)}\right].$$

In such a way, as a result we obtain the Cauchy problem in respect to the unknown functions $u_0^{(i)}(t)$, $w_0^{(i)}(t)$, $i=1,\ldots,N$. The numerical integration to this problem is performed by Runge-Kutta fourth-order method. Then from the displacements, found for nodal points, deformations, stresses and tractions are calculated.

Convergence of the calculation scheme was verified through varying the number of nodal points N. In all the calculated variants, the displacement and stress values at $N=299$ and $N=399$ differ by no more than 5%.

As an example, let us consider a shell of orthotropic material (six-ply graphite/epoxy composite) having the following elastic characteristics:[93]

$$\begin{aligned} &E_1=11.95\cdot10^{10}\,\mathrm{N/m^2}; \quad E_2=0.95\cdot10^{10}\,\mathrm{N/m^2};\\ &G_{12}=0.457\cdot10^{10}\,\mathrm{N/m^2}; \quad \nu_{12}=0.3; \quad \rho=1.5\cdot10^3\,\mathrm{kg/m^3}. \end{aligned} \tag{5.19}$$

The geometric parameters of the shell are as follows: $R=1$ m, $L/R=2$, $R/h=200$. Let us assume the initial axisymmetric deflection w_0^0 to be zero.

The axial compressive traction acting on both end faces of a shell we consider as linearly increasing in time:

$$P(\tau)=V_P\tau P^*, \tag{5.20}$$

where V_P is the dimensionless loading rate;

$$\tau=\frac{tc}{2L}; \quad c=[E_1/\rho(1-\nu_{12}\nu_{21})]^{1/2}; \tag{5.21}$$

P^* is the critical static traction for a shell with the orientation of the material axis 1 along x-axis; $P^*=0.265\cdot10^6$ N/m. Henceforth, if not otherwise stated, it is assumed that $V_P=5$.

Let us consider the results of integration for momentless eqns (5.4) under boundary conditions (5.11) imposed on N_x and initial conditions

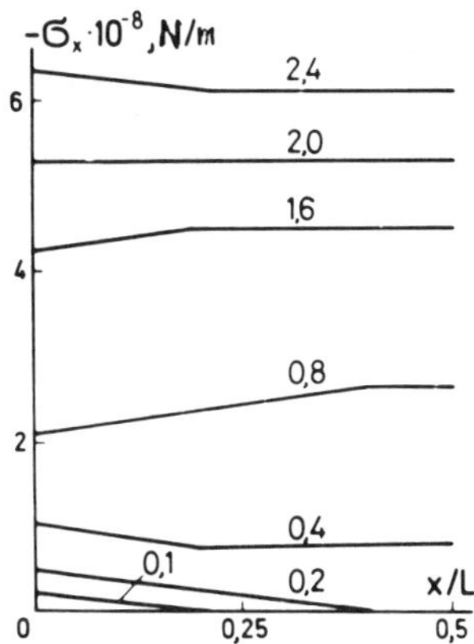

Fig. 5.1 Axial stress dependencies on the x coordinate, corresponding to several consecutive time instants (τ-values are indicated on the curves).

(5.18). In Fig. 5.1, the dependence of axial stress on x coordinate at several time moments is shown. Firstly, let us note that the numerical analysis yields the velocity of disturbances propagation which exactly coincides with the earlier introduced value c. Let us also mention the following result which will be more important further: shortly after the loading waves, moving symmetrically from both end faces, have met (at $\tau = 0.25$) in the middle of the shell, the stress σ_x will very slightly depend on the x coordinate. Thus the maximum difference of this stress from the applied value of stress at the end face is: at $\tau = 0.4$–25%, at $\tau = 0.8$–20%, at $\tau = 1.6$–5%, at $\tau = 2.4$–4%. In such a way, this result allows to neglect the inhomogeneity of the stress state in a shell produced due to the process of loading wave propagation, already at $\tau = 1.6$ for the loading law under consideration.

Let us set about discussing the edge effect having arisen under axial impact. In Fig. 5.2, for boundary conditions (5.11) the deflection dependencies on x coordinate, obtained from eqns (5.6), (5.8) and (5.10), are presented. As is evident, the calculations for the edge zone, with the non-linear term $(N_x w_0')'$ taken into account in the equations of motion and without this term, differ essentially. Taking into account the non-linearity in the strain-displacement formulas reduces somewhat the deflection values, but does not qualitatively affect the results. Thus we can conclude that the solution of the axial impact problem by using the linear moment equations, at least for simply supported end faces, yields a qualitatively incorrect characterization of the shell deformation in the edge zone. Let us also note that the curves 2 and 3 correspond qualitatively to

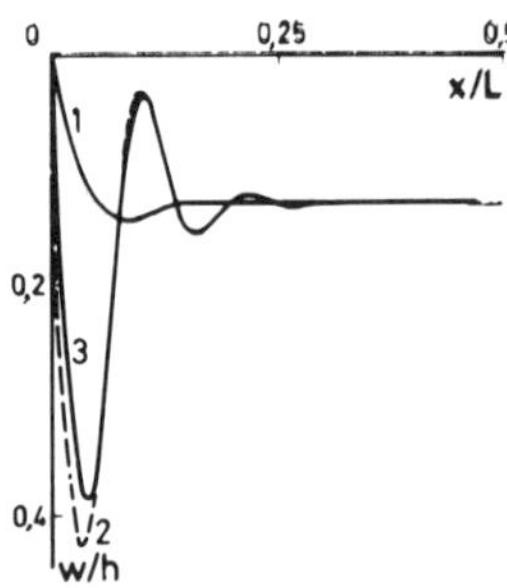

Fig. 5.2. Deflection dependencies on the x coordinate at $\tau = 1.8$ in solving three versions of equations of motion: 1 — (5.6); 2 — (5.8); 3 — (5.10).

the picture of axisymmetric buckling observed in experiments described in Refs 201 and 249, as well as to the results of numerical calculations in Refs 83, 130, 131 and 201.

Figure 5.3 presents the deflection dependencies on the axial coordinate for three boundary conditions (5.11), (5.12) and (5.13) being symmetrical relative to the middle of the shell. For all three cases $V_P = 5$, the value P^*

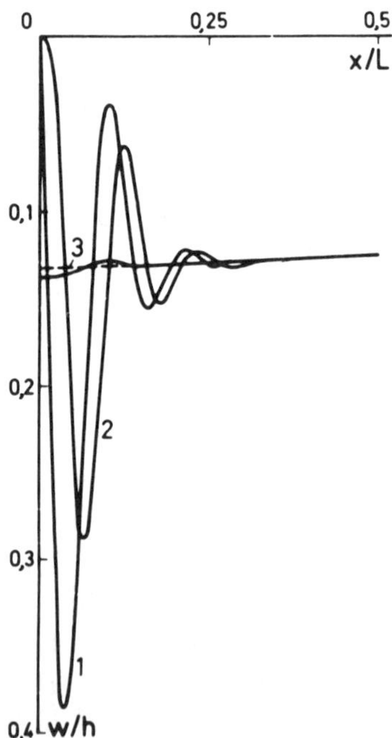

Fig. 5.3. Deflection dependencies on the x coordinate at $\tau = 1.8$ for three types of symmetrical boundary conditions: 1 — (5.11); 2 — (5.12); 3 — (5.13).

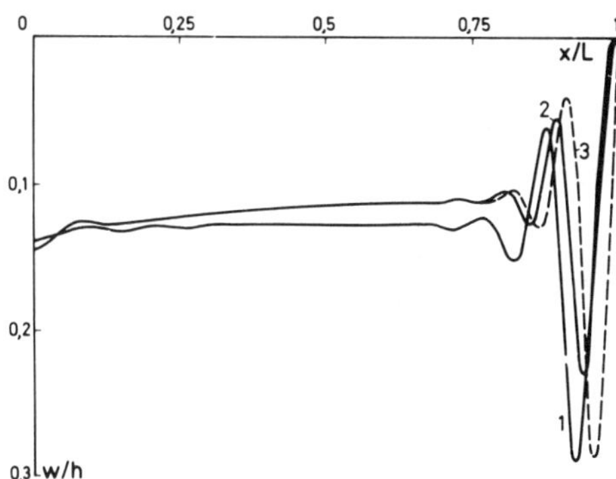

Fig. 5.4. Deflection dependencies on the x coordinate at $\tau = 1.8$ for three types of non-symmetrical boundary conditions: 1 — (5.14); 2 — (5.15); 3 — (5.16).

in (5.20) is the same and corresponds to the simple support conditions. As is shown, the edge effect is the severest under conditions of simple support, less pronounced — under clamping conditions; under completely free in the radial direction edge, the edge effect is practically absent. Let us note that under all three variants of edge conditions the results obtained in Fig. 5.3 coincide at a distance $x \approx 0.3L$ from the edge surface. This figure also shows the solution (marked by a dashed line) of linear moment equations under the conditions of free edge. It differs insignificantly from the solution of the equations of non-linear dynamic edge effect for the same type of boundary conditions.

The deflection dependencies on the axial coordinate for three types of boundary conditions (5.14), (5.15) and (5.16), being non-symmetric relative to the middle of the shell, are given in Fig. 5.4. In all cases the left edge is free in the radial direction. As is evident, the maximum deflection in the edge zone is somewhat greater for the case of axial traction being assigned at the end face than that under the condition of rigid clamping of the edge in the axial direction. From all variants of boundary conditions studied, the deflection is the greatest in the edge zone under simple support conditions (curve 1 of Fig. 5.3).

Further we shall consider the influence of loading rate on the non-linear dynamic edge effect. In Fig. 5.5, under the conditions of simple support the deflection dependencies on x coordinate at three V_P values are shown. The results pertain to the time moments $\tau = 0.25$, 1.8 and 4.4, respectively. As it can be seen, a decrease in loading rate leads to the expansion of edge effect zone, to the increase in width and number of circular dents and bulges.

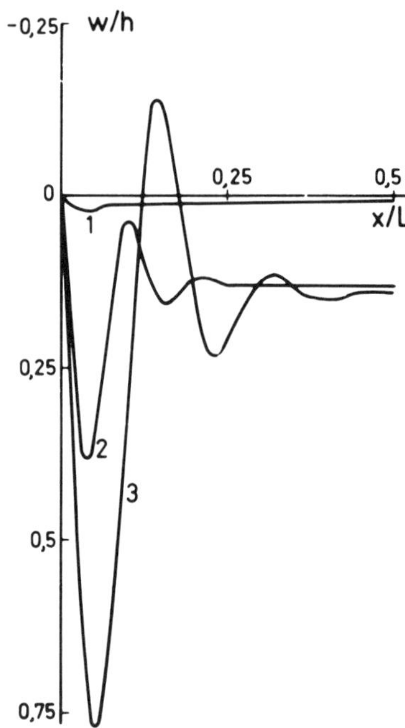

Fig. 5.5. Deflection dependencies on the x coordinate at three magnitudes of loading rate: $V_P = 25$ (1); 5 (2); 1 (3).

In Fig. 5.6, for the boundary conditions of simple support, the deflection and stress σ_x (on the inner shell surface) versus time are plotted. The curves correspond to those magnitudes of coordinate ($x = 0.05L$ for simple support and $x = 0$ for free edge) at which the deflection has its

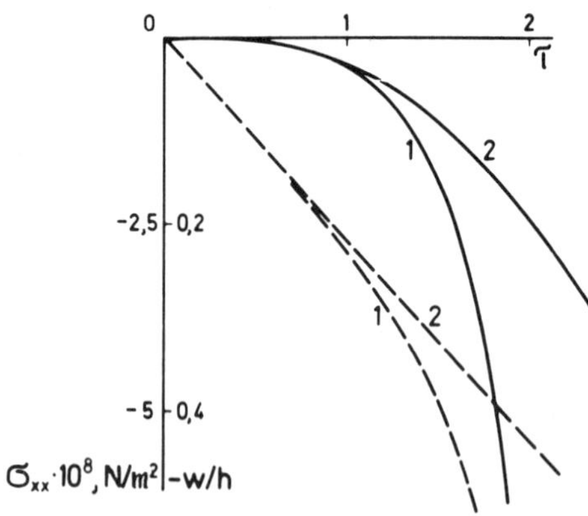

Fig. 5.6. Deflection (———) and axial stress (– – –) dependencies on time for two types of boundary conditions: 1 — (5.11); 2 — (5.13).

maximum. As is evident, starting from the moment $\tau \approx 1$, the axial stress in a point $0.05L$ deviates significantly from the assigned stress at the end face (which increases linearly with time). This is about the time moment from which intensive growth of deflection begins.

The analysis has shown that under restrained possibility of end face motion in the radial direction, flexural deformations undergo intensive growth adjacent to the end face part of a shell. This results in essential inhomogeneity along the x coordinate of the axisymmetric stress–strain state. The stress level in the edge effect zone is considerably higher than in the middle part of a shell.

Finally, we shall dwell upon the question about the effect of initial axisymmetric imperfections on the dynamic buckling of a shell. Let us assume for clarity that the initial deflection is, other than zero, only for one fixed harmonic:

$$w_0{}^0 = 0{,}02h \sin \frac{\pi m}{L} x. \tag{5.22}$$

Figure 5.7 gives the calculated $w_0(x)$ dependencies at the time instant $\tau = 2.0$ and under the boundary conditions (5.11) and (5.13) for several m values. As is seen, in the presence of comparatively small imperfections corresponding to high m numbers the deformation pattern changes qualitatively: along its entire length the shell divides into alternating bands of ring-type dents and bulges. Their depth at fixed τ increases with increasing m from 1 to some m^* (in the case under consideration, $m^* = 17$), whereas at $m > m^*$ it decreases. The value m^*, in its turn, grows with increasing loading rate and decreasing of relative shell thickness h/R.

Let us also note that due to interaction of the buckling process in the edge zone, caused by Poisson's effect, with the process of initial imperfection evolution, the magnitude of deflection in the vicinity of the end face can be essentially different from the case of an ideal shell. For example, at $m = 1$ and $m = 5$, the depth of the bulges first from the left and the right end faces proved to be practically the same as for an ideal shell, whereas at $m = 10$, the first bulge from the left end face was two times deeper, but the first bulge from the right end face practically disappeared.

As it follows from the results presented, under dynamic longitudinal compression of a shell, the small, as to their amplitude and length, initial imperfections can be the reason for strong inhomogeneity of the axisymmetric stress–strain state.

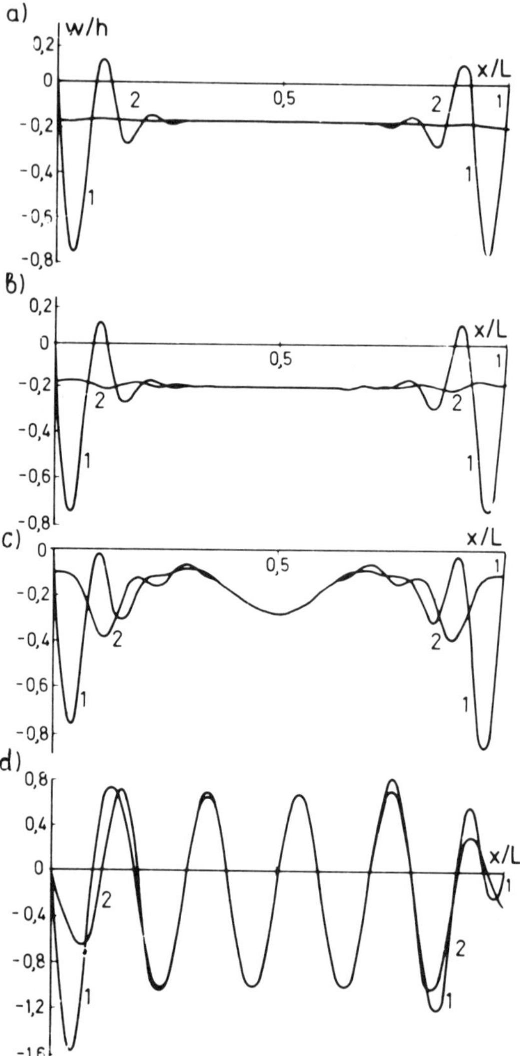

Fig. 5.7. Deflection dependencies on x coordinate at $\tau = 2.0$ for perfect shell (a) and shell with initial deflection (5.22) at $m = 1$ (b), 5 (c), 10 (d). The curves 1 and 2 correspond to boundary conditions (5.11) and (5.13).

5.2 LONGITUDINAL IMPACT BY A RIGID MASS ON THE CYLINDRICAL SHELL

The experimental study of the behavior of thin steel cylindrical shells under longitudinal rigid mass impact was undertaken by V. I. Borisenko and V. T. Voloshin.[82] In accordance with the conditions of experiment, both end faces of a shell can be considered as rigidly clamped in the radial direction. At the lower (out of load) end face the axial displacement u_0 is zero, at the upper (loaded) end face the axial traction $N_x(t)$ is assigned. Therefore, the boundary conditions take the form

$$\begin{aligned} w_0 = w'_0 = 0, \quad N_x = -P(t) \text{ при } x = 0; \\ w_0 = w'_0 = 0, \quad u_0 = 0 \text{ при } x = L. \end{aligned} \tag{5.23}$$

In performing the experiment in Ref. 82 the contact force between the falling weight and the edge face of a shell was not measured. To evaluate $P(t)$ function the experimental data about contact forces produced during the impact interaction between two bodies[125] were used. The experimental $P(t)$ dependence was compared with the curve

$$f(t) = A \sin\left(\pi \frac{t}{t_c}\right),$$

which was utilized in approximation of the load at the shell end face. The duration of impact t_c, found from St Venant's solution, in Ref. 82 was raised by 20% in accordance with experimental data. With the introduced dimensionless time $\tau = (ct/R)$, where $c = \sqrt{(E)/(\rho(1-\nu^2))}$ and for the ratio of striker's mass to shell mass used in the experiment, it was found that $\tau_c = 16.36$. The amplitude A was evaluated from the law of conservation momentum.

Finally on the basis of the data presented in Ref. 82 we obtain

$$P(\tau) = \begin{cases} 1.57 \dfrac{Eh}{1-\nu^2} V^* \dfrac{M_1 + M_2}{M_1} \sin\left(\pi \dfrac{\tau}{\tau_c}\right) & \text{at } \tau \le \tau_c; \\ 0 & \text{at } \tau > \tau_c, \end{cases} \tag{5.24}$$

where M_1 is the striker's mass; M_2 is the mass of a ring installed on the upper end face; V^* is the dimensionless impact velocity. The numerical values of parameters: $E = 20.6 \cdot 10^{10}\,\text{N/m}^2$; $\nu = 0.3$; $\rho = 7.8 \cdot 10^3\,\text{kg/m}^3$; $R = 5\,\text{cm}$; $L = 12\,\text{cm}$; $M_1 = 69.62\,\text{g}$; $M_2 = 37.1\,\text{g}$. The thickness of a shell h underwent variation.

Let us consider the calculation results for a shell under boundary conditions (5.23) and the load on the upper end face (5.24), obtained by the procedure, described in paragraph 5.1. Figure 5.8 reveals the $w_0(x)$ functions for the variant $h = 0.1\,\text{mm}$, $V^* = 0.725 \cdot 10^{-3}$ at three time instants. As is evident, at each end face a sufficiently narrow zone of edge effect is formed which extends with increasing τ and at $\tau = 10$ already consisting of five well pronounced ring-type folds of maximum depth of

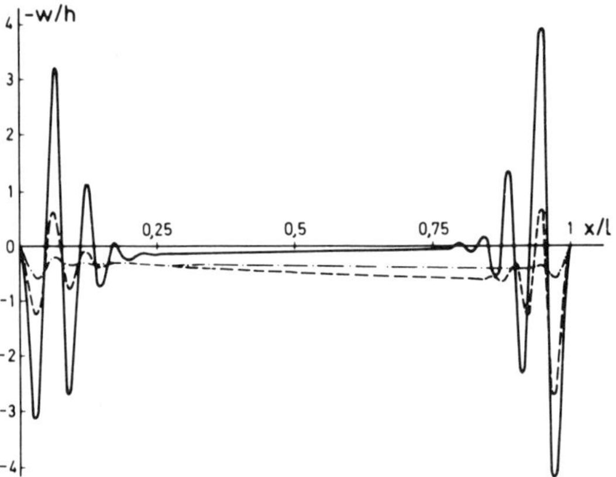

Fig. 5.8. Deflection dependencies on x coordinate at $\tau = 6.0$ (— · —), 8.0 (— — —), 10.0 (———) for $h = 0.1$ mm, $V^* = 0.725 \cdot 10^{-3}$.

order 4 h. In the middle part of the shell, the deflection depends slightly on the coordinate (analogous result was shown in Figs 5.3 and 5.4).

Further let us introduce into consideration the intensity of tangential stresses

$$T = [(\sigma_{xx}^2 + \sigma_{yy}^2 - \sigma_{xx}\sigma_{yy} + 3\sigma_{xy}^2)/3]^{1/2} \tag{5.25}$$

(in the axisymmetric problem under investigation, $\sigma_{xy} \equiv 0$). Figure 5.9 shows the $T(\tau)$ function for a shell with $h = 0.1$ mm at $V^* = 0.5 \cdot 10^{-3}$. The maximum value $T = 860$ MPa is reached at $\tau = 10.3$ on the inner surface in the vicinity of the lower end face*. In such a way, in the case of using von Mises yield criterion $T = (\sigma_s/\sqrt{3}) = T_i$ and starting from the typical values of yield strength σ_s for the material under consideration (steel 45), we can state that at the given loading rate in the shell there will be initiated local zones of plastic deformations. However, this does not mean that the shell will exhibit a notable residual deflection. The value T abruptly reduces, both when passing from the side surfaces deep into the shell thickness and going out of the x-sections, corresponding to the crest of bulges and dents. As a comprehensive analysis of $T(x, z)$ field over single ring-type bulge and fold shows (the assumed number of nodes $N = 299$ allows this to be done), at the mentioned impact velocity, over the entire volume of that shell part, over which the deepest bulges and dents are formed, the condition $T < T_i$ is not violated.

In the experiment of Ref. 82 the dynamic instability was assessed

*On the lower end face the shell deflection is also maximum: $\max_{\tau \in [0,20]} w_0 \approx 1.8h$.

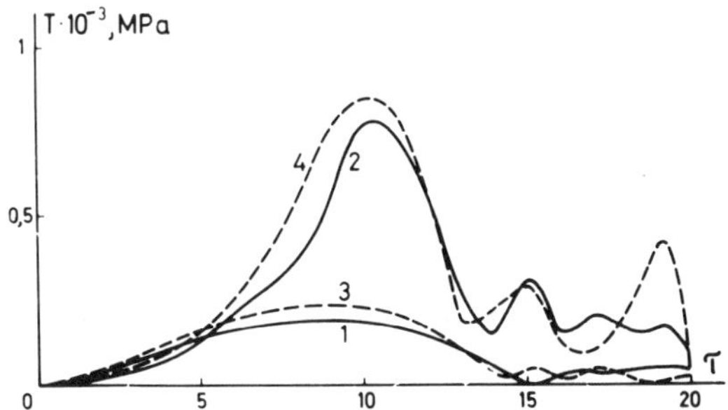

Fig. 5.9. The dependencies of T-function on time for a shell with $h = 10^{-2}$ cm on the outer (———) and inner (– – –) surfaces at $x = 0.06L$ (1); $0.94L$ (2); $0.03L$ (3); $0.97L$ (4); $V^* = 0.5 \cdot 10^{-3}$.

judging from two criteria based on the phenomena observed:

(1) generation of transverse deformations during the process of impact, notable on cine film;
(2) the presence of the residual deflection after the impact.

Taking into account smallness of the deflection (at $V^* = 0.5 \cdot 10^{-3}$, the maximum magnitude of $w_0 \approx 0.2$ mm was obtained in our calculations), it seems natural that the first criterion did not allow to assess experimentally the dynamic instability. The residual deflection was not also observed in the experiment (in view of the above, this does not contradict our calculations).

For a shell having $h = 0.1$ mm, the notable residual deflection (on the lower end face) was observed experimentally at the velocity $V^* = 0.61 \cdot 10^{-3}$. The calculations have shown that the T-function reaches its maximum values (1600 and 1560 MPa, respectively) on the inner and outer shell surfaces in the vicinity of the lower end face. Let us add that practically over the entire material volume within the region of the first three ring-type dents from the face of this end at the impact velocity, named the yield law, is fulfilled. In the vicinity of the upper (under impact) end face during the time interval $\tau \in [0, 20]$ and coordinate range $x \in [0, L/2]$, $z \in [-(h/2), (h/2)]$, the maximum value of the T-function is 540 MPa (i.e. there should be no residual deflection). This was really substantiated in the experiment. As is evident from the calculations provided, at other h and V^* values, the maximum magnitude of the T-function for a shell is increasing according to the law close to the linear one, with rising V^* and falling h.

Let us determine as the critical value of impact velocity the value $V_{cr} = V^*$, at which over the specified intervals $\tau \in [0, \tau_c]$, $x \in [0, L]$,

$z \in [-(h/2), (h/2)]$ the condition

$$\max T(\tau, x, z) = T_i \tag{5.26}$$

is fulfilled. In such a way, V_{cr} depends on T_i as on a parameter. The $V_{cr}(h)$ function at various T_i, at least over the range of velocities studied, $V^* = 0.4 \cdot 10^{-3}$–$0.7 \cdot 10^{-3}$ and thicknesses range $h = 0.09$–0.13 mm, are very well approximated by the linear functions (in Fig. 5.10 they are shown by solid lines). The dashed line corresponds to theoretical function, obtained in the paper of V. I. Borisenko and V. T. Voloshin,[82] where the stresses in undisturbed state of a shell were calculated on the basis of St Venant's solution and subsequent analysis of motion stability was carried out on the basis of linearized equations. Some experimental results presented in Ref. 82 are shown by circles. At the velocity of impact, marked by an empty circle, the dynamic instability was not observed, whereas at the velocity, marked by a dark circle, the dynamic instability was established in accordance with the experimental criteria used in Ref. 82.

The fact that our criterion of dynamic instability (5.26) gives an underrated value of critical velocity, can be fairly well interpreted. This criterion is of local character and allows to determine the value of impact velocity, at which at the sites of most intensive shell buckling, the first local plastic zones are formed. Except those situations when this criterion cannot be applied at all (in particular, if the dynamic buckling process occurs under

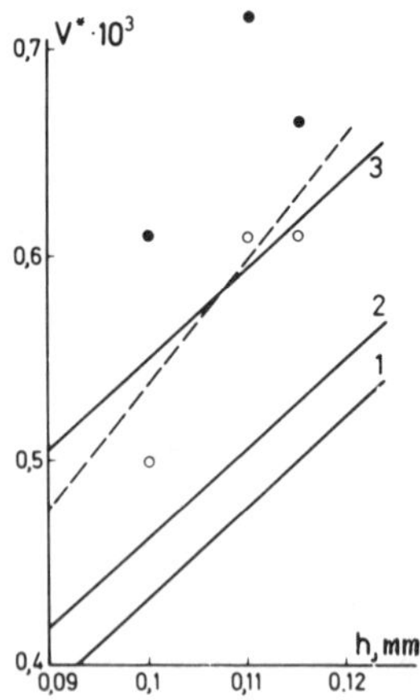

Fig. 5.10. Critical impact velocity dependencies on shell thickness. Solid lines correspond to: $T_i = 400$ MPa (1), 600 (2); 1200 (3). A dashed line is the calculated dependence from Ref. 82; dots indicate the experimental data from Ref. 82. Other designations in the text.

the conditions of elastic material behavior), it naturally leads to lower values of V_{cr} compared to the experimental ones, at which it is possible to establish the change in shell shape visually. Let us add that by choosing the T_i-value it is possible to fairly well correlate the obtained theoretical results with the experimental data. This conventional 'yield point', exceeding the real yield strength of the material, is related to the loading conditions, under which the ring-type fold completely passes from elastic to plastic state.

In the conclusion let us note that the procedure evolved in paragraph (5.1), does not allow to describe all the diversity of experimental results obtained in Ref. 82. In particular, from among 17 test samples, 10 showed instability on the lower end face and 7 on the upper one. We did not succeed in calculating the magnitudes of the T-function, higher on the upper end face than on the lower one. Besides, on one of the samples in Ref. 82 formation of a series of rhombic bulges in the vicinity of the lower end face was established, as well as appearance of a single fold in the middle part of a shell was stated. This fold shifted toward the upper end face. This is a certification of the process of non-axisymmetric dynamic buckling. To describe this phenomenon, a more general statement of the theoretical problem and special methods for its solving should be employed.

5.3 CALCULATION OF NON-AXISYMMETRIC DEFORMATION UNDER LONGITUDINAL IMPACT

The first attempt to evolve the calculation procedure, taking into consideration the processes of stress wave propagation in the middle surface, axisymmetric and non-axisymmetric dynamic buckling, was undertaken in Ref. 37. The problem was solved by using the Bubnov-Galerkin method along the circumferential coordinate, the finite-difference method along the longitudinal coordinate and, in the long run, the primary system of the equations of motion was reduced to the system of non-linear algebraic and ordinary differential equations. The numerical results obtained, in fact, for a semi-infinite shell have shown that during some time interval after the load application, along with stress wave propagation in the middle surface, it is the axisymmetric form of buckling which develops in the vicinity of the end face under impact. When the load reaches a certain magnitude, the axisymmetric form is gradually transformed into non-axisymmetric one. The procedure outlined in Ref. 37, was further developed in our paper[59] for the case of a shell of finite length made up of an orthotropic material.

Let us assume the equations of motion for a shell in the form (2.36) by putting

$$X^*_1 = X^*_2 = X^*_3 = F^*_1 = F^*_2 = \Phi^*_1 = \Phi^*_2 = 0; \quad P^*_1 = -\mu\frac{\partial^2 u}{\partial t^2};$$

$$P^*_2 = -\mu\frac{\partial^2 v}{\partial t^2}; \quad P^*_3 = -\mu\frac{\partial^2 w}{\partial t^2}$$

and introducing the notations: $T^*_{11} = N_x$, $T^*_{21} = T^*_{12} = T$, $T^*_{22} = N_y$. With (2.41) and (2.45) taken into account, these equations are transformed in the form

$$\frac{\partial N_x}{\partial x} + \frac{\partial T}{\partial y} = \mu\frac{\partial^2 u}{\partial t^2}; \tag{5.27}$$

$$\frac{\partial T}{\partial x} + \frac{\partial N_y}{\partial y} = \mu\frac{\partial^2 v}{\partial t^2}; \tag{5.28}$$

$$\begin{aligned} -\frac{N_y}{R} - D_{11}\frac{\partial^4(w-w^0)}{\partial x^4} - 2(D_{12}+2D_{66})\frac{\partial^4(w-w^0)}{\partial x^2\partial y^2} - \\ -D_{22}\frac{\partial^4(w-w^0)}{\partial y^4} + \frac{\partial}{\partial x}\left(N_x\frac{\partial w}{\partial x} + T\frac{\partial w}{\partial y}\right) + \\ + \frac{\partial}{\partial y}\left(T\frac{\partial w}{\partial x} + N_y\frac{\partial w}{\partial y}\right) = \mu\frac{\partial^2 w}{\partial t^2}. \end{aligned} \tag{5.29}$$

According to (2.41) and (2.45) the following relations take place

$$\begin{aligned} N_x &= C_{11}\left[\frac{\partial u}{\partial x} + \frac{1}{2}\left(\frac{\partial w}{\partial x}\right)^2 - \frac{1}{2}\left(\frac{\partial w^0}{\partial x}\right)^2\right] + \\ &+ C_{12}\left[\frac{\partial v}{\partial y} + \frac{w-w^0}{R} + \frac{1}{2}\left(\frac{\partial w}{\partial y}\right)^2 - \frac{1}{2}\left(\frac{\partial w^0}{\partial y}\right)^2\right]; \\ N_y &= C_{12}\left[\frac{\partial u}{\partial x} + \frac{1}{2}\left(\frac{\partial w}{\partial x}\right)^2 - \frac{1}{2}\left(\frac{\partial w^0}{\partial x}\right)^2\right] + \\ &+ C_{22}\left[\frac{\partial v}{\partial y} + \frac{w-w^0}{R} + \frac{1}{2}\left(\frac{\partial w}{\partial y}\right)^2 - \frac{1}{2}\left(\frac{\partial w^0}{\partial y}\right)^2\right]; \\ T &= C_{66}\left(\frac{\partial u}{\partial y} + \frac{\partial v}{\partial x} + \frac{\partial w}{\partial x}\frac{\partial w}{\partial y} - \frac{\partial w^0}{\partial x}\frac{\partial w^0}{\partial y}\right). \end{aligned} \tag{5.30}$$

In the general case, the initial deflection w^0 can be expressed in terms of Fourier series in respect of the circumferential coordinate:

$$w^0(x, y) = w_0^0(x) + \sum_{n=1}^{\infty} [w_n^0(x)\cos\beta_n y + \tilde{w}_n^0(x)\sin\beta_n y], \tag{5.31}$$

where $\beta_n = n/R$. To make the results more vivid, we shall assume that in (5.31), only the axisymmetric term $w_0^0(x)$ and one of the Fourier coefficients $w_n^0(x)$, corresponding to certain fixed number n of the circumferential harmonic are other than zero (analogous assumption was made in Ref. 37; the case when a finite number of Fourier coefficients in the expansion (5.31) are other than zero, will be thoroughly studied in paragraph 5.4). Thus the initial deflection of the shell can be written in the form

$$w^0(x, y) = w_0^0(x) + w_n^0(x) \cos \beta_n y. \tag{5.32}$$

In the approximation of deflection in respect to the circumferential coordinate, let us also restrict ourselves to axisymmetric component and one harmonic of Fourier series, corresponding to the same n-value, as in (5.32):

$$w(x, y, t) = w_0(x, t) + w_n(x, t) \cos \beta_n y. \tag{5.33}$$

Using (5.32) and (5.33), the following dependencies of u and v on the y coordinate can be obtained:

$$\begin{aligned} u(x, y\ t) &= u_0(x, t) + u_n(x, t) \cos \beta_n y + u_{2n}(x, t) \cos 2\beta_n y; \\ v(x, y, t) &= v_n(x, t) \sin \beta_n y + v_{2n}(x, t) \sin 2\beta_n y, \end{aligned} \tag{5.34}$$

where u_0, u_n, u_{2n}, v_n, v_{2n}, are the unknown functions. Substituting (5.32)–(5.34) into (5.30) and separating axisymmetric and non-axisymmetric parts, we obtain for tractions:

$$N_x = N_x{}^0 + N_x{}^n; \quad N_y = N_y{}^0 + N_y{}^n; \quad T = T^n, \tag{5.35}$$

where the axisymmetric terms are

$$\begin{aligned} N_x{}^0 = C_{11} & \left[\frac{\partial u_0}{\partial x} + \frac{1}{2}\left(\frac{\partial w_0}{\partial x}\right)^2 + \frac{1}{4}\left(\frac{\partial w_n}{\partial x}\right)^2 - \frac{1}{2}\left(\frac{\partial w_0{}^0}{\partial x}\right)^2 - \right. \\ & \left. - \frac{1}{4}\left(\frac{\partial w_n{}^0}{\partial x}\right)^2 \right] + C_{12}\left[- \frac{w_0{}^0 - w_0}{R} + \frac{1}{4}\beta_n{}^2 (w_n{}^2 - w_n{}^{02}) \right]; \\ N_y{}^0 = C_{12} & \left[\frac{\partial u_0}{\partial x} + \frac{1}{2}\left(\frac{\partial w_0}{\partial x}\right)^2 + \frac{1}{4}\left(\frac{\partial w_n}{\partial x}\right)^2 - \frac{1}{2}\left(\frac{\partial w_0{}^0}{\partial x}\right)^2 - \right. \\ & \left. - \frac{1}{4}\left(\frac{\partial w_n{}^0}{\partial x}\right)^2 \right] + C_{22}\left[- \frac{w_0{}^0 - w_0}{R} + \frac{1}{4}\beta_n{}^2 (w_n{}^2 - w_n{}^{02}) \right], \end{aligned} \tag{5.36}$$

while the non-axisymmetric ones have the form

$$N_x{}^n = N'_x \cos \beta_n y + N''_x \cos 2\beta_n y; \quad N_y{}^n = N'_y \cos \beta_n y + N''_y \cos 2\beta_n y;$$
$$T = T' \sin \beta_n y + T'' \sin 2\beta_n y \tag{5.37}$$

Let us disregard further the inertia terms:

$$\frac{\partial^2 u_n}{\partial t^2}, \quad \frac{\partial^2 u_{2n}}{\partial t^2}, \quad \frac{\partial^2 v_n}{\partial t^2}, \quad \frac{\partial^2 v_{2n}}{\partial t^2}$$

connected with non-axisymmetric components of tangential displacements in eqns (5.27) and (5.28). Then with (5.35) taken into account, we obtain from (5.27) the following equation

$$\frac{\partial N_x{}^0}{\partial x} = \mu \frac{\partial^2 u_0}{\partial t^2}. \tag{5.38}$$

For non-axisymmetric components of tractions, it is possible to introduce the function $\Phi_n(x, y, t)$ in such a way that

$$N_x^n = \frac{\partial^2 \Phi_n}{\partial y^2}, \quad N_y^n = \frac{\partial^2 \Phi_n}{\partial x^2}, \quad T = -\frac{\partial^2 \Phi_n}{\partial x \partial y},$$

and then employ the known equation of deformation compatibility, which with (5.32), (5.33) and (5.36) taken into account, assumes the form

$$A_{11} \frac{\partial^4 \Phi_n}{\partial y^4} + (A_{66} + 2A_{12}) \frac{\partial^4 \Phi_n}{\partial x^2 \partial y^2} + A_{22} \frac{\partial^4 \Phi_n}{\partial x^4} =$$
$$= \left[\beta_n{}^2 \left(w_n \frac{\partial^2 w_0}{\partial x^2} - w_n{}^0 \frac{\partial^2 w_0{}^0}{\partial x^2} \right) + \frac{1}{R} \frac{\partial^2 (w_n - w_n{}^0)}{\partial x^2} \right] \cos \beta_n y +$$
$$+ \frac{1}{2} \beta_n{}^2 \left[w_n \frac{\partial^2 w_n}{\partial x^2} - w_n{}^0 \frac{\partial^2 w_n{}^0}{\partial x^2} - \left(\frac{\partial w_n}{\partial x} \right)^2 + \right.$$
$$\left. + \left(\frac{\partial w_n{}^0}{\partial x} \right)^2 \right] \cos 2\beta_n y, \tag{5.39}$$

where A_{ij} are compliances of an orthotropic material related to C_{ij} in accordance with the well-known formulas.[24] Representing Φ_n accordingly with the form of the right-hand part of (5.39) in the form $\Phi_n(x, y, t) = \varphi_n(x, t) \cos \beta_n y + \varphi_{2n}(x, t) \cos 2\beta_n y$, we obtain two equations:

$$A_{22} \frac{\partial^4 \varphi_n}{\partial x^4} - \beta_n{}^2 (A_{66} + 2A_{12}) \frac{\partial^2 \varphi_n}{\partial x^2} + \beta_n{}^4 A_{11} \varphi_n =$$
$$= -\frac{1}{R} \frac{\partial^2 (w_n - w_n{}^0)}{\partial x^2} + \beta_n{}^2 \left(w_n \frac{\partial^2 w_0}{\partial x^2} - w_n{}^0 \frac{\partial^2 w_0{}^0}{\partial x^2} \right).$$
$$A_{22} \frac{\partial^4 \varphi_{2n}}{\partial x^4} - 4\beta_n{}^2 (A_{66} + 2A_{12}) \frac{\partial^2 \varphi_{2n}}{\partial x^2} + 16\beta_n{}^4 A_{11} \varphi_{2n} = \tag{5.40}$$
$$= \frac{1}{2} \beta_n{}^2 \left[w_n \frac{\partial^2 w_n}{\partial x^2} - w_n{}^0 \frac{\partial^2 w_n{}^0}{\partial x^2} - \left(\frac{\partial w_n}{\partial x} \right)^2 + \left(\frac{\partial w_n{}^0}{\partial x} \right)^2 \right].$$

The non-axisymmetric components of tractions assume the form:

$$N_x{}^n = -\beta_n{}^2\varphi_n \cos\beta_n y - 4\beta_n{}^2\varphi_{2n} \cos 2\beta_n y;$$
$$N_y{}^n = \frac{\partial^2\varphi_n}{\partial x^2}\cos\beta_n y + \frac{\partial^2\varphi_{2n}}{\partial x^2}\cos 2\beta_n y; \qquad (5.41)$$
$$T=\beta_n\left(\frac{\partial\varphi_n}{\partial x}\sin\beta_n y+2\frac{\partial\varphi_{2n}}{\partial x}\sin 2\beta_n y\right).$$

Finally, substitution of (5.32)–(5.34), (5.35), (5.41) into (5.29) and the use of orthogonalization procedure lead to two equations:

$$\mu\frac{\partial^2 w_0}{\partial t^2}=-\frac{N_y{}^0}{R}-D_{11}\frac{\partial^4(w_0-w_0{}^0)}{\partial x^4}+\frac{\partial}{\partial x}\times$$
$$\times\left[N_x{}^0\frac{\partial w_0}{\partial x}-\frac{1}{2}\beta_n{}^2\left(\varphi_n\frac{\partial w_n}{\partial x}+w_n\frac{\partial\varphi_n}{\partial x}\right)\right];$$
$$\mu\frac{\partial^2 w_n}{\partial t^2}=-\frac{1}{R}\frac{\partial^2\varphi_n}{\partial x^2}-D_{11}\frac{\partial^4(w_n-w_n{}^0)}{\partial x^4}+2\beta_n{}^2(D_{12}+2D_{66})\times \qquad (5.42)$$
$$\times\frac{\partial^2(w_n-w_n{}^0)}{\partial x^2}-D_{22}\beta_n{}^4(w_n-w_n{}^0)+\frac{\partial}{\partial x}\left(N_x{}^0\frac{\partial w_n}{\partial x}\right)-$$
$$-\beta_n{}^2\left[N_y{}^0 w_n+\varphi_n\frac{\partial^2 w_0}{\partial x^2}+2\frac{\partial}{\partial x}\left(\varphi_{2n}\frac{\partial w_n}{\partial x}\right)+\frac{1}{2}w_n\frac{\partial^2\varphi_{2n}}{\partial x^2}\right].$$

In such a way, the primary problem is reduced to solving the system of five equations (5.38), (5.40) and (5.42), containing unknown functions u_0, w_0, w_n, φ_n, φ_{2n}. Let us remark that the introduction of the function of tractions is not principal, but allows merely to reduce the number of final equations from seven to five and write them in more compact form.

Let us represent the conditions at the end faces of a shell, for which the subsequent numerical realizations will be done. Assuming that in all cases the equals on both end faces uniformly distributed compressive forces $P(t)$ have been applied, for components of axial traction we have

$$N_x{}^0\,|_{x=0,L}=-P(t); \qquad (5.43)$$
$$N_x{}^n\,|_{x=0,L}=0, \qquad (5.44)$$

thereby (5.44) assumes the shape

$$\varphi_n\,|_{x=0,L}=\varphi_{2n}\,|_{x=0,L}=0. \qquad (5.45)$$

Let us assume that the circumferential displacement v at the end faces is zero; this condition is formulated in the form

$$\left.\frac{\partial^2\varphi_n}{\partial x^2}\right|_{x=0,L}=\left.\frac{\partial^2\varphi_{2n}}{\partial x^2}\right|_{x=0,L}=0. \qquad (5.46)$$

For deflection, we shall specify one of the three following types of boundary conditions:

$$1)\ w_0\Big|_{x=0,L}=\frac{\partial^2 w_0}{\partial x^2}\Big|_{x=0,L}=w_n\Big|_{x=0,L}=\frac{\partial^2 w_n}{\partial x^2}\Big|_{x=0,L}=0; \tag{5.47}$$

$$2)\ w_0\Big|_{x=0,L}=\frac{\partial w_0}{\partial x}\Big|_{x=0,L}=w_n\Big|_{x=0,L}=\frac{\partial w_n}{\partial x}\Big|_{x=0,L}=0; \tag{5.48}$$

$$3)\ \frac{\partial w_0}{\partial x}\Big|_{x=0,L}=\frac{\partial^3 w_0}{\partial x^3}\Big|_{x=0,L}=\frac{\partial w_n}{\partial x}\Big|_{x=0,L}=\frac{\partial^3 w_n}{\partial x^3}\Big|_{x=0,L}=0. \tag{5.49}$$

The initial conditions are specified in the form

$$w_0\Big|_{t=0}=w_0{}^0;\quad \frac{\partial w_0}{\partial t}\Big|_{t=0}=0;\quad w_n\Big|_{t=0}=w_n{}^0;\quad \frac{\partial w_n}{\partial t}\Big|_{t=0}=0. \tag{5.50}$$

One important feature of the problem stated should be mentioned here. In the linear approach this problem is separated into two independent ones. The axisymmetric deformation is characterized by the system of equations

$$\mu\frac{\partial^2 u_0}{\partial t^2}=\frac{\partial}{\partial x}\left(C_{11}\frac{\partial u_0}{\partial x}+C_{12}\frac{w_0-w_0{}^0}{R}\right);$$
$$\mu\frac{\partial^2 w_0}{\partial t^2}=-\frac{1}{R}\left(C_{12}\frac{\partial u_0}{\partial x}+C_{22}\frac{w_0-w_0{}^0}{R}\right)-D_{11}\frac{\partial^4(w_0-w_0{}^0)}{\partial x^4} \tag{5.51}$$

with inhomogeneous boundary conditions at the end faces (5.43). The non-axisymmetric deformation is described by the equation

$$\mu\frac{\partial^2 w_n}{\partial t^2}=-\frac{1}{R}\frac{\partial^2 \varphi_n}{\partial x^2}-D_{11}\frac{\partial^4(w_n-w_n{}^0)}{\partial x^4}+2\beta_n{}^2(D_{12}+2D_{66})\times$$
$$\times\frac{\partial^2(w_n-w_n{}^0)}{\partial x^2}-D_{22}\beta_n{}^4(w_n-w_n{}^0), \tag{5.52}$$

the evident solution of which for any version of axisymmetric restraint of end faces and under initial conditions (5.50) is as follows:

$$\varphi_n(x,t)\equiv 0;\quad w_n(x,t)\equiv w_n{}^0(x). \tag{5.53}$$

Consequently, the information about axisymmetric loading of the end faces comes to non-axisymmetric forms of deformation only through the group of non-linear terms

$$\frac{\partial}{\partial x}\left(N_x{}^0\frac{\partial w_n}{\partial x}\right)-\beta_n{}^2N_y{}^0w_n-\beta_n{}^2\varphi_n\frac{\partial^2 w_0}{\partial x^2},$$

entering the eqn (5.42). Therefore, it is clear that in providing the numerical integration of the problem these terms must be calculated with special accuracy. Even small errors in the calculation, for example, of traction N_x^0 can greatly misrepresent the results of calculating the process of non-axisymmetric deformation of a shell.

The numerical solution of the mixed boundary problem is performed, just as in paragraph 5.1, using the longitudinal scheme of a straight lines method. The finite-difference approximation of deflection and its derivatives was described in paragraph 5.1. Let us note here that under boundary conditions (5.45) and (5.46), the following relationships are valid:

$$\varphi_k{}^0=\varphi_k{}^{(N+1)}=0;\quad \varphi_k{}^{(-1)}=-\varphi_k{}^{(1)};\quad \varphi_k{}^{(N+2)}=-\varphi_k{}^{(N)};\quad k=n,\,2n,$$

while the non-uniform boundary condition (5.43) is written, after (5.36) taken into account, in the form

$$\left\{C_{11}\left[\frac{\partial u_0}{\partial x}+\frac{1}{2}\left(\frac{\partial w_0}{\partial x}\right)^2+\frac{1}{4}\left(\frac{\partial w_n}{\partial x}\right)^2-\frac{1}{2}\left(\frac{\partial^2 w_0{}^0}{\partial x}\right)^2-\right.\right.$$
$$\left.-\frac{1}{4}\left(\frac{\partial w_n{}^0}{\partial x}\right)^2\right]-C_{12}\left[\frac{w_0-w_0{}^0}{R}-\frac{\beta_n{}^2}{4}\times\right.$$
$$\left.\left.\times\,(w_n{}^2-w_n{}^{0^2})\right]\right\}_{x=0,L}=-P(t). \tag{5.54}$$

The system of finite-difference equations, having been eventually obtained, is integrated in time in the following succession.

(1) In each time step, two linear in respect to φ_n, φ_{2n} systems of N algebraic equations, which have been obtained after the finite-difference approximation of eqns (5.40), are being solved. For this sake the method of five-diagonal run is employed; the values of functions φ_n and φ_{2n} in the nodes of the grid are expressed in terms of w_0 and w_n:

$$\varphi_k{}^{(i)}=\varphi_k{}^{(i)}[w_0{}^{(1)},\,w_0{}^{(2)},\,\ldots,\,w_0{}^{(N)},\,w_n{}^{(1)},\,\ldots,\,w_n{}^{(N)}]; \tag{5.55}$$
$$k=n,\,2n;\quad i=1,\,\ldots,\,N.$$

(2) The expressions (5.55) are substituted into the system of $3N$ non-linear ordinary differential second-order equations, having been obtained in the finite-difference approximation of eqns (5.38) and (5.42). The formulated Cauchy's problem is integrated by the fourth-order Runge-Kutta method. As a result, the functions $u_0^{(i)}(t)$, $w_0^{(i)}(t)$ and $w_n^{(i)}(t)$ are found.

(3) Using expressions (5.55) as well as the formulas (5.35), (5.36) and

(5.41) the tractions and further, if necessary, also stresses and strains in each shell point in an arbitrary time instant are calculated.

Let us consider the shell having the same parameters as in paragraph 5.1 and assume the axisymmetric component of the initial deflection w_0^0 as equal to zero, whereas the non-axisymmetric component w_n^0 at fixed n specify by means of Fourier series

$$w_n{}^0(x)=\sum_{m=1}^{\infty} W_{mn}{}^0 \sin\frac{\pi m}{L}x, \tag{5.56}$$

Let us assume the distribution of Fourier coefficients W_{mn}^0 in the form

$$W_{mn}{}^0=0{,}2h\frac{(-1)^l}{m^2}, \tag{5.57}$$

where $l=(m/2)$, provided m is even and $l=(m+1)/2$ for odd m. To perform calculations according to the procedure above, it is necessary to calculate first the $w_n^0(x)$ function by summing the series (5.56). We assume that the axial compressive force at both end faces is linearly increasing in time (5.20) just like in paragraph 5.1. Further, if not otherwise specified, $V_P=5$.

Let us treat some results of numerical calculations having been performed for the simple support boundary conditions:

$$w_0=\frac{\partial^2 w_0}{\partial x^2}=w_n=\frac{\partial^2 w_n}{\partial x^2}=0 \quad \text{at} \quad x=0, L. \tag{5.58}$$

It is assumed that the initial deflection is other than zero only for the circumferential harmonic $n=3$.

The convergence of the results of calculating the deflection was checked by varying the number of nodes N in the finite-difference scheme. In Fig.

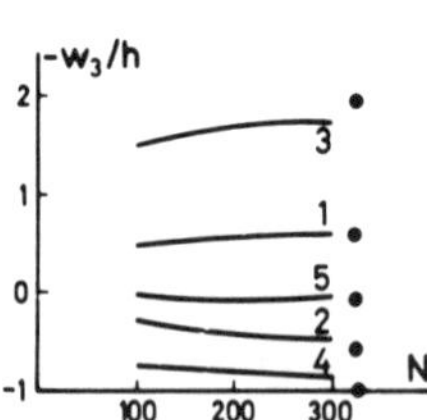

Fig. 5.11. An illustration of convergency for a finite difference scheme (N is the number of nodes). Numbers on curves correspond to x-values: 1 — $0.36L$; 2 — $0.42L$; 3 — $0.48L$; 4 — $0.54L$; 5 — $0.66L$.

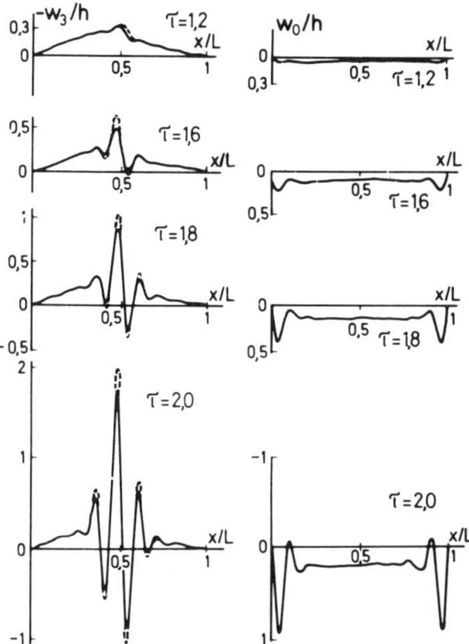

Fig. 5.12. The process of development of the axisymmetric and non-axisymmetric deflection components in time. Dashed lines correspond to the case of solving the non-axisymmetric dynamic buckling problem by the Bubnov-Galerkin method.

5.11, the variance of w_3 values in several shell sections with increasing N from 99 to 299 is shown (the results indicated by dots, will be discussed in paragraph 5.5). All the numerical results to be published in this paragraph have been obtained at $N = 299$.

In Fig. 5.12, the $w_0(x)$ and $w_3(x)$ functions are presented in several successive time instants. As is seen, two processes are taking place at the same time: the axisymmetric buckling in the vicinities of edges and non-axisymmetric buckling in the middle part of a shell. The intensive development of the axisymmetric flexural deformations in the immediate vicinity of end face is explainable by the restriction imposed in the radial direction. In its turn, the main factor which defines the location of the zone of intensive non-axisymmetric strain development is the type of $w_n^0(x)$ function, specifying the initial imperfections.

The effect of the end conditions on the dependence of non-axisymmetric deflection component on the x coordinate is illustrated in Fig. 5.13. Here, alongside with boundary conditions of simple support (5.58) also the clamping conditions

$$w_0 = \frac{\partial w_0}{\partial x} = w_n = \frac{\partial w_n}{\partial x} = 0, \quad \text{at} \quad x = 0, L \tag{5.59}$$

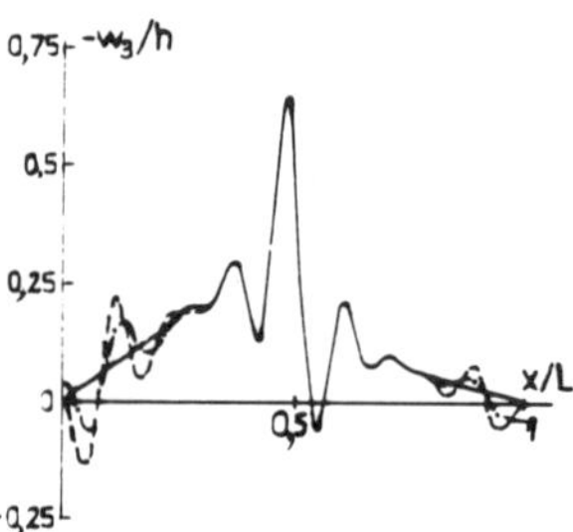

Fig. 5.13. Non-axisymmetric deflection component dependence on the x coordinate at $\tau = 1.7$ and boundary conditions: (5.37) (———), (5.38) (– – –), (5.39) (–·–). The initial imperfections are specified by the formulas (5.35) and (5.36).

and the free edge conditions

$$\frac{\partial w_0}{\partial x} = \frac{\partial^3 w_0}{\partial x^3} = \frac{\partial w_n}{\partial x} = \frac{\partial^3 w_n}{\partial x^3} = 0, \quad \text{at} \quad x = 0, L. \qquad (5.60)$$

are considered. As is seen from the figure, the $w_3(x)$ dependencies for all three types of boundary conditions practically coincide in the middle part of a shell, whereas in fairly extended zones adjacent to the end faces, the character of shell buckling changes essentially when substituting (5.58) by (5.59) or (5.60). Therefore, we can assume that at such $w_n^0(x)$ functions which lead to intensive non-axisymmetric bulging just in the vicinity of shell ends, the conditions of edge restraint affect significantly the maximum values not only of axisymmetric, but also non-axisymmetric components of

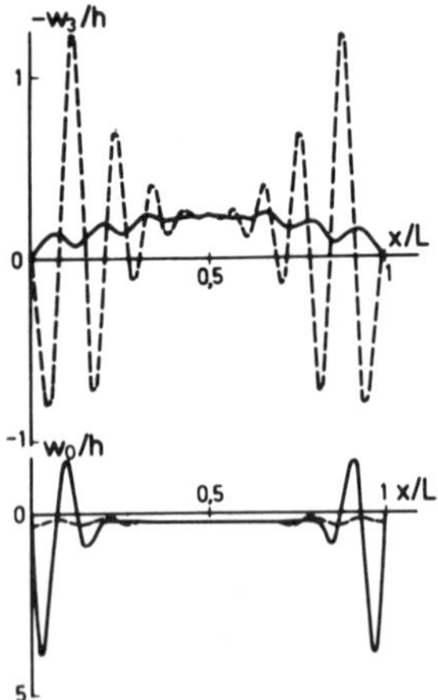

Fig. 5.14. Non-axisymmetric (a) and axisymmetric (b) deflection component dependence on the x coordinate at $\tau = 2.3$ and boundary conditions: (5.37) (———), (5.39) (– – –). The initial imperfections are specified by (5.40).

deflection. It is obvious also that in such a situation the non-linear effect of the interaction of axisymmetric and non-axisymmetric modes of bulging will be strongly pronounced.

As an illustration of the above let us consider the results of calculation, presented in Fig. 5.14, at parabolic initial deflection dependence on the axial coordinate

$$w_n{}^0(x) = 0{,}2h \frac{4}{L^2} x(L-x). \tag{5.61}$$

Under the conditions (5.60), the intensive non-axisymmetric bulging occurs in the zones adjacent to the end faces of a shell (let us note that the axisymmetric edge effect at such boundary conditions is absent). Under the boundary conditions (5.58), for which the axisymmetric edge effect is well pronounced, the non-axisymmetric deflection component at the same time instant is small over the entire shell surface.

In such a way, in the presence of initial imperfections (5.61) a certain type of 'suppression' of non-axisymmetric form of dynamic bulging takes place. It is possible to explain this effect through the following considerations. As was mentioned, the information about axisymmetric loading of the shell end faces is passed on to the non-axisymmetric forms only by the group of three non-linear terms, enclosed in the second equation (5.42):

$$\frac{\partial}{\partial x}\left(N_x{}^0 \frac{\partial w_n}{\partial x} \right) - \beta_n{}^2 N_y{}^0 w_n - \beta_n{}^2 \varphi_n \frac{\partial^2 w_0}{\partial x^2},$$

besides, the first one plays the main role in initiating the process of non-axisymmetric bulging. As the analysis of $N_x^0(x)$ functions shows, the compressive longitudinal traction in the zones of ring-type dent formation is considerably less than in the middle part of a shell, where this traction is close to the magnitude, having been specified at the end faces. Thus, as a result of axisymmetric bulging the edge zones are unloaded from compressive longitudinal tractions. It is natural that in such a situation the process of intensive evolution of non-axisymmetric initial imperfections, under otherwise equal conditions, will start later in the edge zones than in the middle part of a shell.

The process of non-axisymmetric bulging, if it occurs in the vicinity of end faces, leaves just the reversed effect on the evolution of axisymmetric edge effect. It is possible because of a group of non-linear terms

$$-\frac{1}{2} \beta_n{}^2 \frac{\partial}{\partial x}\left(\varphi_n \frac{\partial w_n}{\partial x} + w_n \frac{\partial \varphi_n}{\partial x} \right),$$

entering the first eqn (5.42).

In completing the discussion of the effect of interaction of the axisymmetric and non-axisymmetric forms of bulging, let us note that this effect is the stronger, the larger w_0 and w_n. The qualitative effect of w_0 on w_n (and *vice versa*) is defined by the relation between these two values. Thus, 'suppression' of the non-axisymmetric component occurs in those cases, when w_0 considerably exceeds w_n.

In this paragraph, the calculation procedure evolved for solving the problem of dynamic bulging of an imperfect cylindrical shell is based on approximation of the deflection over the circumferential coordinate by two terms (5.33). In principle, its generalization to the case of deflection approximation, with an arbitrary finite number of circumferential harmonics taken into account, does not cause any difficulties. Owing to the non-linearity of the problem, the axisymmetric component of the deflection and all Fourier coefficients for the non-axisymmetric harmonics turn out to be interrelated within a united system of non-linear equations through particular derivatives. The numerical solution of such a system is of certain technical difficulty, even with a small number of circumferential harmonics taken into account. Besides, the computer-aided realization of this solution requires a considerable amount of computer time. Therefore, it is reasonable to study the effect of interconnection of circumferential harmonics in the problems of non-linear dynamic bulging of cylindrical shells in a simplified statement of the problem.

5.4 CALCULATION OF NON-AXISYMMETRIC DYNAMIC BUCKLING USING BUBNOV-GALERKIN'S METHOD

The method of solution, evolved in paragraph 5.3, was based on a two-term approximation of the deflection over the circumferential coordinate: the axisymmetric component and one harmonic of Fourier series, corresponding to some fixed number n, were taken into account. Further we shall evolve the solution, in which the deflection will be approximated by a finite sum of Fourier series, containing an arbitrary number of circumferential harmonics.

Let us employ the simplified, in comparison to paragraph 5.3, statement of the problem: disregard the longitudinal inertia of a shell. The eqn (5.38) then will take the form:

$$\frac{\partial N_x{}^0}{\partial x} = 0. \tag{5.62}$$

It follows from (5.62) that N_x^0 does not depend on the x coordinate, whereas with the boundary conditions (5.43) specified at the end faces, we shall obtain the equality $N_x^0 = -P(t)$. The introduced simplification means that the process of stress wave propagation from the impacted end faces of the shell is not treated, but momentary established as uniform, in respect to x'axisymmetric stress state, is assumed. As was shown in paragraph 5.1, under the load linearly increasing in time (see Fig. 5.1) such an assumption can be employed already after 3–4 times of longitudinal compressive wave passage over the shell length.

In such a simplified approach the equations of motion for a shell compressed along the axis coincide with (4.43) and (4.44), if the vibrational load $P_0 + P_t \cos \Theta t$ is substituted by an impulsive one $P(t)$. The equations of motion can also be obtained from (2.51) and (2.52), if we put

$$\begin{gathered} F^*{}_1 = F^*{}_2 = \Phi^*{}_1 = \Phi^*{}_2 = 0, \quad P^*{}_3 = -\mu \frac{\partial^2 w}{\partial t^2}; \\ X^*{}_3 = -P(t) \frac{\partial^2 w}{\partial x^2} - Rq(t) \left(\frac{\partial^2 w}{\partial y^2} + \frac{w}{R^2} \right). \end{gathered} \tag{5.63}$$

The last term in (5.63) is present if the pressure $q(t)$, which is uniformly distributed over the outer side surface, acts on the shell. In such a way, the equations of motion are written in the form

$$\begin{gathered} D_{11} \frac{\partial^4 (w-w^0)}{\partial x^4} + 2(D_{12}+2D_{66}) \frac{\partial^4 (w-w^0)}{\partial x^2 \partial y^2} + D_{22} \frac{\partial^4 (w-w^0)}{\partial y^4} + \\ + \frac{1}{R} \frac{\partial^2 \Phi}{\partial x^2} - \frac{\partial^2 w}{\partial x^2} \frac{\partial^2 \Phi}{\partial y^2} - \frac{\partial^2 w}{\partial y^2} \frac{\partial^2 \Phi}{\partial x^2} + 2 \frac{\partial^2 w}{\partial x \partial y} \frac{\partial^2 \Phi}{\partial x \partial y} + \\ + P(t) \frac{\partial^2 w}{\partial x^2} + Rq(t) \left(\frac{\partial^2 w}{\partial y^2} + \frac{w}{R^2} \right) + \mu \frac{\partial^2 w}{\partial t^2} = 0; \end{gathered} \tag{5.64}$$

$$\begin{gathered} A_{22} \frac{\partial^4 \Phi}{\partial x^4} + (2A_{12}+A_{66}) \frac{\partial^4 \Phi}{\partial x^2 \partial y^2} + A_{11} \frac{\partial^4 \Phi}{\partial y^4} = \frac{1}{R} \frac{\partial^2 (w-w^0)}{\partial x^2} + \\ + \left(\frac{\partial^2 w}{\partial x \partial y} \right)^2 - \left(\frac{\partial^2 w^0}{\partial x \partial y} \right)^2 - \frac{\partial^2 w}{\partial x^2} \frac{\partial^2 w}{\partial y^2} + \frac{\partial^2 w^0}{\partial x^2} \frac{\partial^2 w^0}{\partial y^2}. \end{gathered} \tag{5.65}$$

Though the statement of two kindred problems — about the non-linear parametric vibrations and non-linear dynamic buckling — differ only by the type of $P(t)$ function in (5.64), essentially different approaches are needed for their solution. The principle for selecting the multi-term approximation of the deflection, proposed in 4.3, is evidently inapplicable to the problems of impulsive loading.

The subject of deflection approximation in solving the non-linear problems of non-axisymmetric dynamic buckling of cylindrical shell was comprehensively treated in our paper.[57] In particular, the two-term approximation (4.51) was considered. It permits to describe the peculiarities of the non-linear interaction of various axial and circumferential harmonics. Based on numerical calculations, the conclusion was drawn that in the region where non-linear terms play an essential role, taking into account the interaction of circumferential harmonics leads to a qualitative change of time dependence of the deflection. In studying the interaction of a group of axial harmonics, characterized by the greatest rate of growth, a definite qualitative effect was established. However, it remained insignificant in comparison to the effect of interaction of circumferential harmonics. On the basis of these results in Ref. 57 it was suggested to solve the problem of non-axisymmetric dynamic buckling of a cylindrical shell loaded in longitudinal compression, proceeding from the following multi-term approximation of deflection

$$w_m(x, y, t) = \sin \alpha_m x \sum_{n=n_0}^{N} W_{mn}(t) \cos \beta_n y, \tag{5.66}$$

where

$$\alpha_m = \frac{\pi m}{L}; \quad m = 1, 2, \ldots; \quad \beta_n = \frac{n}{R}.$$

The initial deflection is assumed in an analogous form

$$w_m{}^0(x, y) = \sin \alpha_m x \sum_{n=n_0}^{N} W_{mn}{}^0 \cos \beta_n y. \tag{5.67}$$

The procedure of solving the system (5.64) and (5.65) is as follows. At first we substitute the expressions (5.66) and (5.67) into (5.65). Making equal the coefficients at the same trigonometric functions in both parts of this equation, we find the function of tractions:

$$\begin{aligned}
\Phi_m(x, y, t) = {} & \sin \alpha_m x \sum_{n=n_0}^{N} C_{mn}{}^{(1)}(W_{mn} - W_{mn}{}^0) \cos \beta_n y + \\
& + \cos 2\alpha_m x \sum_{n=n_0}^{N} C_{mn}{}^{(2)}(W_{mn}{}^2 - W_{mn}{}^{0^2}) + \\
& + \sum_{n=n_0}^{N} C_{mn}{}^{(3)}(W_{mn}{}^2 - W_{mn}{}^{0^2}) \cos 2\beta_n y + \\
& + \sum_{i=n_0+1}^{N} \sum_{j=n_0}^{i-1} [(C_{mij}{}^{(4)} + C_{mij}{}^{(6)} \cos 2\alpha_m x) \cos(\beta_i - \beta_j) y + \\
& + (C_{mij}{}^{(5)} + C_{mij}{}^{(7)} \cos 2\alpha_m x) \cos(\beta_i + \beta_j) y] (W_{mi} W_{mj} - W_{mi}{}^0 W_{mj}{}^0),
\end{aligned} \tag{5.68}$$

where

$$
\begin{gathered}
C_{mn}{}^{(1)} = -\frac{\alpha_m{}^2}{R[\alpha_m{}^4 A_{22} + (2A_{12} + A_{66})\alpha_m{}^2 \beta_n{}^2 + A_{11}\beta_n{}^4]};\\
C_{mn}{}^{(2)} = \frac{\beta_n{}^2}{32\alpha_m{}^2 A_{22}}; \quad C_{mn}{}^{(3)} = -\frac{\alpha_m{}^2}{32\beta_n{}^2 A_{11}};\\
C_{mij}{}^{(4)} = -\frac{\alpha_m{}^2}{4A_{11}(\beta_i - \beta_j)^2}; \quad C_{mij}{}^{(5)} = -\frac{\alpha_m{}^2}{4A_{11}(\beta_i + \beta_j)^2};\\
C_{mij}{}^{(6)} = \frac{\alpha_m{}^2(\beta_i + \beta_j)^2}{4[16A_{22}\alpha_m{}^4 + 4(A_{66} + 2A_{12})\alpha_m{}^2(\beta_i - \beta_j)^2 + A_{11}(\beta_i - \beta_j)^4]};\\
C_{mij}{}^{(7)} = \frac{\alpha_m{}^2(\beta_i - \beta_j)^2}{4[16A_{22}\alpha_m{}^4 + 4(A_{66} + 2A_{12})\alpha_m{}^2(\beta_i + \beta_j)^2 + A_{11}(\beta_i + \beta_j)^4]}.
\end{gathered}
\tag{5.69}
$$

The subsequent substitution of (5.66) and (5.67) into (5.64) and the use of orthogonalization procedure lead to the system of $N - n_0 + 1$ non-linear ordinary differential equations (m is fixed, $k = n_0, \ldots, N$):

$$
\begin{gathered}
\frac{d^2 W_{mk}}{dt^2} + \omega_{mk}{}^2 \left\{ W_{mk} \left[1 - \frac{P(t)}{P^*{}_{mk}} - \frac{q(t)}{q^*{}_{mk}} \right] - W_{mk}{}^0 \right\} +\\
+ \frac{1}{\mu} \left[\beta_k{}^2 W_{mk} \sum_{i=n_0}^{N} B_i{}^{(1)} (W_{mi}{}^2 - W_{mi}{}^{0^2}) + \alpha_m{}^4 \sum_{i=n_0}^{N} \sum_{j=n_0}^{N} B_{ijk}{}^{(2)} W_{mj} \times \right.\\
\times (W_{mi}{}^2 - W_{mi}{}^{0^2}) - \alpha_m{}^2 \sum_{n=n_0}^{N} \sum_{i=n_0+1}^{N} \sum_{j=n_0}^{i-1} B^{(3)}_{mijnk} W_{mn} \times\\
\left. \times (W_{mi} W_{mj} - W_{mi}{}^0 W_{mj}{}^0) \right] = 0.
\end{gathered}
\tag{5.70}
$$

Here,

$$
\begin{gathered}
B_i{}^{(1)} = \frac{\beta_i{}^2}{16A_{22}}; \quad B_{ijk}{}^{(2)} = \frac{1}{16A_{11}} (\delta_{2i+j-k} + \delta_{2i-j+k} + \delta_{2i-j-k});\\
B^{(3)}_{mijnk} = \left\{ \frac{1}{2} C_{mij}{}^{(4)} (\beta_i - \beta_j)^2 - C_{mij}{}^{(6)} \left[\frac{(\beta_i - \beta_j)^2}{4} + \beta_n{}^2 \right] \right\} \times\\
\times (\delta_{i-j+n-k} - \delta_{i-j-n+k} + \delta_{i-j-n-k}) + \left\{ \frac{1}{2} C_{mij}{}^{(5)} (\beta_i + \beta_j)^2 - C_{mij}{}^{(7)} \times \right.\\
\left. \times \left[\frac{(\beta_i + \beta_j)^2}{4} + \beta_n{}^2 \right] \right\} (\delta_{i+j+n-k} + \delta_{i+j-n+k} + \delta_{i+j-n-k}) + \beta_n \times\\
\times [C_{mij}{}^{(6)} (\beta_i - \beta_j)(\delta_{i-j-n+k} + \delta_{i-j-n-k} - \delta_{i-j+n-k}) + C_{mij}{}^{(7)} (\beta_i + \beta_j) \times\\
\times (\delta_{i+j-n+k} + \delta_{i+j-n-k} - \delta_{i+j+n-k})],\\
\delta_l = \begin{cases} 1 \text{ при } l = 0, \\ 0 \text{ при } l \neq 0. \end{cases}
\end{gathered}
\tag{5.71}
$$

The system (5.70) is supplemented by the initial conditions:

$$W_{mk}\Big|_{t=0}=W_{mk}{}^{0};\qquad \frac{dW_{mk}}{dt}\Big|_{t=0}=0. \tag{5.72}$$

To solve the Cauchy's problem (5.70) and (5.72) the standard computer programs for numerical integration of the systems of non-linear ordinary differential equations can be employed. For this purpose, we have used the program, utilizing the Runge-Kutta fourth-order accuracy method.

After finding Fourier coefficients $W_{mk}(t)$ the $w_m(x,y,t)$ function can be calculated according to formula (5.66), then the $\Phi_m(x,y,t)$ function can be determined according to formula (5.68). Further the deflection $w(x,y,t)$ and traction function $\Phi(x,y,t)$ can be calculated by summing in respect to m the series:

$$w(x,y,t)=\sum_{m=m_0}^{M} w_m(x,y,t)=\sum_{m=m_0}^{M}\sin\alpha_m x\sum_{n=n_0}^{N} W_{mn}(t)\cos\beta_n y;$$
$$\Phi(x,y,t)=\sum_{m=m_0}^{M}\Phi_m(x,y,t). \tag{5.73}$$

The boundaries of summation m_0, M, n_0, N in (5.73) in solving each particular problem are defined, proceeding from the accuracy needed in calculating the characteristics of stress–strain state.

Henceforth the numerical results which have been obtained on the basis of multi-term approximation (5.66) will be compared to the results based on one-term approximation

$$w_{mn}(x,y,t)=W_{mn}(t)\sin\alpha_m x\cos\beta_n y. \tag{5.74}$$

The use of (5.74) implies that the non-linear terms, arising due to interconnection of different axial and circumferential harmonics, have been disregarded in the solution. In this case, for each $W_{mn}(t)$ function we obtain from the system (5.64) and (5.65) the non-linear ordinary differential equation

$$\frac{d^2W_{mn}}{dt^2}+\omega_{mn}{}^2\left\{W_{mn}\left[1-\frac{P(t)}{P^*_{mn}}-\frac{q(t)}{q^*_{mn}}\right]-W_{mn}{}^0\right\}+$$
$$+d_{mn}W_{mn}(W_{mn}{}^2-W_{mn}{}^{0^2})=0, \tag{5.75}$$

in which d_{mn} is determined according to (4.71).

The Cauchy's problem (5.75) and (5.72) is also numerically integrated by the Runge-Kutta method. The deflection of a shell and traction function are calculated by summing Fourier series:

$$w(x, y, t) = \sum_{m=m_0}^{M} \sum_{n=n_0}^{N} W_{mn}(t) \sin \alpha_m x \cos \beta_n y; \tag{5.76}$$

$$\Phi(x, y, t) = \sum_{m=m_0}^{M} \sum_{n=n_0}^{N} [\Phi_{mn}^{(1)} \sin \alpha_m x \cos \beta_n y + \Phi_{mn}^{(2)} \cos 2\alpha_m x + + \Phi_{mn}^{(3)} \cos 2\beta_n y], \tag{5.77}$$

where

$$\Phi_{mn}^{(1)} = C_{mn}^{(1)} (W_{mn} - W_{mn}^0); \quad \Phi_{mn}^{(2)} = C_{mn}^{(2)} (W_{mn}^2 - W_{mn}^{0^2}); \quad \Phi_{mn}^{(3)} = C_{mn}^{(3)} (W_{mn}^2 - W_{mn}^{0^2}). \tag{5.78}$$

Let us pass over to evaluation of the results of numerical calculations. Consider a shell loaded with linearly increasing in time longitudinal compressive forces (5.20) applied to the end faces. The shell parameters are assumed to be: $R = 1$ m; $L/R = 2$; $R/h = 200$; $E = 2 \cdot 10^{11}$ N/m^2; $\nu = 0.3$; $\rho = 7.7 \cdot 10^3$ kg/m^3. Let us specify the distribution of Fourier coefficients W_{mn}^0 of the initial deflection in the form

$$W_{mn}^0 = 0{,}2h \frac{(-1)^{l+n}}{m^2} e^{-|n-3|}, \tag{5.79}$$

where $l = m/2$ at even m and $l = (m+1)/2$ at odd m.

In Fig. 5.15, the results of numerical integration of the system (5.70) at $m = 19$, $n_0 = 1$ and $N = 5$ are presented. The results, marked by a dashed line, are obtained by solving the eqn (5.75). As is seen, in the case of one-term approximation, the values W_{mn} for all circumferential harmonics over the investigated time interval reaches magnitudes of the order $2h$. In the case of taking the interconnection into account, the dominant form $n = 3$ which 'suppresses' all the other forms is marked out. Let us note that the maximum of distribution (5.79) corresponds precisely to $n = 3$. The effect of 'suppression' is observed in the cases, when the rate of growth of one or several forms of bulging is considerably higher than that of the remaining forms. For example, it is seen from comparison of the dependencies of Fourier coefficients on time (marked by dashed lines in Fig. 5.15), corresponding to different n, that the rate of $W_{19.3}$ growth over the time interval $\tau \in [3.0; 3.5]$ is considerably higher than that of the remaining Fourier coefficients. Let us add that the effect of 'suppression' evidently follows from the form of non-linear terms of the system (5.70).

Let us consider the calculation of a shell, loaded with external pressure

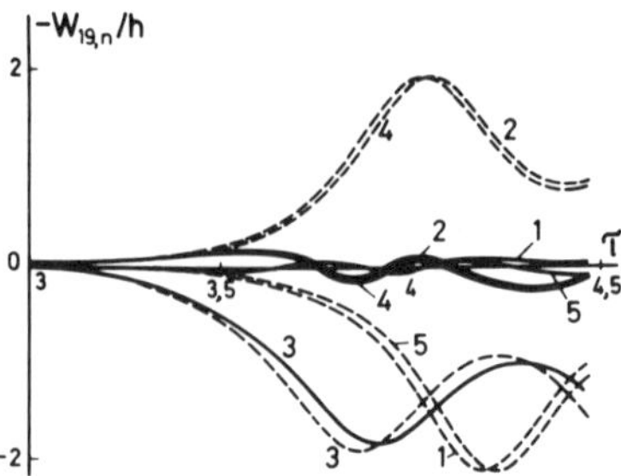

Fig. 5.15. Fourier coefficients dependence on time for $m = 19$, with (———) and without (– – –) the interrelation of circumferential harmonics taken into account. The initial imperfections are specified by (5.55). Numbers on curves correspond to n-values.

$q(t) = at$, linearly increasing in time. The material of the shell is duralumin; $L = 0.2\,\text{m}$; $R = 0.09\,\text{m}$; $h = 8 \cdot 10^{-4}\,\text{m}$. Such shells were tested at several velocities of pressure front growth in Ref. 102. The information about the initial imperfections of the shell shapes, necessary for calculations in accordance with the proposed method, is absent in Ref. 102. Because of that we restrict ourselves to an illustrative example by specifying the distribution W_{mn}^0 in the form

$$W_{mn}{}^0 = 0{,}02h\,(-1)^{l+n} \exp\left[-\frac{(m-1)^2}{10} - \frac{(n-4)^2}{10} \right], \tag{5.80}$$

where $l = m/2$ at even m and $l = (m+1)/2$ at odd m.

The results of numerical integration of the system (5.70) at the loading rate $a = 6500\,\text{atm/s}$ for $m = 1$, $n_0 = 5$ and the two magnitudes of N are given in Fig. 5.16 (a dashed line shows the results of numerical integration of eqn (5.75) at $m = 1$, $n = 5$–11). As is seen, on adding the harmonic with $n = 11$ to approximation (5.66) over the studied time interval, the $W_{1n}(\tau)$ dependencies for $n = 5$–10 change insignificantly; the $W_{1,11}(\tau)$ function itself at $0 < \tau < 16$ is essentially smaller than the other ones. The additional calculations showed that taking into account harmonics with $n = 4$ and $n = 12$ in the deflection approximation (5.66) does not cause notable changes in $W_{1n}(\tau)$ dependencies presented in Fig. 5.16(b), at $\tau \leq 20$. In such a way, to calculate Fourier coefficients over the time interval $\tau \in [0, 20]$ it is enough to retain, in the deflection approximation, the circumferential harmonics with $n = 5$–11.

The comparison of results (see Fig. 5.16(b)), obtained through multi-term and one-term approximations of deflection, shows that at $\tau < 11$ the corresponding curves practically coincide. At greater τ values, taking into account the interconnection of circumferential forms for all values of n under investigation results in a notable decrease in $|W_{1n}(\tau)|$.

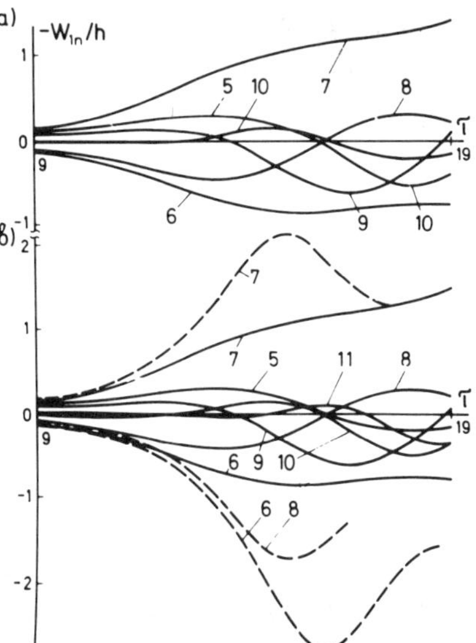

Fig. 5.16. Fourier coefficients dependence on time at $m = 1$, with (———) and without (– – –) the interrelation of circumferential harmonics taken into account; $n_0 = 5$, $N = 10$ (a); $n_0 = 5$, $N = 11$ (b). The initial imperfections are specified by (5.56). The numbers on curves correspond to n-values.

On the basis of the above, we can draw a conclusion that the time dependencies of the most intensively growing Fourier coefficients can essentially differ in the cases of multi-term and one-term approximations of deflection. In the case of one-term approximation the values of Fourier coefficients during the initial stage of bulging are over-rated. The revealed effects are the stronger, the higher is the absolute values of Fourier coefficients.

The examples treated above allow us to assume that both under axial dynamic compression and dynamic external pressure the appearance of the effects, that can be attributed to the interconnection of circumferential harmonics are, to a considerable extent, defined by the field of initial imperfections.

5.5 PECULIARITIES OF NON-AXISYMMETRIC DEFORMATION UNDER LONGITUDINAL DYNAMIC COMPRESSION

The methodology of calculating the deflection w and function of tractions Φ, outlined in the preceding paragraph, allows us to solve the following problem: to determine strains and stresses being generated in a shell at any moment during the given loading program with axial force $P(\tau)$ and external pressure $q(\tau)$.

The stresses in a homogeneous orthotropic cylindrical shell are calculated in accordance with the formulas (4.76) taking the following relations into account

$$T_{11}{}^{0}=-P(\tau);\quad T_{22}{}^{0}=-Rq(\tau). \tag{5.81}$$

The deformations being found from the above relations for stresses, with the generalized Hooke's law for an orthotropic material taken into account, have the form

$$\begin{aligned}
\varepsilon_{xx}(x,y,z,\tau) &= -\left[A_{11}P(\tau)+A_{12}Rq(\tau)\right]+\left[A_{11}\frac{\partial^2\Phi}{\partial y^2}+A_{12}\frac{\partial^2\Phi}{\partial x^2}\right]-\\
&\quad -z\frac{\partial^2(w-w^0)}{\partial x^2};\\
\varepsilon_{yy}(x,y,z,\tau) &= -\left[A_{12}P(\tau)+A_{22}Rq(\tau)\right]+\left[A_{22}\frac{\partial^2\Phi}{\partial x^2}+A_{12}\frac{\partial^2\Phi}{\partial y^2}\right]-\\
&\quad -z\frac{\partial^2(w-w^0)}{\partial y^2};\\
\varepsilon_{xy}(x,y,z,\tau) &= -A_{66}\frac{\partial^2\Phi}{\partial x\partial y}-2z\frac{\partial^2(w-w^0)}{\partial x\partial y}.
\end{aligned} \tag{5.82}$$

Eventually the deformations and stresses in a shell are determined by summing up Fourier series.

In solving each particular problem the first stage involves finding out of Fourier coefficients $W_{mn}(\tau)$ and studying their dependencies on m and n. Several typical dependencies of Fourier coefficients in the expansion of additional deflection $\tilde{W}_{mn}=W_{mn}-W_{mn}^{0}$ on m and n are presented in Fig. 5.17.

As the results of numerous calculations have revealed, the distributions of W_{mn} in respect to n for any m-values both at axial dynamic compression and dynamic external pressure have only one maximum. The value $n=n^*$ to which this maximum corresponds, is mainly defined by the loading rate, by the type of Fourier coefficients W_{mn}^{0} distribution in respect to n and by the time instant τ. Under dynamic external pressure, the distribution of W_{mn} in respect to m has always only one maximum corresponding to $m=1$.

The main specific feature of the problem about the axial dynamic compression consists in that the distribution W_{mn} in respect to m has, as a rule, two maximums. One of them corresponds to $m=1$, the other to an essentially greater value, $m=m^*$. As it was established, m^* depends mainly

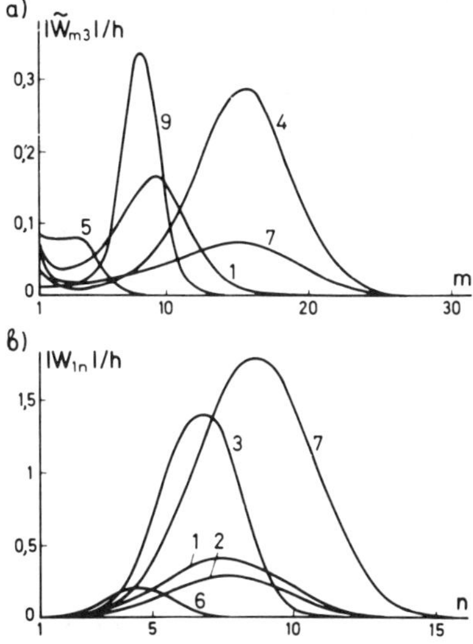

Fig. 5.17.

on the loading rate V_P and shell geometry. The indicated specific feature of the problem leads to the requirement in order to escape coarse errors in determination of the boundaries for summation of the series, that a preliminary analysis of distribution of Fourier coefficients in respect to m should be made within the time interval of interest.

After the values of m and n, corresponding to all maximums of Fourier coefficient distribution, have been established, summation can be made in the following way. Let the maximum with respect to n correspond to $n = n_1$, but the maximum with respect to m — to $m = m_1 = 1$ and $m = m_2$. As the first we choose the term in Fourier series with numbers (m_1, n_1). To this we consecutively add the terms with numbers $(m_1, n_1 - 1)$, $(m_1, n_1 + 1), \ldots, (m_1, n_0), \ldots, (m_1, N)$. As a result we calculate the sum $S^{m_1}_{n_0,N}$. After this the m_1-value increases by a unity, we repeat all the procedure, determine the sum $S^{m_1+1}_{n_0,N}$, then $S^{m_1+2}_{n_0,N}$, etc. until the condition

$$\frac{\sum\limits_{i=1}^{k_0} S^{m_1+k+i}_{n_0,N}}{\sum\limits_{j=0}^{k} S^{m_1+j}_{n_0,N}} < \varepsilon, \qquad (5.83)$$

is fulfilled. Here, ε is the specified allowable error; k_0 is the specified in advance integer value. The resulting double sum is denoted by $S^{(1)}$.

Henceforth the relation between m_1, k, entering (5.83), and m_2 is of essential significance. If $m_1+k\geq m_2$, the summation is over. If $m_1+k<m_2$, the summation should be continued by adding the terms $S_{n_0,N}^{m_2}, S_{n_0,N}^{m_2-1},\ldots,S_{n_0,N}^{m_2-p}$ to $S^{(1)}$. The summation from m_2 toward decreasing m is due to be stopped either after the condition $p=m_2-m_1-k$ has been fulfilled or the specified accuracy achieved:

$$\frac{\sum\limits_{i=1}^{k_0} S_{n_0,N}^{m_2-p-i}}{S^{(1)}+\sum\limits_{r=0}^{p} S_{n_0,N}^{m_2-r}}<\varepsilon. \tag{5.84}$$

Let us denote the resulting sum by $S^{(2)}$. The last stage involves summing from m_2 toward increasing m. Thereby the terms $S_{n_0,N}^{m_2+1}, S_{n_0,N}^{m_2+2},\ldots,S_{n_0,N}^{m_2+q}$ are added to $S^{(2)}$. The process is stopped after the condition

$$\frac{\sum\limits_{i=1}^{k_0} S_{n_0,N}^{m_2+q+i}}{S^{(2)}+\sum\limits_{s=1}^{q} S_{n_0,N}^{m_2+s}}<\varepsilon. \tag{5.85}$$

has been fulfilled. Let us denote the resulting sum by S. In such a way, we eventually have

$$S=\sum_{j=0}^{k} S_{n_0,N}^{m_1+j}+\sum_{r=0}^{p} S_{n_0,N}^{m_2-r}+\sum_{s=1}^{q} S_{n_0,N}^{m_2+s}=$$

$$=\sum_{j=0}^{k}\sum_{n=n_0}^{N} S_n^{m_1+j}+\sum_{r=0}^{p}\sum_{n=n_0}^{N} S_n^{m_2-r}+\sum_{s=1}^{q}\sum_{n=n_0}^{N} S_n^{m_2+s}; \tag{5.86}$$

$$m_2-p<m_1+k.$$

The boundaries of summation k, p, q are determined provided that the addition of any group consisting of k_0 terms, which have been disregarded in (5.86), would lead to correction which does not exceed the specified ε-value.

The procedure of summing can be performed in reverse order by making the n-value fixed and starting calculations by summing in respect to m. The result can be presented in the form

$$S = \sum_{n=n_0}^{N} \left(\sum_{j=0}^{k} S_n^{m_1+j} + \sum_{r=0}^{p} S_n^{m_2-r} + \sum_{s=1}^{q} S_n^{m_2+s} \right). \tag{5.87}$$

In solving the problem by multi-term approximation of deflection, n_0 and N are specified in advance. Therefore, the accuracy could be evaluated only through repeated calculations and comparison of the magnitudes of sums (5.86) or (5.87), calculated at various limiting numbers n_0, N. In solving the problem by one-term approximation of deflection the limits for summation n_0, N are not specified in advance; the summation in respect to n is continued until the required accuracy is reached.

It should be added that depending on $\{x, y\}$ coordinates of the point on shell surface, at which the series are summed, and the time moment τ, the number of terms which must be taken into account in order to reach the specified accuracy, varies greatly. As is revealed from calculations, the series converge very fast in those $\{x, y\}$ points, where the searched-for functions (deflection, deformations, stresses) reach their maxima. Taking this circumstance into account, it is possible to save considerably the computer time necessary for summing of the series, since it is of no practical necessity to calculate small values with a high degree of accuracy.

Let us consider some results of summing the series for deflection in the case of a shell loaded with linearly increasing in time axial compressive force $P(\tau)$. The shell parameters are assumed the same as in the numerical example for axial loading described in paragraph 5.4. The distribution of Fourier coefficients W_{mn}^0 is specified in the form of (5.79); the loading rate $V_P = 0.5$. The limits for summing $m_0 = 1$, $M = 30$, $n_0 = 1$ and $N = 9$ were predetermined by the condition that on adding each following circumferential harmonic and each next group of four axial harmonics the sum might not change more than 1%.

In Fig. 5.18, the $w(x)$ function in the section $y = \pi R$ is shown for two time instants. As is seen, in the given example the non-axisymmetric dynamic bulging is characterized by formation and development of several bands of dents and bulges in the part of a shell. Such a location of the most intensive bulging is caused by the given dependence of the initial deflection on x. In Fig. 5.19, the dependencies of deflection on circumferential coordinate at $\tau = 3.4$ in the sections $x = 0.485L$ (curves 1 and 2) and $x = 0.54L$ (curve 3) are presented. As was noted, in using the distribution (5.79) the circumferential harmonic with $n = 3$ is the dominant; this explains the existence of six nodal points of the $w(y)$ function. The deflection reaches its highest value at $y = \pi R$. As to the effect of interconnection of

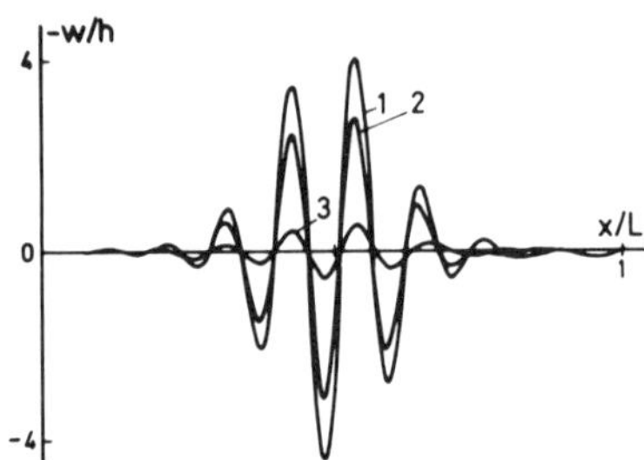

Fig. 5.18. Deflection dependencies on the x coordinate in the $y = \pi R$ section. The curve 1 corresponds to one-term approximation solution at $\tau = 3.4$; curves 2 and 3 correspond to the multi-term approximation solution at $\tau = 3.4$ and 3.0.

circumferential harmonics, it leads to a notable decrease of deflection as seen from Figs 5.18 and 5.19. Since this effect is stipulated by taking into account the geometric non-linearity, it is more pronounced in those parts of the shell where the magnitude of deflection is greater.

Further, we shall consider how the characteristic parameters entering the equations of motion — the loading rate V_P, R/h and L/R relations — affect the shape of a shell during the process of dynamic bulging. The distribution W^0_{mn} is maintained in the form (5.79).

In Figs 5.20a, b, c, the dependencies of deflection on axial coordinate at three loading rates are presented. Let us underline two most characteristic features. Firstly, the number of dents and bulges formed along the shell axis increases with decreasing loading rate; secondly, the dimensions of separate dents and bulges decrease with growing V_P. The latter result was also observed in the experiments of S. A. Uteshev;[249] it may be interpreted

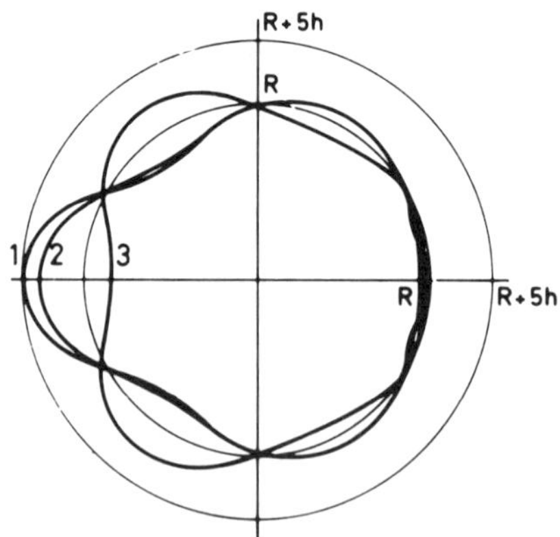

Fig. 5.19. Deflection dependencies on y coordinate at $\tau = 3.4$ in the $x = 0.485L$ (curves 1 and 2) and $x = 0.54L$ (3) sections. The curve 1 corresponds to one-term approximation solution, 2 and 3 correspond to multi-term approximation solution.

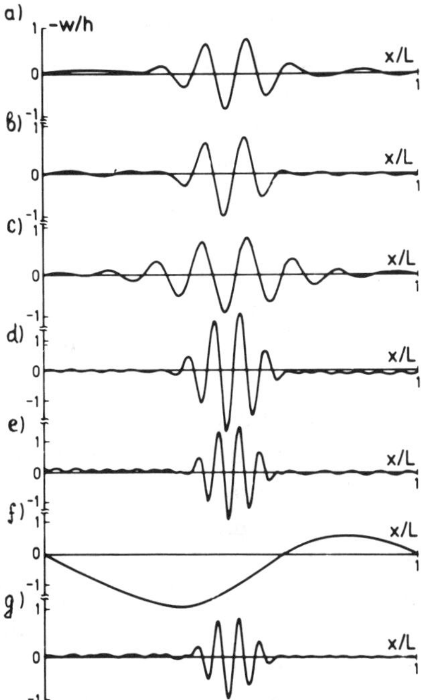

Fig. 5.20. Deflection dependencies on x coordinate at $V_P = 0.5$ (a, d, e, f, g), 1 (b), 0.125 (c); $R/h = 200$ (a, b, c, f, g), 500 (d), 800 (e); $L/R = 2$ (a, b, c, d, e), 0.25 (f), 4 (g).

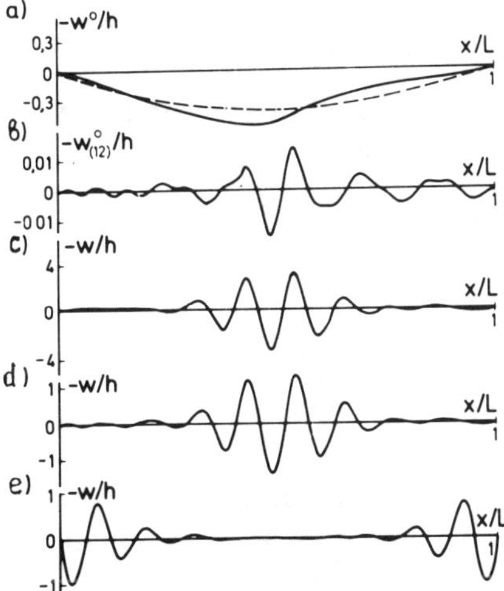

Fig. 5.21. Initial (a, b) and overall (c, d, e) deflection dependencies on the x coordinate: (a) solid line corresponds to (5.64) with $m_0 = 1$, dashed line to (5.65); (b) corresponds to formula (5.64) with $m_0 = 12$, (c) to initial deflection (5.64) and $\tau = 3.4$; (d) to initial deflection (5.66) and $\tau = 3.8$; (e) to initial deflection (5.65) and $\tau = 3.4$.

by the increasing number of characteristic axial harmonic m^*, corresponding to the maximum of $\tilde{W}_{mn}(m)$ function. This result has been established in our calculations (see Fig. 5.17a).

Using the results shown in Figs 5.20(a, d, e), the $w(x)$ functions can be compared at three R/h relations. As can be seen, the size of dents and bulges along the generatrix decreases with increasing R/h and the localization of the zone of intensive bulging in the middle part of a shell becomes more pronounced. This can be interpreted through the increase in m^*-value (see Fig. 5.17(a)) just as in the case of the increase in V_P.

From the results presented in Figs 5.20(a, f, g), the $w(x)$ functions can be compared at three L/R relations. Let us note that the number m^* increases with increasing L/R. This results in the reduction of dimensions of dents and bulges along the generatrix and more pronounced localization of intensive bulging in the middle part of the shell. In such a way, qualitatively the increase of parameters V_P, R/h, L/R similarly affect the character of $w(x)$ function.

Further let us consider carefully the effect of initial imperfections on the peculiarities of dynamic buckling. In the problem under consideration, this aspect seems to be the most important.

In Fig. 5.21(a), the dependencies of the $w^0(x, y)$ function on x in the $y = \pi R$ section are presented for two cases. A solid line corresponds to the sum of Fourier series with coefficients (5.79):

$$w^0(x, y) = 0{,}2h \sum_{m=1}^{30} \frac{(-1)^l}{m^2} \sin \alpha_m x \sum_{n=1}^{10} (-1)^n e^{-|n-3|} \cos \beta_n y, \tag{5.88}$$

while the dashed line — to parabolic dependence on x:

$$w^0(x, y) = 0{,}2h \frac{4x(x-L)}{L^2} \sum_{n=1}^{10} (-1)^n e^{-|n-3|} \cos \beta_n y. \tag{5.89}$$

As can be seen, the curves of Fig. 5.21(a) differ insignificantly. However, presented in Figs 5.21(c, e), $w(x)$ functions, corresponding to these forms of initial deflection, show that the intensive bulging occurs in different shell parts. In the case (5.88) the deepest dents and bulges are formed in the middle part, whereas for imperfections (5.89) they are formed at both end faces. Let us note particularly that both $w(x)$ curves (Figs 5.21(c, e)) are completely dissimilar to corresponding $w^0(x)$ curves (Fig. 5.21(a)).

In such a way, uniform development of the given form of initial imperfections in time is not characteristic for the process of shell bulging under axial dynamic loading. The highest harmonics (see Fig. 5.17(a)) which determine mainly the deformed shell surface, play the principal role in this process. This is confirmed by comparing $w(x)$ function (see Fig. 5.21(c)) and $w^0_{(12)}(x)$ function (Fig. 5.21(b)); the latter was obtained through summing (5.88) with the lower limit $m_0 = 12$.

In Fig. 5.21(d), the $w(x)$ function, corresponding to the initial deflection with more rapid than in (5.88) decrease in Fourier coefficients

$$w^0(x, y) = 0,2h \sum_{m=1}^{30} \frac{(-1)^l}{m^3} \sin \alpha_m x \sum_{n=1}^{10} (-1)^n e^{-|n-3|} \cos \beta_n y \tag{5.90}$$

is presented. As is seen, the curves in Figs 5.21(c, d), are qualitatively similar, only in the case of imperfections (5.66) the intensive development of deflection begins somewhat later. Let us add to the results discussed that uniform increase (or decrease) in all W^0_{mn} coefficients on m leads to proportional increase (decrease) of deflection in all points of a shell.

Further let us dwell on the following important aspect: comparison of the results of calculating the deflection of a shell loaded with axial dynamic forces in accordance with two procedures outlined in paragraphs 5.3 and 5.4. Since the procedure of 5.3 disregards the interconnection of circumferential harmonics, such a comparison is possible only with a version of one-term approximation, described in 5.4.

Let us consider a graphite/epoxy shell having the following geometric parameters: $R = 1$ m, $L/R = 2$, $R/h = 200$ and material characteristics of (5.19). The loading rate is $V_P = 5$; the initial imperfections are assumed other than zero only for circumferential harmonic $n = 3$. Let us specify them through formulas (5.56) and (5.57). The results of solving the problem in accordance with the procedure described in paragraphs 5.3 and 5.4 for this calculation variant are presented in Fig. 5.12. Besides, in Fig. 5.11, the magnitudes of deflection for the corresponding x-section are indicated (marked by dots). They have been calculated in accordance with procedure 5.4. In these calculations the $W_{mn}(t)$ coefficients were found by integrating eqns (5.75), while the summation of Fourier series in respect to m was made within the range $m = 1$, $M = 30$.

It should be once more especially emphasized that the procedures outlined in paragraphs 5.3 and 5.4, differ in essence both the statement of the problem and the methods of solving used. In the first case, the system of equations of motion do not comprise the external load $P(t)$, which enters only the non-uniform boundary conditions (5.43). In the second case, $P(t)$ enters the equation of motion (5.64) as a coefficient, while all the boundary conditions are uniform. In the first case, it is necessary to calculate the $w^0_3(x, y)$ function in advance by summing Fourier series (5.56) in respect to m. In the subsequent solution procedure the values of this function are used only in nodal points $x = x_i$. In the second case the Fourier series for $w_3^0(x, y)$ is nowhere summed up and the W_{m3}^0 coefficients are used only in solving the Cauchy's problem (5.75) and (5.72). And, finally, in the first procedure the finite-difference method, in respect to the x coordinate, is used, whereas in the second the Bubnov-Galerkin method, with one-term approximation of deflection, is used.

Bearing in mind the above, the coincidence of the results of calculating $w_3(x)$ dependencies according to two procedures can be considered as very good. Practically those x-values coincide, at which deflection has maximums. And only the magnitudes of maximums themselves, calculated in accordance with the procedure of paragraph 5.4, are slightly greater. This insignificant disagreement of the results is likely to be explained by disregarding, in the procedure of paragraph 5.4, the effect of interconnection of axial harmonics, which in the procedure of paragraph 5.3 is automatically taken into account. As is shown in Ref. 57, with this effect taken into account, Fourier coefficients W_{mn} decrease. Consequently, the magnitude of deflection must also reduce in those shell regions where deflection has its maximums. The explanation proposed is substantiated, moreover, by the fact that, on the one hand, the relative decrease in deflection, with the effect of interconnection of axial harmonics taken into account, grows with the increase of deflection itself, but, on the other hand, the $w_3(x)$ values, having been calculated according to two procedures, differ more considerably (Fig. 5.12), if the value of $w_3(x)$ is greater.

The given example shows that in those situations when intensive non-axisymmetric bulging occurs at a sufficient distance from the shell end faces (outside the zone of axisymmetric edge effect), the non-axisymmetric deflection component can be fairly reliably established using the procedure outlined in paragraph 5.4. Thereby it should be borne in mind that this procedure is valid only at limited (from above) loading rates, since the respective statement of the problem presupposes the condition of homogeneity of the axisymmetric stress state outside the zones of edge effect. This condition is being disturbed under comparatively short-time loads. In those cases when the region of intensive non-axisymmetric buckling is outside the zones of edge effect, the procedure outlined in paragraph 5.1 can be used for calculating the axisymmetric component of deflection.

In such a way, under the condition that the processes of axisymmetric and non-axisymmetric dynamic buckling occur in different coordinate regions, the general problem, outlined in paragraph 5.3, can be divided into two, essentially simpler ones, studied in paragraphs 5.1 and 5.4. The solution to the first problem yields the axisymmetric component of deflection, while the solution to the second one — the non-axisymmetric components (which can be calculated with the interconnection of arbitrary finite number of circumferential harmonics which are taken into account). The total shell deflection can be evaluated by superposition of these two components. Let us note that the computer time for solving the two named particular problems is 4–5 times less than for solving the general problem.

The formulated principle of superposition for the non-linear problem is, of course, of an approximate character and its applicability is restricted. It is inapplicable in those situations when calculations made in accordance to the procedure of paragraph 5.4 indicate that the region of non-axisymmetric dynamic buckling is located within the zones of axisymmetric edge effect (see, for example, Fig. 5.21(e)). Really, in calculations performed in paragraph 5.3 with the parabolic dependence on x of the non-axisymmetric component of the initial deflection it was established (see Fig. 5.14(a)) that under the conditions of simple support, the evolution of non-axisymmetric component of deflection is slightly pronounced. The character of the $w(x)$ function presented in Fig. 5.21(e), contradicts this result. Therefore, in the given case correct magnitudes of deflection can be obtained only with the interconnection of the axisymmetric and non-axisymmetric components taken into account, i.e. by using the procedure outlined in paragraph 5.3.

The examples treated refer to the calculation of the shell deflection. Deformations and stresses, as stated earlier, can be calculated by substituting the series (5.73) into the formulas (5.81) and (4.76). Of course, in summing the series for deformations and stresses it is necessary to ensure that the required accuracy be reached. For this purpose, the boundaries for summation should be established according to the procedure described earlier in this paragraph. As a rule, the series for deformations and stresses converge more slowly than the series for deflection. This requires a greater number of terms both in respect to m and n to be retained.

The numerical examples, illustrating the stress dependence on the x coordinate and time for various isotropic and orthotropic shells were presented in Refs 55 and 57, where the loading with linearly increasing in time axial compressive forces was treated. Without dwelling on the illustrations of these results, let us note only one, the most essential to our mind, conclusion.

It was emphasized in Chapter 2 that all the employed versions of the equations of motion had been based on the model of linearly elastic behavior of the material. Therefore, strictly speaking, the outlined calculation procedures become inapplicable, provided the stresses even in one shell point have passed out of the region into stress space eliminated by the yield surface (or strength surface).

In this connection, there arises a question: whether the above stated effects of geometrical non-linearity are manifested and what their extent is in the case of linearly elastic material behavior? This subject has been comprehensively studied for a steel shell at various magnitudes of R/h, L/R, V_P parameters. The dependencies of intensity of tangential stresses, function $T(\tau)$ (5.25), obtained from the solutions of geometrically non-linear problem, with and without interconnection of circumferential harmonics taken into account, as well as from the solution of the linearized equations of motion have been compared. It was established that right up to the moment $\tau = \tau^*$, corresponding to the initiation of plastic deformations these dependencies differ slightly, even for shells with $R/h = 500$.

One of the ways of getting the most notable effects, associated with the geometric non-linearity in the region of physically linear material behavior, can be the study of shells made up of the materials with higher yield strength (or proportionality limit for non-linear-elastic materials). One of such materials is MAYLAR having the following mechanical characteristics: $E = 5 \cdot 10^9\ \mathrm{N/m^2}$; $\nu = 0.38$; $\rho = 1.4 \cdot 10^3\ \mathrm{kg/m^3}$; $T_i = 9 \cdot 10^7\ \mathrm{N/m^2}$. The $T(\tau)$ dependencies (Fig. 5.22) have been calculated in accordance with the three named approaches for shells having $R/h = 800$, $L/R = 2$, at loading rate $V_P = 0.5$ and initial imperfections (5.79). As is seen, at the time moment $\tau = \tau^*$, which corresponds to the $T(\tau^*) = T_i$ condition, T-values, calculated with and without the interconnection of circumferential harmonics taken into account, differ appreciably. The difference between T-values found from the solution of the linearized equations and from

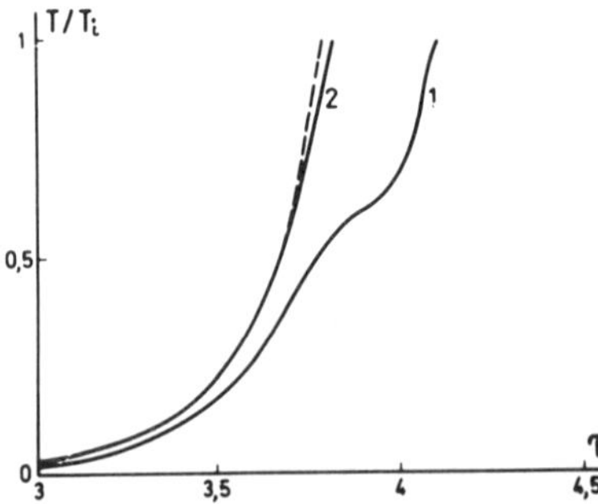

Fig. 5.22. Shear stress intensity dependence on time for Mylar shell calculated according to linearized (curve 1) and non-linear (2, 3) equations; the curve 2 corresponds to the solution by one-term approximation and 3 by multi-term approximation; $x = 0.49L$, $y = \pi R$.

solution of the non-linear equations with one-term approximation, is considerably weaker.

A similar analysis was performed for shells of several composite materials (fiberglass-, carbon- and boron-plastics). Thereby, instead of von Mises yield condition, the strength condition postulated for an orthotropic layer (expressed in terms of a quadratic polynomial of the stress tensor components), was employed. This problem is treated comprehensively in Chapter 6. Let us emphasize here only that failure of the named composite materials in a shell starts usually when the effect of geometric non-linearity is still poorly pronounced, both for axial dynamic compression and external dynamic pressure loadings.

The explanation for this result can be based on sufficiently simple considerations. Really, owing to the high density of the spectra of natural frequencies and critical static axial forces, a large number of axial and circumferential harmonics take part in the process of dynamic buckling of a cylindrical shell. In such a way, each of the characteristics of stress–strain state in a shell is determined as a sum of a large number of Fourier series terms. Thereby, each single term in the sum is sufficiently small compared to the magnitude of the total sum. For example, $|W_{mn}| \ll \max |w(x,y)|$ for $x \in [0, L]$, $y \in [0, 2\pi R]$ and all the m and n values. From the relative smallness of Fourier coefficients, in their turn, follows that they can be calculated with sufficiently high accuracy, without taking the geometric non-linearity into account. This result is of paramount significance in solving the dynamic problems for more complicated cylindrical structural elements (for example, stiffened shells and shells with an inner core). In many cases this result can permit to rid oneself of the necessity of constructing extremely awkward non-linear solutions, because right up to the moment of the onset of plastic deformations (or the moment of initial failure), it is possible to calculate the characteristics of stress—strain state on the basis of linearized equations of motion.

5.6 PECULIARITIES OF NON-AXISYMMETRIC DEFORMATION UNDER DYNAMIC LATERAL PRESSURE

Let us consider a shell loaded with pressure $q(\tau)$ uniformly distributed over the outer surface. Deformations and stresses in a homogeneous orthotropic shell are specified for this case by the formulas (5.81) and (4.76) at $P(\tau) \equiv 0$ and, eventually, are calculated by summing the double Fourier series. From the standpoint of constructing the procedure for summation of the series, the dynamic external pressure problem differs from the problem of longitudinal dynamic compression only in that for the former case the distribution of Fourier coefficients W_{mn} in respect to m always has only one maximum, which usually corresponds to the lowest harmonic $m = 1$.

Let us dwell on some of the most characteristic results of calculating the non-axisymmetric deformation of a cylindrical shell under linearly increasing in time lateral pressure

$$q(\tau) = V_q q^* \tau. \tag{5.91}$$

Here, q^* is the critical static external pressure; τ is the dimensionless time, determined according to the formula (5.21); V_q is the dimensionless loading rate.

The dependence of Fourier coefficients on time for a duralumin shell having the parameters $L = 0.2\,\mathrm{m}$, $R = 0.09\,\mathrm{m}$, $h = 8 \cdot 10^{-4}\,\mathrm{m}$ at initial imperfections (5.80) and loading rate $V_q = 0.2$ (6500 atm/s) are presented in Fig. 5.16. As was noted in paragraph 5.4, shell deflection can be calculated for this case with sufficient accuracy, taking into account the circumferential harmonics with numbers $n = 4$–12. The effect of interconnection of circumferential harmonics influences essentially the $W_{mn}(\tau)$ functions. The results of calculating the deflection are presented in Fig. 5.23 (the series were summated at $m_0 = 1$, $M = 3$, $n_0 = 4$, $N = 12$). As is seen, shell deflection develops intensively in several strictly definite zones on the shell surface: the location of these zones depends on the specified field of initial imperfections $w^0(x, y)$. Just in these zones the difference in magnitudes of deflection, calculated on the basis of multi-term and one-term approximations is the most pronounced. An analogous result was outlined in discussing the problem in longitudinal dynamic compression (see Fig. 5.18). Obviously, this result is characteristic for all non-linear problems: the magnitudes of deflection, deformations, stresses, having been obtained on the basis of multi-term and one-term approximations, differ the stronger, the greater are their values. It can also be noted that the solution by one-term approximation yields over-rated magnitudes of deflection in the zones of the most intensive deflection growth.

The transformation of the deformed shell surface in time is illustrated in Fig. 5.23(b). The dependencies show that the maximum value of deflection changes slightly in time, while the points of maximum deflection keep shifting continually along the circumference. Let us add that the solution based on one-term approximation indicates monotonous growth of deflection at the points of maximums with practically unchanging location of these points on the circumference.

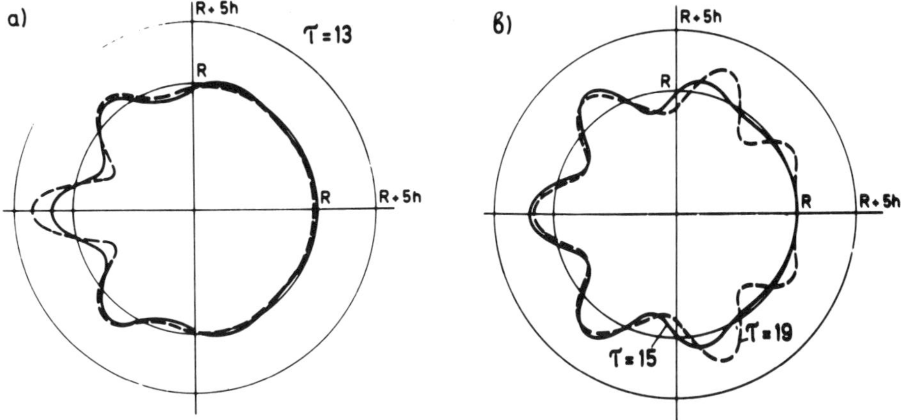

Fig. 5.23. Deflection dependencies on the y coordinate, corresponding to solutions by one-term (– – –) and multi-term (———) approximations at $\tau = 13$ (a) and to solution by multi-term approximation at $\tau = 15$ (———) and 19 (– – –) (b).

Further we shall treat the results of calculating the intensity of tangential stresses (5.25) at loading rate 6500 atm/s. In Fig. 5.24, the $T(\tau)$ dependencies are plotted for $x = L/2$ section and some characteristic points on the circumference. As is seen, right up to the moment $\tau = 11$, at which the $W_{mn}(\tau)$ dependencies, calculated with and without interconnection (see Fig. 5.16(b)) taken into account, were close, the respective $T(\tau)$ curves also differ slightly. At later time moments, the difference between these curves becomes of a qualitative character: the solution with one-term approximation leads to monotonous $T(\tau)$ growth, while the solution

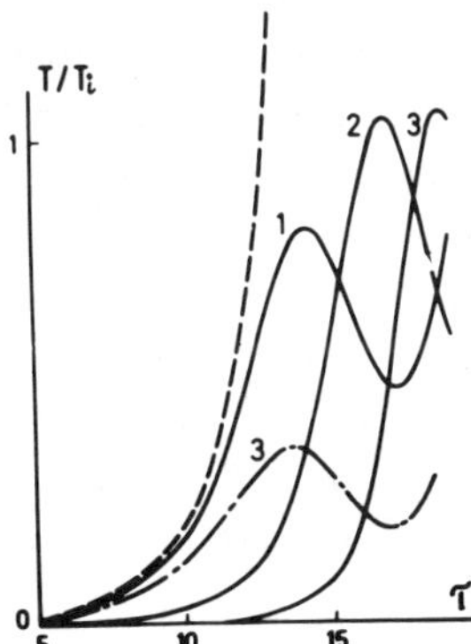

Fig. 5.24. Dependencies of T-function on time in the $x = 0.5L$ section for inner (curves 1–3, 5) and outer (4) shell surfaces. The curves 1–4 correspond to the solution by multi-term, 5 — by one-term approximations. The y-values are: $0.85\pi R$ (1), $0.544\pi R$ (2), $0.272\pi R$ (3, 4), $0.886\pi R$ (5).

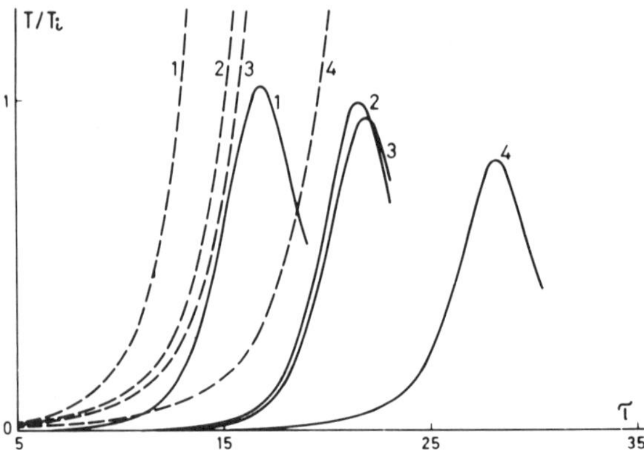

Fig. 5.25. Dependencies of T-function on time in the multi-term (———) and one-term (– – –) approximation solutions. The numbers on curves correspond to V_q-values: 1 — 0.2 (6500 at/s); 2 — 0.154 (5000 at/s); 3 — 0.145 (4700 at/s); 4 — 0.103 (3340 at/s).

with multi-term approximation reveals non-monotonous character of $T(\tau)$ function which can be explained by local bulgings in different parts of a shell which alternate with reverse bulgings. As is seen from Fig. 5.24, each of the curves has its own characteristic instant when it reaches the first maximum. The magnitudes of these maximums depend essentially on y and z coordinates.

The $T(\tau)$ dependencies calculated at four rates of external pressure loading, used in experiments by A. S. Vol'mir and V. E. Mineev,[102] are presented in Fig. 5.25. These dependencies correspond to the inside surface, $x = L/2$ section and to individual for each dependence point on the circumference, in which the first maximum of T-function on τ reaches its greatest value. As is seen, the effect of interconnection of circumferential harmonics increases with decreasing loading rate. Besides, let us note that the magnitude of the first maximum of T-function on τ diminishes with decreasing V_q.

The results treated show that the criterion of dynamic instability, used in paragraph 5.2, and consisting in fulfilling the $T = T_i$ condition, at least in one point of a shell, can also be used in the problem under consideration here. In order to find such a point $\{x^*, y^*, z^*\}$ in which at a fixed loading rate the yield condition is fulfilled earliest of all, a special algorithm was evolved.

Let us note, by the way, that the yield condition was, obviously, first employed for evaluation of the critical magnitude of dynamic external pressure by Yu. P. Kadashevich and A. K. Pertsev,[162] and later in a paper by E. I. Grigolyuk and A. I. Srebovskii,[139] where the moment in which the earliest plastic deformations initiated in a shell was assessed from the hoop stress values calculated separately for each circumferential harmonic. Such an approach can provide a highly over-rated value of critical pressure, since as a result of superposition of several circumferential harmonics having, possibly, the same growth rate, the zones with much higher stresses than the stresses for individual harmonic are formed in a shell. Let us emphasize that in our procedure, in order to calculate the T-function, a definite number of terms of Fourier double series were used in respect to the accuracy required, while the Fourier coefficients were calculated, with the interrelation of all circumferential harmonics taken into account.

As the calculations based on multi-term approximation of the deflection (the value $T_i = 3 \cdot 10^8\,\mathrm{N/m^2}$ was used) have shown that at loading rate $V_q = 0.2$ (6500 atm/s) the formation of plastic deformations starts at the moment $\tau^* = 17.0$ in the $\{0.5L; 0.54\pi R; -0.5h\}$ point. When using one-term approximation, $\tau^* = 12.8$ and coordinates $\{0.5L; 0.87\pi R; -0.5h\}$ were obtained, respectively. At $V_q = 0.154$ (5000 atm/s) we find $\tau^* = 21.4$ from the solution by multi-term approximation and $\tau^* = 16.5$ — by one-term approximation. For the other two magnitudes of V_q — 0.145 (4700 atm/s) and 0.103 (3340 atm/s) the yield criterion was not fulfilled in the interval $\tau \in [0, 35]$ (the solution with one-term approximation allows to evaluate τ^*-value also for these rates: 17.2 at $V_q = 0.145$ and 21.4 at $V_q = 0.103$) in any point of the shell in the case of multi-term approximation. Consequently, it is possible to employ the yield condition as a criterion of dynamic instability only within a limited from below range of external pressure loading rates: $V_q > V_q^0$. As is seen from Fig. 5.25, at $V_q \leq V_q^0$, after the characteristics of stress–strain state have reached their first maximum in various parts of the shell, they start to fall abruptly: the entire process of alternating 'bulgings' and 'reverse bulgings' occurs without infringement on the condition of elastic behavior of the material.

Thus, within different ranges of loading rates it is necessary to use various criteria of dynamic instability. The outlined procedure of calculation permits the following criteria to be used. Firstly, this is the condition for generation of plastic deformations in a shell, treated earlier (let us designate it Criterion 1). Secondly, the condition for the deflection to exceed the specified ultimately allowable w^* level at least in one point of the shell (Criterion 2). Thirdly, assessment of the critical time instant τ^* in accordance with fulfilment of the condition $(\partial w(x, y, \tau))/\partial \tau = 0$, at least in one point of the shell (Criterion 3) which is the mathematical formulation of the deflection transition through the first maximum. Finally, more complicated procedure can be employed. Proceeding from $w(\tau, y)$ functions, to evaluate at first for each y such $\tau^*(y)$ value, at which the deflection reaches its first maximum. Further, to choose from all magnitudes of $w^*(y) = |w[\tau^*(y)]|$ the highest one and the corresponding τ^* value to name it as the critical one (Criterion 4). It should be emphasized that all these criteria of dynamic instability presuppose the presence of non-axisymmetric imperfections, which are necessary for calculating the process of non-axisymmetric shell deformation.

Let us also note that these criteria are equally well applicable to problems of external pressure loading and axial compression loading. The advantage of these criteria consists in that they have a strict mathematical formulation and, consequently, can be realized on computers. Of course, the approaches enumerated do not exhaust all the variety of possible criteria of dynamic instability for imperfect shells.

Let us compare the obtained theoretical results and the experimental data of A. S. Vol'mir and V. E. Mineev.[102] Let us introduce the dynamicity coefficient $K_d = (q(\tau^*))/q^* = V_q\tau^*$. In Table 5.1, the experimental magnitudes of K_d from Ref. 102 and two groups of theoretical values, calculated on the basis of multi- and one-term approximations, are listed.

On evaluating the theoretical and experimental data correlation it should be borne in mind that the field of initial imperfections $w^0(x, y)$, assumed in calculations, was introduced through Fourier coefficients (5.80) and the assumed yield strength value T_i can significantly differ from the respective characteristics of actual samples tested in Ref. 102. However, at all four loading rates the calculated magnitudes of K_d, obtained through multi-term approximation, are fairly close to experimental ones. At the same time, the solution obtained through one-term approximation at the same initial parameters yield highly over-rated K_d magnitudes.

TABLE 5.1

Theoretical and Experimental Values of K_d-Coefficient of a Shell Loaded by Dynamic External Pressure

V_q *(atm/s)*	*Solution by one-term approximation*	*Solution by multi-term approximation*	*Experimental data*[102]
0.20 (6500)	2.56[a]	3.40[a]	3.98
0.154 (5000)	2.31[a]	3.30[a]	3.54
0.145 (4700)	2.23[a]	3.15[b]	3.19
0.103 (3340)	2.00[a]	2.88[b]	2.80

[a]Criterion 1.
[b]Criterion 4.

The fact that the K_d magnitudes, obtained by using Criterion 1, are lower than the experimental ones, has its explanation analogous to that given in 5.2 when comparing the theoretical and experimental results of the axial impact problem. The formation of the zones of plastic deformations in a shell is only the 'precursor' of notable residual deflection, detectable in the experiment.

As to orthotropic composite shells, any of the above mentioned criteria of dynamic instability is applicable in the same manner. Here let us restrict ourselves to an illustrative example of calculating the deflection in graphite/epoxy composite shells having various layer stacking sequence.

Let us consider a shell having the following geometric parameters: $R = 1$ m, $L/R = 2$, $R/h = 200$ (h is the total thickness of a package) and characteristics of a monolayer (5.19). Let us designate the angle between the shell generatrix and axis 1 of the i-th layer by Θ_i. Let us specify the initial imperfections in the form (5.80) by assuming instead of coefficient $0.02h$ the following values: $10^{-2}h$ for $V_q = 0.5$ and $0.1h$ and $10^{-3}h$ for rate $V_q = 2$.

In Fig. 5.26, the deflection dependencies on time in the $\{0.5L, \pi R\}$ point at three loading rates are presented (the results were obtained by using multi-term approximation). Let us outline the abrupt increase in the sensitivity of $w(\tau)$ function to the structure of a package with decreasing V_q.

The $w(\tau)$ dependencies, plotted for the same point at $V_q = 0.5$ (Fig. 5.27), allow to demonstrate the effect of interrelation of circumferential harmonics once more. As is seen, the deflection, calculated on the basis of one-term approximation, for all structures of a package is highly overrated. The calculations have also shown that the difference in results, obtained by multi-term and one-term approximations, increases with decreasing V_q.

We can conclude from the results of this paragraph that in the case of loading with dynamic external pressure, in contrast to the case of

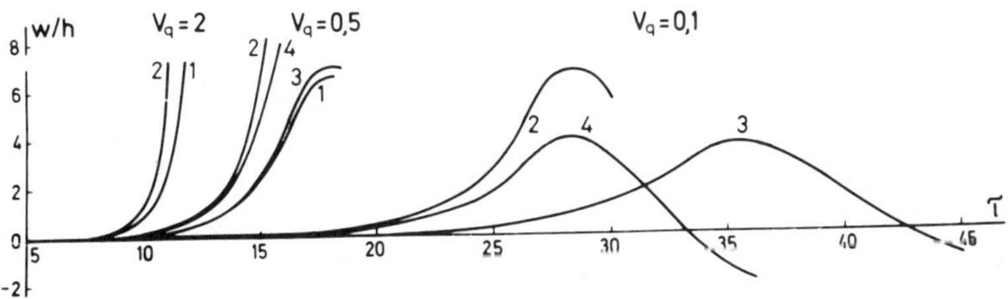

Fig. 5.26. Deflection dependencies on time for three rates of external pressure loading. Lamina layup angles: 90° (curves 1); 0° (2); $[90°, 0°]_S$ (3); $[0°, 90°]_S$ (4).

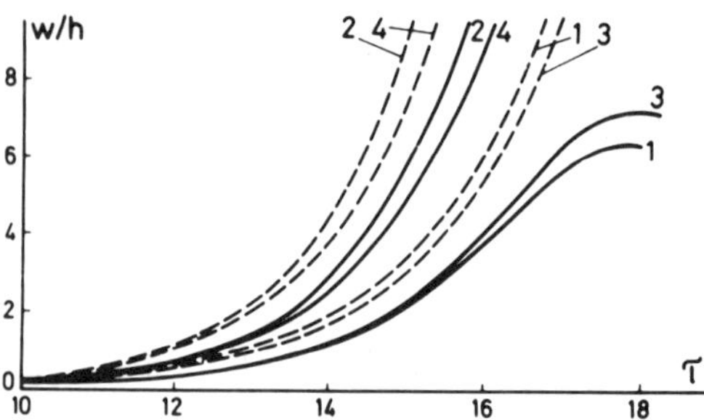

Fig. 5.27. Deflection dependencies on time at $V_q = 0.5$. The numbers on curves correspond to the same reinforcement angles as in Fig. 5.26. Solid lines correspond to multi-term approximation solution and dashed ones to one-term approximation solution.

axial dynamic compression, treated in 5.5, the influence of geometrical non-linearity on the calculated characteristics of stress–strain state, even in the region of linearly elastic material behavior, is significant. The effects, caused by geometric non-linearity, become weaker with increasing loading rate.

5.7 DEFORMATION UNDER JOINT DYNAMIC LONGITUDINAL COMPRESSION AND LATERAL PRESSURE

On the basis of the calculation method, worked out in 5.4, we can consider the problem of simultaneous effect of uniformly distributed over the end faces axial compressive tractions $P(t)$ and uniformly distributed over the outer side surface pressure $q(t)$. Fourier coefficients $W_{mn}(t)$ are found by numerical integration of the system (5.70) with initial conditions (5.72). The deflection and other characteristics of stress–strain state in a shell are determined in accordance with the procedure, described in 5.5, by summing up double Fourier series.

The possibility of realizing fairly complex buckling modes is the specific feature of the combined effect of the two different dynamic loads. In fact, under dynamic external pressure the most intensive buckling is accompanied, as a rule, by formation of one half-wave along the generatrix, whereas under longitudinal dynamic compression, a group of rather high axial harmonics prevail. The numbers of the dominant circumferential modes of buckling also differ essentially for the two loading types. This leads to the result that during the deformation process, the shell surface can have the most diverse configurations, depending on the ratio of loading rates.

Let us assume that both loads are the linear time functions (5.20) and (5.91) and settle the total dimensionless loading rate $V = V_P + V_q = 1/3$ fixed. Let us consider a six-ply graphite/epoxy shell having $R/h = 188$, $L/R = 2$. The monolayer characteristics are specified in accordance to (5.19). The angles between axis 1 of each layer and the shell generatrix are as follows: 45°, −45°, 90°, 90°, 45°, −45°. Let us introduce the initial imperfections of the shell shape by means of Fourier series (5.67) with coefficients

$$W_{mn}{}^{0} = 0{,}02h \frac{(-1)^{l+n}}{m^2} \exp\left[-\frac{(n-4)^2}{4} \right], \tag{5.92}$$

where $l = (m/2)$ at even m and $l = (m+1)/2$ at odd m.

In Fig. 5.28, the deflection dependence on the x coordinate at five V_P/V_q

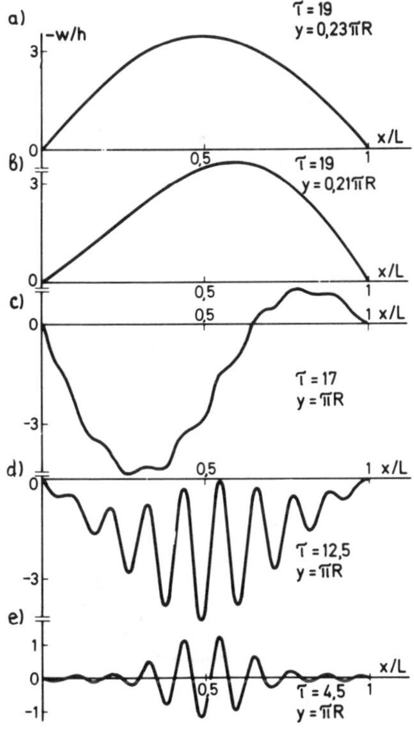

Fig. 5.28. Deflection dependencies on the x coordinate at $V_p = 0$ (a), $V_q = 0$ (e) and $V_p/V_q = 1/9$ (b), 1/4 (c), 3/7 (d).

ratios is shown. For each of the variants under consideration, these dependencies have been plotted for those sections y, where the deflection has reached its maximum. As is seen from comparison of the curves of Fig. 5.28(a, b), addition of relatively slowly increasing longitudinal load (for variant b, at $\tau = 19$ it is equal to $0.78P^*$) to the more rapidly increasing external pressure leads to the maximum of $w(x)$ function being shifted. This effect is the result of considerably earlier (compared to the case of pure external pressure) development of $m = 2$ mode and its notable contribution to the process of dynamic shell buckling. However, since at $\tau = 19$ the axial load is less than the static critical load, the intensive buckling by high m harmonics, has not yet started and they do not practically affect the shape of the deformed shell surface.

For the variant c, the axial loading is equal to $1.41P^*$ at $\tau = 17$; the harmonics, corresponding to high m, exceed the prescribed initial imperfections and, as a result, affect appreciably the $w(x)$ function, although the deflection, being caused by axial compression, is still considerably smaller than that caused by external pressure. In the variant d, at $\tau = 12.5$, the axial load is equal to $1.55P^*$ and the deflection components, defined by the mode $m = 1$ and by a group of high axial harmonics, are roughly the same. It results in the characteristic form of $w(x)$ dependence (Fig. 5.28(d)) reminding of the superposition of two $w(x)$ functions, corresponding to single loadings with external pressure and axial compression. At ratios $V_P/V_q > 2/3$, the deformed shell surface has the shape, typical for the case of pure axial dynamic loading (Fig. 5.28(e)).

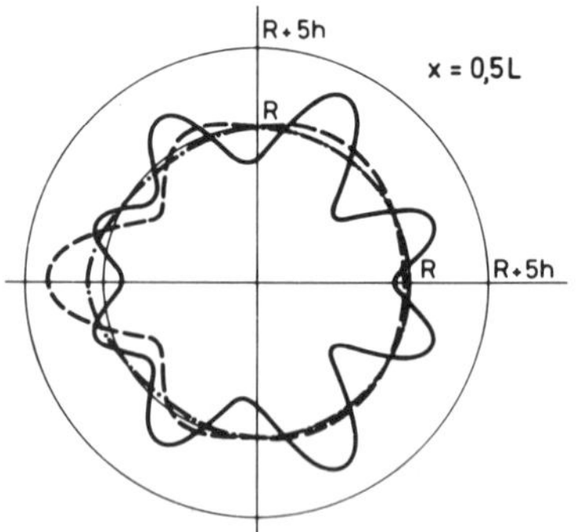

Fig. 5.29. Deflection dependencies on the y coordinate at $V_p = 0$ (———) and $V_p/V_q = 1/4$ (– – –), 3/7 (–· – ·–).

In Fig. 5.29, the deflection dependence on the circumferential coordinate in the $x = L/2$ section is shown for three V_P/V_q ratios. Let us note that under longitudinal compression the circumferential mode $n = 4$, corresponding to the maximum of distribution (5.92), is prevailing, whereas under external pressure at the same initial imperfection distribution the circumferential harmonics $n = 7$ and $n = 8$ are prevailing. In the case $V_P/V_q = 3/7$ the shell takes a configuration, which combines certain specific features of buckling modes, having been realized for the loadings under consideration, applied separately.

As is seen from the results of Fig. 5.28, a sharp change in the character of shell deformation occurs within the range $V_P/V_q = 1/4$–$3/7$. In Fig. 5.30 the results show in detail, how the $w(x)$ dependence is transformed with increasing relative longitudinal loading rate. Let us note that at comparatively small V_P/V_q increase, the contribution of the group of high

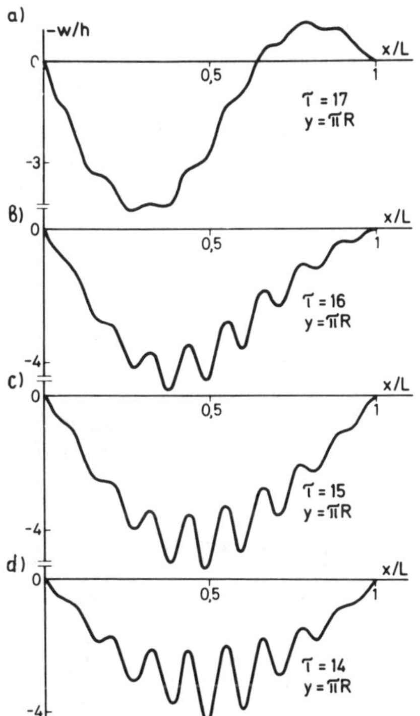

Fig. 5.30. Deflection dependencies on the x coordinate at $V_p/V_q = 0.25$ (a), 0.275 (b), 0.3 (c), 0.325 (d).

axial harmonics becomes very strong: the presence of external pressure is revealed in the characteristic 'sagging' of $w(x)$ curve, which becomes less notable with increasing V_P/V_q.

The time dependencies of axial σ_{xx} and circumferential σ_{yy} stresses generated on the inner shell surface are shown in Fig. 5.31 for three V_P/V_q ratios. The results of each variant refer to a certain $\{x, y\}$ point of shell surface, at which deflection reaches its maximum. Let us pay attention to the fact that at $V_P = 0$, the stress σ_{yy} considerably exceeds σ_{xx} (which is tensile). At $V_P/V_q = 1/4$, the stress σ_{xx} becomes compressive, while at $\tau < 17$, σ_{yy} remains the dominant one. But in the case $V_P/V_q = 3/7$ an intensive growth of σ_{xx} at $\tau > 11$ is observed; the axial stress becomes the dominant one.

The example treated here, shows that the problem of optimal design (in its simplest version, finding the optimal angles of ply layup) of fibrous composite shells under combined effect of two dynamic loads under consideration, is highly complicated. Within different V_P/V_q ranges, it is necessary to guarantee high resistance either to axial, circumferential, or both stresses simultaneously. It should also be emphasized that, depending on the V_P/V_q ratio, the boundaries for summing up Fourier series, defining the characteristics of stress–strain state in a shell, do change essentially.

Let us add that in those situations, when the deflection and stresses initiated by dynamic external pressure play an essential role, the interaction of circumferential modes of buckling must be taken into account.

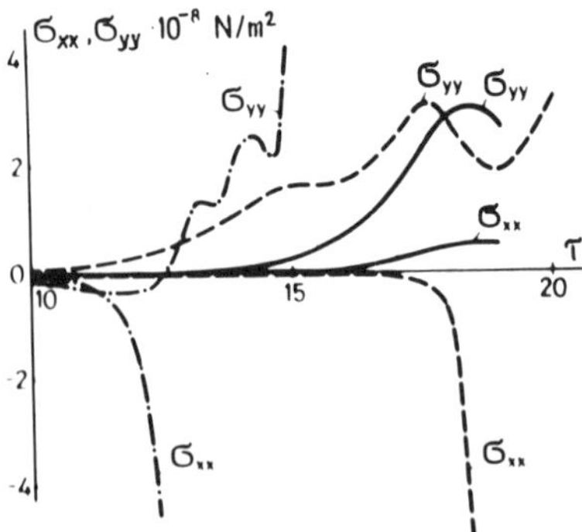

Fig. 5.31. Stress dependencies on time at $V_p = 0$ (———) and $V_p/V_q = 1/4$ (– – –), 3/7 (–· – ·–).

5.8 DEFORMATION UNDER COMBINED EFFECT OF STATIC AND DYNAMIC LOADS

The method worked out in 5.4 can be generalized also to the case, when prior to the instant in which dynamic loads were applied, the shell has been loaded with static axial compression, static external or internal pressure. Thereby deformations and stresses in the shell are defined by both static and dynamic loads, acting over the specified time interval $[0, T]$.

Static preloading results in two effects: load increase on a shell in each time moment $t \in [0, T]$, and growth of initial imperfections. Owing to this, equations of motion (5.64) and (5.65) should be modified: constant values of P_0 and q_0 ($-q_0$ in case of internal pressure) are to be added to $P(t)$ and $q(t)$, as well as $w^0(x, y)$ substituted by $w^{\text{st}}(x, y)$. The latter is determined from the solution to the problem of buckling of imperfect cylindrical shell under corresponding static loads. (The action of internal static pressure leads to the decrease in initial deflection $w^0(x, y)$. Calculations of $w^{\text{st}}(x, y)$ in this case is simple, because it can be performed in geometrically linear statement.) Correspondingly, when solving the Cauchy's problem (5.70) and (5.72), the Fourier coefficients W_{mn}^{st} in Fourier series for the $w^{\text{st}}(x, y)$ function should be utilized as the initial value $W_{mn}|_{t=0}$.

When solving the problem of static buckling in geometrically linear statement, the magnitudes of W_{mn}^{st} can be calculated by the formula

$$W_{mn}{}^{st} = \frac{W_{mn}{}^{0}}{1 - P_0/P^*{}_{mn} - q_0/q^*{}_{mn}}, \tag{5.93}$$

where P_{mn}^* and q_{mn}^* are critical values of static loads under axial compression and external pressure for the mode (m, n). The expression (5.93) is inapplicable at those P_0 and q_0 values, which are close to P_{mn}^* and q_{mn}^*, because for them geometrical non-linearity must be taken into account.

Without having discussed the complex independent problem of static buckling of imperfect cylindrical shells in detail, let us consider the simplest solution, based on one-term deflection approximation:

$$w_{mn}{}^{st}(x, y) = W_{mn}{}^{st} \sin \alpha_m x \cos \beta_n y. \tag{5.94}$$

The solution by the Bubnov-Galerkin method of the static version of equations (5.64) and (5.65) with approximation (5.94) leads to the following non-linear algebraic equation in respect to W_{mn}^{st}:

$$W_{mn}{}^{st}\left(1 - \frac{P_0}{P^*{}_{mn}} - \frac{q_0}{q^*{}_{mn}}\right) + \\ + d_{mn} W_{mn}{}^{st}\left(W_{mn}{}^{st^2} - W_{mn}{}^{0^2}\right) = W_{mn}{}^{0}, \tag{5.95}$$

where d_{mn} is determined according to (4.71).

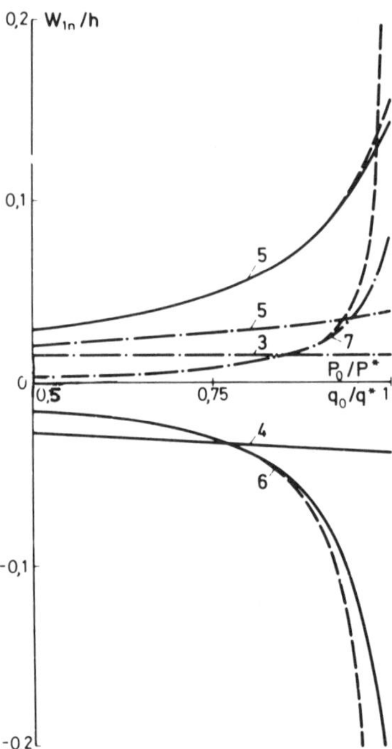

Fig. 5.32. Initial deflection Fourier coefficients dependence on static load magnitude calculated in accordance with eqn (5.95) for the cases of axial compression (———) and external pressure (–· – ·–) and according to formula (5.93) (– – –). The numbers on curves correspond to n-values.

In Fig. 5.32, the dependencies of W_{mn}^{st} on the magnitude of applied static loads is shown. Calculations have been performed for a six-ply graphite/epoxy shell, described in 5.7. As we see, under external pressure the greatest difference in the results obtained according to (5.95) and formula (5.93) is observed for circumferential harmonic $n = 7$ (corresponding critical load $q_{1,7}^*$ coincides with q^*), while under axial compression — for harmonics $n = 5$ and 6 (the values of $P_{1,5}^*$ and $P_{1,6}^*$ are very close to P^*). Within the range $P_0/P^* \leq 0.9$ and $q_0/q^* \leq 0.9$, the acceptable results were obtained from the solution to the problem of static buckling in geometrically linear statement.

As follows from Fig. 5.33, the results obtained according to eqn (5.95) and formula (5.93) for a number of high axial modes, being of great interest for the problem of axial dynamic compression, practically coincide. This can be accounted for by the fact that the values of P_{mn}^* at high m exceed appreciably P^*. The presence of external static pressure q_0/q^* does not practically affect W_{mn}^{st} at high m. It should be emphasized that these estimates have been obtained for distribution (5.92); concrete

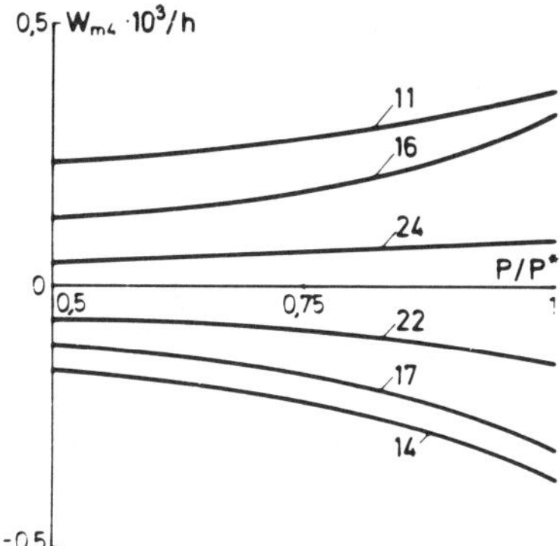

Fig. 5.33. Initial deflection Fourier coefficient dependencies on the magnitudes of axial static load. The numbers on curves correspond to n-values.

types of initial imperfections affect the range of applicability of the linear theory to static buckling problems, to a considerable extent.

From the results of Figs 5.32 and 5.33, we can conclude that W_{mn}^{st}-values, corresponding to various m and n, change very non-uniformly with increasing static loads. Therefore, static pre-loading can lead to notable distinction between the 'imperfections' w^{st}, existing in the shell prior to the moment of application of dynamic loads, and initial imperfections w^0, and thus affect essentially the process of dynamic buckling.

The dynamic problems were solved for the following versions of combined loading: (a) $P_0 + P(t)$; (b) $q_0 + P(t)$; (c) $P_0 + q(t)$; (d) $q_0 + q(t)$. The loads $P(t)$ and $q(t)$ were assumed as linearly increasing in time. From among the obtained results let us note the fact that in the case (b), the value q_0 at $0 \leq q_0 \leq 0.9q^*$ does not practically affect the stress–strain state in a shell, having arisen under the effect of dynamic loading. This can be easily explained by taking into account that high axial harmonics are prevailing under longitudinal dynamic compression, at the same time, as was stated, these harmonics are insensitive in the case of static external pressure. In other cases, under sufficiently high levels of preliminary static loading, there are notable changes in the character of deflection and stress distribution over the shell surface in a fixed time instant and also in their time dependencies. The strongest effect was assessed under combined loading with static and dynamic external pressure. Figure 5.34 shows the time dependencies of stresses at the inner shell surface under this variant of combined loading. These dependencies were obtained for a six-ply shell having the following angles between the axis 1 of the layer and generatrix of the shell: 90°, 90°, 80°, −80°, 90° and 90°. The characteristics of a

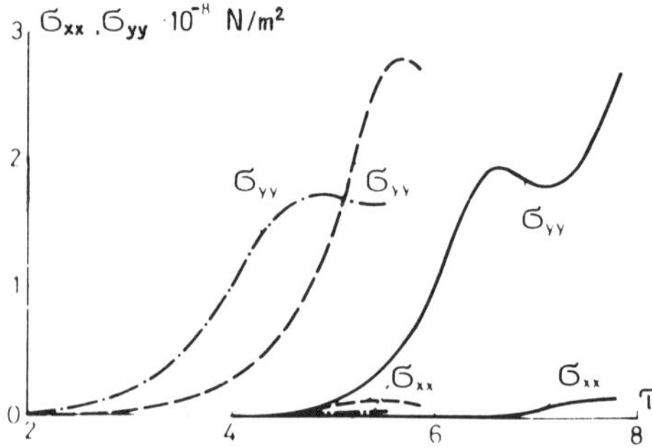

Fig. 5.34. Stress dependencies on time in the case of combined effect of static and dynamic components of external pressure for $q_0 = 0$ (———) and $q_0/q^* = 0.5$ (– – –), 0.9 (–· – ·–).

monolayer and geometric parameters of a shell are assumed as in 5.7. As is clear from Fig. 5.34, the increase in the level of static loading leads to significant shortening of the time interval prior to the start of intensive dynamic buckling. An analogous, but quantitatively weaker effect takes place under the variants of combined loading (a) and (c). The calculation results also show that preliminary static loading changes the coordinates of those points on the shell surface, at which the deflection and stresses reach their maximums. This effect is most pronounced in variants (c) and (d). In certain cases the reduction of stress maximums in the initial stage of dynamic buckling was detected (this result is illustrated in Fig. 5.34). In such a way, preliminary static compression of a shell can prolong the time interval till the start of a failure process.

Let us emphasize also that the case of combined loading (a) offers no difficulty for generalizing the calculation procedure, outlined earlier in 5.3 for the case of axial dynamic compression only.

5.9 THE EFFECT OF TRANSVERSE SHEAR DEFORMATIONS ON DYNAMIC BUCKLING OF ORTHOTROPIC CYLINDRICAL SHELLS

The calculation procedures, elaborated in this chapter, for calculating the characteristics of stress–strain state in orthotropic cylindrical shells under dynamic compression loads are based on non-linear equations of motion, obtained by using the Kirchhoff-Love kinematic model. As it follows from the previous numerical results, the coefficients of variability of stress–strain state in a shell for the problems under investigation can be sufficiently high. Besides, resistance to shear deformations of fibrous composites is comparatively weak. These circumstances put forth a natural question: how strongly do the results of solving the above treated problems change when the transverse shear deformations are additionally taken into account?

Let us treat the problem of dynamic buckling of orthotropic cylindrical shells on the basis of linearized equations of motion of Timoshenko type, presented in 3.3, by supplementing the left-hand part of the third equation of the system (3.43) by 'loading' term (5.63).

Assuming the conditions of free support at the end faces of a shell, let us represent the unknown functions as series (3.5) and (3.44). Then in respect to Fourier coefficients, we obtain the system (3.45), supplemented by the term

$$\left[P(t)\alpha_m^2 + q(t)R\left(\beta_n^2 - \frac{1}{R^2}\right)\right]W_{mn}$$

in the left-hand part of the third equation. A further procedure of solution, analogous to that outlined in 3.3, leads to a linear ordinary differential equation with a time-dependent coefficient

$$\frac{d^2W_{mn}}{dt^2} + \omega_{mn}^2\left\{\left[1 - \frac{P(t)}{P^*_{mn}} - \frac{q(t)}{q^*_{mn}}\right]W_{mn} - W_{mn}^0\right\} = 0, \tag{5.96}$$

This equation differs from the respective equation, obtained on the basis of the Kirchhoff-Love model, only by the expressions for coefficients ω_{mn}^2, P^*_{mn}, q^*_{mn}. In such a way, the accuracy of solution to the problem of dynamic buckling in geometrically linear statement is fully determined by the accuracy of determination of the natural frequencies and critical static loads. The effect of transverse shear deformations on these characteristics of a shell was studied in detail in 3.3.

Let us consider several illustrative examples for a cylindrical shell made of a unidirectionally reinforced graphite/epoxy composite having the characteristics of (3.23) in the case of axial dynamic compression. The axis 1 of the material is directed along the x-axis of the shell. The load is assumed as linearly increasing in time (5.20). Let us assign the distribution of Fourier coefficients of the initial deflection in the form

$$W_{mn} = 0{,}2h\frac{(-1)^{l+n}}{m^2}\exp\left[-\frac{(n-4)^2}{4}\right], \tag{5.97}$$

where $l = m/2$ at even m and $l = (m+1)/2$ at odd m.

In Fig. 5.35 the $w(\tau)$ functions are presented in the $y = \pi R$ section at such x_0-values, which correspond to the most intensive deflection

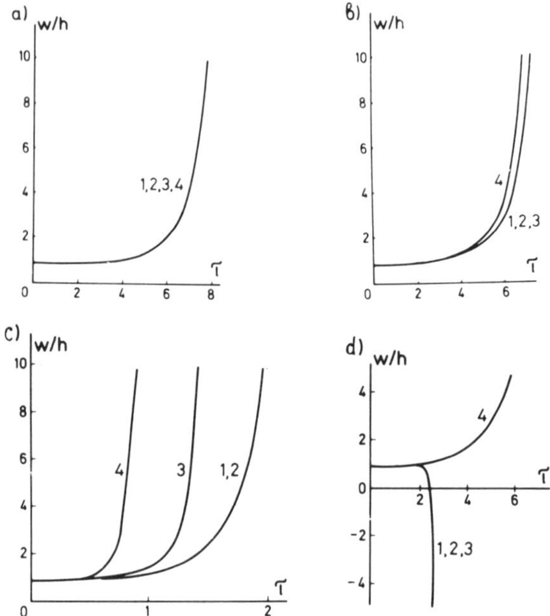

Fig. 5.35. Deflection dependencies on time, calculated in accordance with the Kirchhoff-Love theory (curves 1) and to the Timoshenko-type theory (2, 3, 4) at: $G_{13} = G_{23} = G_{12}$ (2), $G_{13} = G_{23} = 0.1G_{12}$ (3) and $G_{13} = G_{23} = 0.01G_{12}$ (4); $V_p = 0.5$ (a, b, d) and 5 (c); $R/h = 500$ (a), 200 (b), 50 (c, d); $x_0 = 0.48L$ (a), $0.47L$ (b), $0.49L$ (c), $0.45L$ (d).

development for each concrete variant. Various V_P-values, R/h and shear moduli ratios $G_{13}/G_{12} = G_{23}/G_{12}$ are considered. Let us note that at $V_P = 0.5$ and $R/h = 500$ (Fig. 5.35(a)) the fact of taking into account the transverse shears changes the deflection (at the magnitude of deflection $w \approx 3h$) by 0.08% at $G_{i3}/G_{12} = 1$ and by 0.6% at $G_{i3}/G_{12} = 0.1$. At $R/h = 200$ (Fig. 5.35(b)) the deflection magnitudes differ by 0.1% at $G_{i3}/G_{12} = 1$ and by 1% at $G_{i3}/G_{12} = 0.1$, while at $R/h = 50$ (Fig. 5.35(d)) — by 0.3% at $G_{i3}/G_{12} = 1$. At the same R/h relation and $V_P = 5$ (Fig. 5.35(c)) the differences in deflection magnitudes, calculated with and without transverse shear deformations taken into account, amounts to 0.1% at $G_{i3}/G_{12} = 1$. It should be emphasized that for realistic values of transverse shear moduli ($G_{i3}/G_{12} = 1$–0.1), even at $R/h = 50$, the correction introduced into w-value, with transverse shear deformations taken into account, in the treated range of V_P does not exceed 1%.

Let us underline the qualitative difference of $w(\tau)$ functions, calculated with and without transverse shear deformations taken into account, at

$G_{i3}/G_{12} = 10^{-2}$ (curve 4 in Fig. 5.35(d)). This result can be interpreted as follows. At very low values of G_{i3}/G_{12} the critical axial tractions P^*_{mn} for intensively developing in time harmonics, being calculated with transverse shears taken into account, prove to be notably smaller than in the case of using the classical theory. Consequently, intensive dynamic shell buckling will start at essentially lower magnitude of load $P(\tau)$ (and at lower magnitude of τ, respectively). As a result, for one and the same τ-value, the shape of the deformed shell surface, calculated according to the classical theory, does not practically differ from the initial one (Fig. 5.36(a)), whereas, with the transverse shears taken into account, it has the appearance characteristic for the developed dynamic buckling state (Fig. 5.36(b)). The deflection maximum is approximately $20h$. It is natural that in a fixed cross section x, the picture of deflection variance in time essentially differs for two theories under consideration.

The results presented give a certain substantiation for the applicability of the Kirchhoff-Love theory in calculations of the deflection of thin-walled orthotropic shells under the effect of dynamic compressive loads. The calculations have also shown that for deformations and stresses the correction introduced by taking transverse shears into account, is somewhat greater than for the deflection, though it also does not exceed several per cent at $G_{i3}/G_{12} \geq 0.1$; $R/h \geq 100$ in the range of loading rates $V_P \leq 10$; $V_q \leq 20$.

It should be mentioned also that in Chapter 2 the non-linear equations of motion for orthotropic cylindrical shells, with transverse shear deformations taken into account, were obtained. The solutions to non-linear problems, treated in Chapters 4 and 5, can analogously be constructed on the basis of these equations. The difficulties arising are of purely technical character. Eventually the systems of non-linear ordinary differential equations will be obtained, in which the coefficients at non-linear terms depend on transverse shear moduli, too.

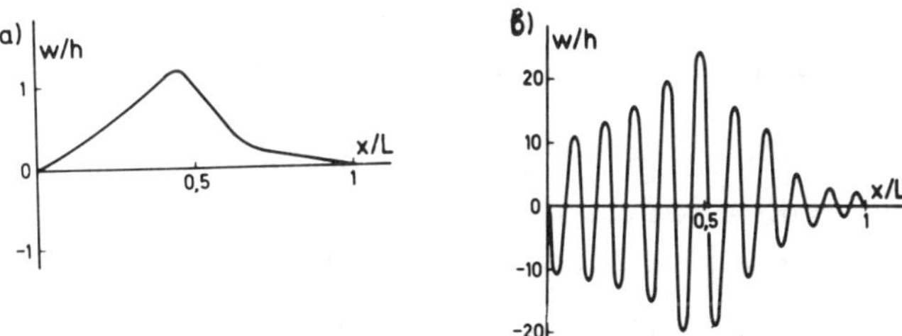

Fig. 5.36. Deflection dependencies on x coordinate at $\tau = 2.8$. Calculation done in accordance to the Kirchhoff-Love theory (a) and the Timoshenko-type theory (b) at $R/h = 50$, $V_p = 5$, $G_{13} = G_{23} = 10^{-2}G_{12}$.

5.10 ON THE EFFECT OF IMPULSE SHAPE ON DYNAMIC STABILITY OF CYLINDRICAL SHELLS

The numerical results, treated in this chapter (except paragraph 5.2), were obtained under linearly increasing in time axial compression and external pressure. Such a character of $P(\tau)$ and $q(\tau)$ dependencies was chosen from the considerations of compactness in discussing the principal results. From this viewpoint, the main advantage of the linear $P(\tau)$ and $q(\tau)$ functions is that only one parameter-loading rate $V_P(V_q)$ is to be specified.

The procedures elaborated in 5.1, 5.3 and 5.4 reduce the calculation of shell deformation to the Cauchy's problem for complex non-linear systems of ordinary second-order differential equations. The external loads enter the equations obtained in 5.4, only as time-dependent coefficients. In the simplest version — the case of one-term approximation of the deflection — it is necessary to solve the single non-linear ordinary differential equation with a variable coefficient. But even in this case the solution cannot be written down in an analytical form. Only in geometrically linear statement, when the primary problem is reduced to integration of the linear ordinary differential equation with a variable coefficient, the solution can be written down through quadrature. However, even for this case, only in particular cases the solution can be defined through special functions. The examples are Mathieu's equation, to which the problem of parametric vibrations is reduced (see Chapter 4) and the equation with linearly increasing in time coefficient, whose solution is expressed through Bessel's function. Even in geometrically linear statement of the problem, calculation of a shell under dynamic loading programs of more general character requires necessarily that numerical integration should be applied.

For this purpose, in all the problems treated in this book the standard program realizing Runge-Kutta method of fourth-order accuracy was used. This program was comprehensively checked by us in solving diverse problems and, as was stated, it allows, in principle, to carry out calculations for any continuous in time external loads. The concrete character of load-time dependence affects only the value of integration step.

As an example, illustrating the possibilities of the procedure used, let us consider the results of calculating the shell loaded with axial compressive tractions, varying in time in accordance with the following laws:

$$\text{(a)} \quad P(\tau) = \begin{cases} \frac{\tau}{2} P^* & \text{at } \tau \leq \tau_0; \\ 0 & \text{at } \tau > \tau_0; \end{cases} \tag{5.98}$$

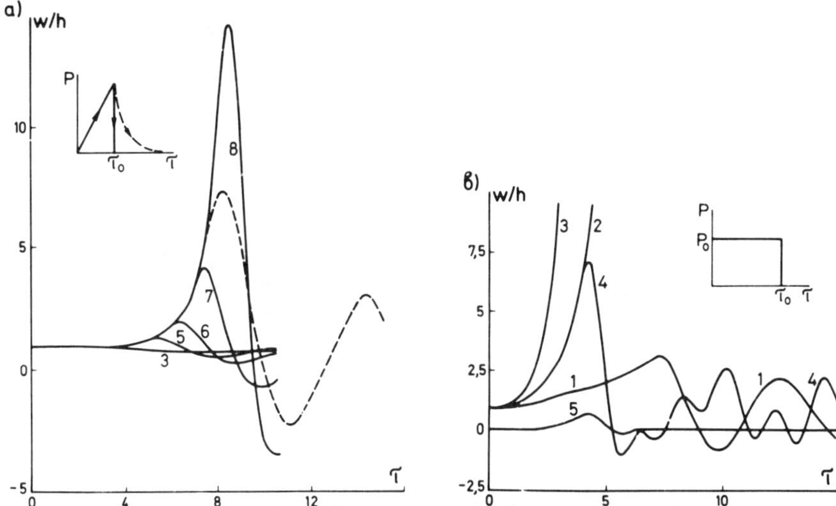

Fig. 5.37. Dependencies of w on time for loading programs (5.98), (5.99) (a) and (5.100) (b). The numbers on curves correspond for Fig. 5.37(a) to values: $\tau_0 = 3, 5, 6, 7, 8$, for Fig. 5.37(b) to values $P_0 = 2$ (1), 3 (2, 4, 5), 4 (3) and $\tau_0 = 7$ (1, 2, 3) and 4 (4, 5). Other explanations in the text.

$$\text{(b)} \quad P(\tau) = \begin{cases} \dfrac{\tau}{2} P^* & \text{at } \tau \leq \tau_0; \\ \dfrac{\tau}{2} P^* e^{-(\tau-\tau_0)} & \text{at } \tau > \tau_0; \end{cases} \tag{5.99}$$

$$\text{(c)} \quad P(\tau) = P_0[H(\tau) - H(\tau - \tau_0)]. \tag{5.100}$$

Calculations were carried out in non-linear statement (procedure 5.4) on the basis of one-term approximation.

Typical $w(\tau)$ dependencies are presented in Fig. 5.37. In Fig. 5.37(a), solid lines correspond to loading program (5.98); numbers at curves indicate the τ_0-value. A dashed line in Fig. 5.37(a) corresponds to (5.99) with $\tau_0 = 7$. The initial deflection at the point of a shell, to which the given results refer, is $w^0 = w|_{t=0} \approx 0.9h$. In Fig. 5.37(b) $w(\tau)$ dependencies for several combinations of P_0 and τ_0 parameters in formula (5.100) are given. The curves 1–4 correspond to the magnitude of initial deflection $w^0 \approx 0.9h$ at the point under consideration, whereas the curve 5 — to $w^0 \approx 0.09h$. Let us note that a tenfold reduction of the amplitude of initial deflection has resulted in maximum deflection magnitude also being reduced ten times. The presented results illustrate, in particular, the fact that the question about dynamic stability of realistic thin-walled structure can hardly be solved without its imperfections being taken into account.

6

Strength Analysis of Laminated Cylindrical Shells under Dynamic Compressive Loads

The calculation procedures, presented in Chapter 5, allow to evaluate the magnitudes of displacements, deformations and stresses in an arbitrary shell point in any time instant under assigned program of dynamic axisymmetric (longitudinal compression and external pressure) load. The following problem can be formulated on this basis: at specified strength characteristics of the shell material to determine the magnitude of load (or that of applied impulse) under which, at least in one shell point, the failure condition of the material is satisfied and to locate the zone of initial failure. In this chapter, the main attention is paid to methods and results of calculating the initial failure of laminated cylindrical shells under the effect of dynamic compressive loads. Besides, a model describing the process of layer-by-layer failure of a shell beyond the first-ply failure moment, is treated. A procedure for calculating transversal stresses in laminated cylindrical shells, subjected to dynamic compressive loads, is proposed. The danger of failure due to transverse and interlaminar stresses prior to the onset of in-plane failure of the layer is analyzed for graphite-epoxy shells.

Let us dwell briefly on several works, pertaining to the above mentioned questions. First of all, let us note that in review papers[71,93,108,215,220,226,237,238,255] considerable attention was paid to various aspects of strength and failure problems for fibrous, laminated composites and filament wound structural elements. One of the most important tasks now is the evolvement of strength criteria for anisotropic and multi-layer composites under complex stress state and complex loading. Fundamental works in this field were made by A. K. Malmeister,[184] S. W. Tsai and E. M. Wu,[443] E. M. Wu[108] and A. H. Puppo and H. A. Evensen.[400] In the last years many works devoted to strength criteria for composites, their experimental check-up and experimental determination of strength characteristics, entering these criteria, were published. In this connection, let us mention a few published papers.[93,241,242,255,340,409,411]

The strength criteria, applicable to single layers, allow determination of only the load and type of 'initial' failure of a multi-layer composite. In this connection, great attention is naturally being paid to experimental study and theoretical characterization of the processes of layer-by-layer failure, starting from the moment of first-ply failure right up to macrofailure of a multi-layer specimen. This problem, first investigated in Ref. 341 was treated in Refs 158, 159, 226, 231, 255 and 394.

In the last years there is a growing interest in deformation and failure of composite materials under high-velocity impact loads, although the number of works published so far is comparatively scarce.[153,275,276,284,299,300,306,325,343,356–358,361,364,365,395,403,408,412,414,426] Tests have been conducted in tension,[299,306,343,356,357,403] compression,[275,276] in-plane shear,[365] inter-laminar shear,[414] as well as three-point bending.[358] The experiments under single[153,408,412,426] and repeated[395] impact loading applied to the surface of composite plates were carried out too. Loading was applied by the use of gas guns,[153,275,412] impact pendulum[299] and falling weight,[365] by generating internal pressure on a ring due to explosion in the liquid contained,[306] according to the Hopkinson's bar technique.[275,276,etc] The materials tested were fiberglass,[275,276,343,356,357,395,403] unidirectional[299,306,343,356,357] and laminated[153,358,408,412] graphite plastics.

Let us especially characterize some results, obtained in the papers listed. In Refs 275 and 276 it was noted that glass-epoxy strength increases with increasing strain rate, independently of the angle between the reinforcement directions and acting compressive load. It was concluded in Ref. 306 for a unidirectional carbon plastic that the effect of strain rate is not essential under the loading applied along the reinforcing fibers but increases abruptly with growing angle between the reinforcing fibers and load direction. It was established in Ref. 343 that the strength and failure mode of a unidirectional carbon plastic does not practically depend on strain rate within a wide range; whereas the modulus and strength of fiberglass fabric composite increased notably with increasing strain rate. It was shown in Refs 356 and 357 that the ultimate stresses and failure strains of fiberglass plastics increase essentially with increasing strain rate, whereas the changes in these values for carbon plastics are barely noticeable. Let us also note that, in accordance with Ref. 153, carbon plastic under impact loading behaves as a brittle material. In Refs 299, 403, 408 and 412, the extent of damageability, character of microdamages, shapes of fracture surfaces as well as specific features of failure mechanisms in carbon and fiberglass plastics under impact loading were investigated.

In the last years, great attention was attracted by the problem of increasing the resistance of composite laminated structures to specific failure modes — delamination on the interfaces between layers and 'peeling' of one of the outer layers when the structural element is under the effect of compressive loads. V. V. Bolotin[73] has investigated the problems of growth of initial local delaminations and peeling in laminated structural elements. In a faultless primary structure these failure modes can be realized if the interlaminar stresses, generated under loading, exceed the respective ultimate strength of the interface layer between the reinforced layers. The method evolved by the author in Ref. 49 makes it possible to evaluate interlaminar stresses in imperfect cylindrical shells under dynamic compressive loads, to determine the loads corresponding to the onset of delamination and location of the zones of initial delaminations.

Few works devoted to the strength analysis for cylindrical shells made of laminated composites (let us mention Refs 27, 155, 221 and 222) refer exceptionally to the simplest kinds of static loads (internal pressure, axisymmetric bending), for which the first stage of solving of the total problem — calculation of stress–strain state in a shell — does not make difficulties. In data[55,56,58,59] the problem of calculating the 'initial' failure of laminated cylindrical shells under dynamic axial compressive loads and external pressure was formulated and solved. For the solutions obtained it was necessary to take into account that essentially non-uniform, in respect to coordinates, stress–strain states would be realized. In Ref. 48, the simplest model was proposed to describe the process of layer-by-layer failure of a cylindrical imperfect shell under the effect of dynamic compressive loads.

Let us also note that the probabilistic strength analysis of multi-layer cylindrical shells under random internal pressure was realized[216] by the Monte Carlo method. In Ref. 62, the procedure for calculating the reliability of a multi-layer cylindrical shell with a random field of initial imperfections under dynamic loads of axial compression and external pressure was developed. Reliability of a shell was determined in terms of outbursts of the stress vector field beyond the strength surface having the form of a three-dimensional rectangular parallelepiped.

On the whole, it can be stated that hitherto only the first steps have been made in solving the strength, failure and reliability problems for thin-walled structural elements made of fibrous and laminated composites.

6.1 ASSESSMENT OF CRITICAL DYNAMIC LOAD IN TERMS OF THE MOMENT OF FIRST-PLY FAILURE

The methods described in Chapter 5 allow us to calculate deformations in an arbitrary point $\{x, y, z\}$ for a shell in any time moment $t \in [0, T]$, where T is the duration of the specified loading program under all the types of loading described.

Let the multi-layered shell consist of N orthotropic layers; we shall determine the principal directions of elasticity for each individual layer by indices 1 and 2. Thus, if in any section z the strains in relation to shell axis are ε_{xx}, ε_{yy}, ε_{xy}, the strains in the principal axes of the i-th layer in the same section, in accordance with the known formulas for tensor components transformation upon rotation of the system of coordinates, take the form

$$\begin{aligned}\varepsilon_{11}{}^{(i)}&=\varepsilon_{xx}\cos^2\Theta_i+\varepsilon_{yy}\sin^2\Theta_i+\varepsilon_{xy}\sin\Theta_i\cos\Theta_i;\\ \varepsilon_{22}{}^{(i)}&=\varepsilon_{xx}\sin^2\Theta_i+\varepsilon_{yy}\cos^2\Theta_i-\varepsilon_{xy}\sin\Theta_i\cos\Theta_i;\\ \varepsilon_{12}{}^{(i)}&=(\varepsilon_{yy}-\varepsilon_{xx})\sin 2\Theta_i+\varepsilon_{xy}\cos 2\Theta_i,\end{aligned} \tag{6.1}$$

where Θ_i is the angle between the principal direction 1 in the i-th layer and x coordinate axis. The strains ε_{xx}, ε_{yy} and ε_{xy}, relative to x, y-axes are linearly distributed through the thickness of a shell. At the same time, at the magnitudes of Θ_i, different from layer to layer, the linear distribution of strains ε_{11}, ε_{22} and ε_{12} through the thickness according to (6.1), will be satisfied only within the limits of each individual layer. Consequently, the strains relative to axes 1 and 2 on the contact surfaces will undergo jumps.

The stresses in the principal axes of the i-th orthotropic layer in accordance with the Hooke's law take the form

$$\begin{aligned}\sigma_{11}{}^{(i)}&=B_{11}{}^{(i)}\varepsilon_{11}{}^{(i)}+B_{12}{}^{(i)}\varepsilon_{22}{}^{(i)};\quad \sigma_{22}{}^{(i)}=B_{12}{}^{(i)}\varepsilon_{11}{}^{(i)}+B_{22}{}^{(i)}\varepsilon_{22}{}^{(i)};\\ \sigma_{12}{}^{(i)}&=B_{66}{}^{(i)}\varepsilon_{12}{}^{(i)},\end{aligned} \tag{6.2}$$

where $B_{jk}^{(i)}$ is the stiffness matrix of the i-th layer.

The formulas of Chapter 2 permit calculation (at known functions u, v, w or w, F) of the stress–strain state in an arbitrary point $\{x, y, z\}$ of a laminated shell. Thereby at the stages of obtaining u, v, w and F functions and determining the strains ε_{xx}, ε_{yy} and ε_{xy}, the laminate may be considered as homogeneous and orthotropic medium; further calculation of strains $\varepsilon_{11}^{(i)}$, $\varepsilon_{22}^{(i)}$ and $\varepsilon_{12}^{(i)}$ and stresses $\sigma_{11}^{(i)}$, $\sigma_{22}^{(i)}$ and $\sigma_{12}^{(i)}$ is performed in terms of individual elastic characteristics of each layer and the orientation of its principal directions relative to the shell axes.

The next stage of investigation involves selection of the strength criterion for a shell. It can be formulated both for an orthotropic shell as a whole (provided that the laminate is conceived as homogeneous) as well as for each individual layer. Further we shall employ the latter approach and calculate the load of initial failure of a shell under the condition that at least in one of the layers the specified strength criterion is satisfied. The magnitude of dynamic load will be further called the critical dynamic load (or the load of initial failure) but the time moment in which the critical dynamic load is reached — the moment of initial failure.

In accordance with the described approach, the strength analysis of a shell can principally be performed by using any of the known phenomenological strength criteria for an anisotropic body. However, in consideration of a composite structure not all of these criteria, as shown in Refs 93, 108, 184, 226, 255, 443, reflect the specific features of composite materials to a sufficient degree. It should also be borne in mind that in solving the complicated dynamic problems the requirement of the mathematical convenience of strength criterion is of significance.

For a phenomenological description of a strength surface in the six-dimensional space of stresses, the following equation was suggested in Ref. 184:

$$p_{\alpha\beta}\sigma_{\alpha\beta}+p_{\alpha\beta\gamma\delta}\sigma_{\alpha\beta}\sigma_{\gamma\delta}+p_{\alpha\beta\gamma\delta\varepsilon\xi}\sigma_{\alpha\beta}\sigma_{\gamma\delta}\sigma_{\varepsilon\xi}+\ldots=1, \qquad (6.3)$$

where $p_{\alpha\beta}$, $p_{\alpha\beta\gamma\delta}$, $p_{\alpha\beta\gamma\delta\varepsilon\xi}, \ldots$ are the strength surface tensors. (At dynamic loading, the values $p_{\alpha\beta}$, $p_{\alpha\beta\gamma\delta}$, $p_{\alpha\beta\gamma\delta\varepsilon\xi}, \ldots$ are the functions of strain rate.) As has been shown in Refs 108 and 184, at present all the practically known phenomenological strength criteria for an anisotropic body can be treated as particular cases of the criterion (6.3) or as its singular case. It was shown in a thorough theoretical study of the properties of a tensor-polynomial criterion[108] that apart from the fact that it is universal and invariant as to selection of the system of coordinates, this criterion possesses 'maximally possible computing flexibility and, at the same time, does not involve extra parameters'.

In assessing the moment of initial failure of the *i*-th layer in the shell, we shall further employ the criterion (6.3) written down for the case of a plane stress state by retaining the terms of the second degree

$$\begin{aligned} F(\sigma_{jk}^{(i)}) = {} & p_{11}^{(i)}\sigma_{11}^{(i)}+p_{22}^{(i)}\sigma_{22}^{(i)}+p_{1111}^{(i)}\sigma_{11}^{(i)2}+p_{2222}^{(i)}\sigma_{22}^{(i)2}+ \\ & +2p_{1122}^{(i)}\sigma_{11}^{(i)}\sigma_{22}^{(i)}+4p_{1212}^{(i)}\sigma_{12}^{(i)2}=1. \end{aligned} \qquad (6.4)$$

For practical use of the criterion (6.4) it is necessary to know the magnitudes of components of the strength surface tensors. They can be calculated from the magnitudes of characteristic strengths, obtained for various simple loading ways, or as a result of approximation of the strength surface by the surface of second order, provided the number of experimental points exceeds the number of known parameters in (6.4). The experimental data in the literature allow us to obtain reliable magnitudes of all the components of strength surface tensors, entering (6.4), except p_{1122}. The difficulties of the assessment of p_{1122} have been analyzed in Refs 108, 226, 242, 255 and 459. It has been noted in particular that the component p_{1122} for certain composite structures has increased sensitivity to the manner of approximation of the set of experimental data by the strength surface of the second order. In Refs 242 and 459 a strong dependence of the accuracy of p_{1122} determination on the stress relation σ_{11}/σ_{22} was revealed.

The analysis, with the aim of assessing the effect of p_{1122} on the initial failure load values for graphite-epoxy cylindrical shells under some types of dynamic loading performed in Ref. 56, permitted to conclude that under axial dynamic compression the magnitude of p_{1122} does not practically affect the results of calculating the moment of initial failure at all reinforcement angles (except reinforcement at small angle, $\Theta < 5°$, to the longitudinal axis of a shell). Under dynamic lateral pressure, the magnitude of p_{1122} affects appreciably the calculated magnitude of the moment of initial failure only within a narrow range of reinforcement angles $\Theta \approx 75–85°$.

6.2 INITIAL FAILURE UNDER LONGITUDINAL DYNAMIC COMPRESSION

Let us consider an orthotropic cylindrical shell loaded with longitudinal compressive traction $P(\tau) = V_P P^* \tau$. Let τ^* be the time instant in which the condition (6.4) is fulfilled, at least in one shell point. The point coordinates $x = x^*$, $y = y^*$, $z = z^*$, in which at $\tau = \tau^*$ the criterion (6.4) is fulfilled, define the location of the zone of initial failure. Henceforth we shall call them the coordinates of initial failure (for a homogeneous shell) or coordinates of first-ply failure (for a multi-layer shell). The load magnitude $P_d^* = V_P P^* \tau^*$ is said to be the critical dynamic load, while the relation $K_d = P_d/P^* = V_P \tau^*$ is said to be the dynamicity coefficient.

All the shell layers are assumed to have equal elastic and strength characteristics, as well as the same thickness. The following magnitudes of characteristic lamina strengths are specified:

$$r_{100}^{(i)} = 9.56 \cdot 10^8\ \mathrm{N/m^2}; \quad r_{020}^{(i)} = 5.76 \cdot 10^7\ \mathrm{N/m^2}; \quad r_{006}^{(i)} = 4.57 \cdot 10^7\ \mathrm{N/m^2};$$
$$r_{100}^{(i)}/r_{\bar{1}00}^{(i)} = 1.5; \quad r_{0\bar{2}0}^{(i)}/r_{020}^{(i)} = 4. \tag{6.5}$$

The strength surface tensor components $p_{11}^{(i)}$, $p_{22}^{(i)}$, $p_{1111}^{(i)}$, $p_{2222}^{(i)}$, $p_{1212}^{(i)}$ are evaluated from characteristic strengths in accordance with the formulas presented in Ref. 184. Let us assume for the component $p_{1122}^{(i)}$:

$$p_{1122}{}^{(i)}=0{,}265\sqrt{p_{1111}{}^{(i)}p_{2222}{}^{(i)}}\,.$$

Let us consider the results obtained from numerical solution of the axisymmetric problem for the case of longitudinal dynamic compression. The axisymmetric initial deflection $w^0 = 0$. The layer axis 1 is directed along the shell generatrix. The geometric shell parameters and elastic characteristics of the material are given in 5.1.

In Table 6.1, the dynamicity coefficient, coordinate of initial failure x^* and maximum deflection value w^* in the instant of initial failure τ^* are presented for six different end conditions, and several loading rates are considered. As is seen, in all the cases the initial failure occurred on the inner shell surface, in the zones of edge effect. However, we failed to assess, under the condition of free edge, the exact values of the x^* coordinate, since the strength criterion is fulfilled practically simultaneously over an extended region in the vicinity of the end face. From among the remaining boundary conditions, the free support gives the closest location of the first defect to the end face, and the minimum value of the critical dynamic load corresponds to these types of end conditions. The maximum value of the deflection in the instant of initial failure is fairly the same for all types of boundary conditions, except (5.15) and (5.16). As to the effect of the loading rate, then upto its increasing 25 times, the critical dynamic load

TABLE 6.1

Dynamicity Coefficient, Coordinates of Initial Failure and Maximum Deflection Magnitude at the Instant of Initial Failure under Various Boundary Condition Versions

V_P	*Boundary condition*	k_D	z^*/h	x^*/L	$\|w_0^*\|/h$
5	(5.11)	8.75	−0.5	0.04; 0.96	0.31
5	(5.12)	9.0	−0.5	0.07; 0.93	0.29
5	(5.13)	12.0	±0.5	0–1	0.30
5	(5.14)	9.0	−0.5	0.93	0.29
5	(5.15)	8.75	−0.5	0.94	0.19
5	(5.16)	8.6	−0.5	0.96	0.21
1	(5.11)	4.44	−0.5	0.05; 0.95	0.82
25	(5.11)	12.5	−0.5	0.025; 0.975	0.02
50	(5.11)	13.0	−0.5	0; 1	0.005

for the conditions of free support increases three times, while the distance from the end face to the section where the failure occurs reduces twice. Let us note that at high loading rates, when no notable formation of flexural deformations takes place prior to the onset of failure, the K_d does not reveal its dependence on loading rate. For the shell under consideration, $K_d = 13$ is, obviously, the limiting value. Thereby, the strength condition is fulfilled practically simultaneously for all z values.

As the next example, we shall consider two-ply shells with lamina layup $\pm\Theta$ between axis 1 of the material and generatrix (total thickness of a package is the same as for a single-layer shell in the previous example). The specifics of this case consists in that, regardless of axisymmetricity of the load and geometric axisymmetricity of the shell, the displacement v in the circumferential direction is not zero. The system of equations of motion is supplemented by one more equation and takes the form

$$N'_x = \mu \frac{\partial^2 u_0}{\partial t^2}; \quad C_{66} v'' - K_{16} w''_0 = \mu \frac{\partial^2 v}{\partial t^2};$$
$$\frac{N_y}{R} - D_{11} w_0{}^{IV} + K_{16} v'' + (N_x w'_0)' = \mu \frac{\partial^2 w_0}{\partial t^2}, \tag{6.6}$$

where K_{16} is the component of membrane-flexural stiffness matrix and this component is equal to zero only at $\Theta = 0$ or $\Theta = \pi/2$. The expressions (5.9) for N_x and N_y remain valid. Let us specify the necessary additional boundary condition at each end face, alternatively, in two variants:

$$v \big|_{x=0,L} = 0 \tag{6.7}$$

or

$$T \big|_{x=0,L} = (C_{66} v' - K_{16} w''_0)_{x=0,L} = 0, \tag{6.8}$$

where T is the shear traction. The numerical integration of the eqns (6.6) was performed according to the scheme described in paragraph 5.1. In Fig. 6.1, the dependencies of the instant of first-ply failure τ^* and maximum deflection value w_0^* in the moment τ^* as functions of layer layup angle Θ are presented. The results refer to the conditions of free support. At all Θ-values, $V_P = 5$ is assumed; dimensionless time $\tau = tc/2L$, static traction P^* and velocity c are defined for the shell having $\Theta = 0$. As follows from the results presented, the maximum value of the critical dynamic load is reached at $\Theta = 0$. The strong dependence of deflection on the Θ-value, reached by the moment τ^*, means that the often used dynamic instability criterion based on reaching the prescribed deflection value, for composite shells becomes even more vague than for shells made of metals. Let us note

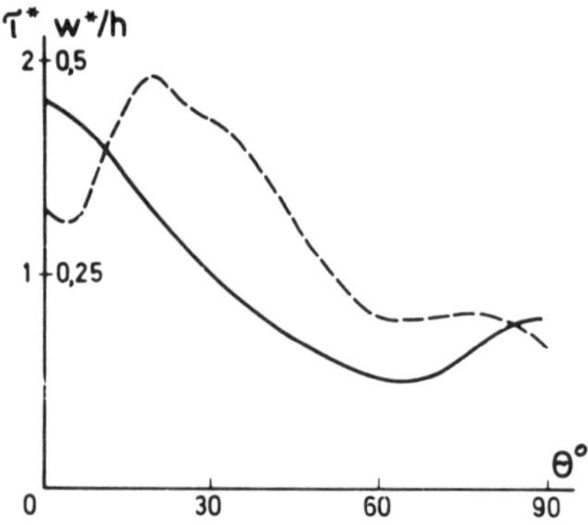

Fig. 6.1. The dependencies on the layer reinforcement angle of the instant of first-ply failure (———) and maximum deflection magnitude at the instant of first-ply failure (– – –) for a two-ply shell.

that in accordance with the data of Table 6.2, the circumferential displacement $v^* = v(\tau^*)$, related to non-symmetricity of the package composition through the thickness, at all Θ is considerably less than w_0^*. Owing to this, the τ^*-values, calculated for boundary conditions (6.7) and (6.8) at all Θ differ less than 1% (maximum difference corresponds to $\Theta \approx 30°$). Disregard of the terms with K_{16} in eqns (6.6) does not practically affect the τ^*-value.

Further we shall consider the results of calculations for the shell made of the same material ($\Theta_i = 0$) in the presence of non-axisymmetric initial shape imperfections (calculation procedure described in 5.3 is used). The initial deflection is assumed, other than zero, only for the circumferential harmonic $n = 3$. The distribution of Fourier coefficients, in respect to m, is

TABLE 6.2

Maximum Value of Circumferential Displacement v at the Instant of First-Ply Failure under Two Versions of Boundary Conditions and Several Angles Θ

Θ, *degrees*	$\|v^*\|/h$		Θ, *degrees*	$\|v^*\|/h$	
	(6.7)	(6.8)		(6.7)	(6.8)
15	0.018	0.030	60	0.005	0.003
30	0.017	0.022	75	0.002	0.002
45	0.005	0.010			

Notes: (6.7) and (6.8) are the corresponding boundary conditions.

specified as

$$W_{m3}{}^{0}=0{,}2h\frac{(-1)^{l}}{m^{2}},$$

where $l = m/2$ at even m and $l = (m+1)/2$ at odd m. The loading rate is $V_P = 5$. The geometrical shell parameters are: $R = 1\,\text{m}$, $L/R = 2$, $R/h = 200$.

For the case of free support of the end faces, in Fig. 6.2 the dependence of the F-function (6.4) on x at two amplitudes of initial imperfections a_0^* is presented. It is seen from Fig. 6.2(a), that both zones of non-axisymmetric and axisymmetric buckling are dangerous on the inner surface. On the outer surface the non-axisymmetric bulge, generated in the middle part of a shell, is the most dangerous. In such a way, at $a_0 = 0.2h$ the shell failure starts at $\tau^* \approx 1.6$ in the middle part, whereby practically simultaneously from both side surfaces. The reduction of a_0 by order, as shown by comparison of Fig. 6.2(a, b), leads to shifting of the site of initial failure into the edge zones. Thereby the instant τ^* somewhat increases. The results of extra calculations showed that softening of restrictions imposed on the end faces in the radial direction, leads to sharp decreasing of the F-function in edge zones and, as a result, to the increase in the τ^*-value.

Figure 6.3 illustrates the effect of loading rate on the character of the $F(x)$-function. As is seen, at sufficiently high loading rate, failure starts already in the stage of momentless deformation and propagates practically simultaneously over spacious regions both along the shell surface and through the thickness. On the whole, the analysis performed allows to

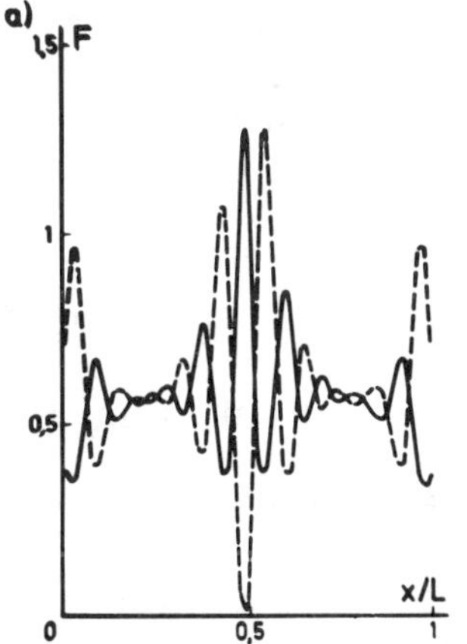

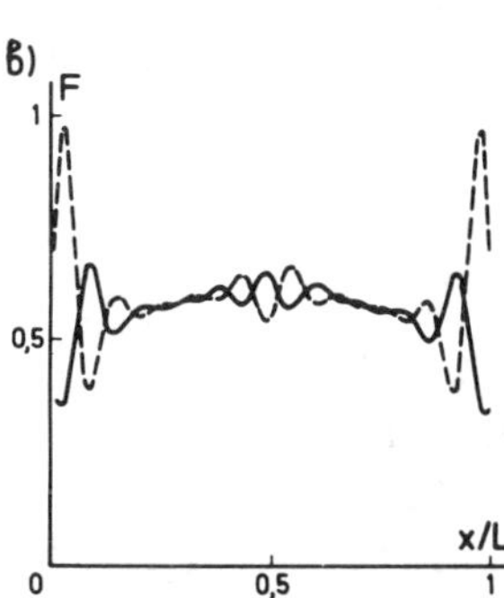

Fig. 6.2. The dependencies of $F(\sigma_{ij})$ function on the x coordinate for the outer (———) and inner (– – –) shell surfaces at $\tau = 1.7$; $a_0 = 0.2h$ (a) and $0.02h$ (b).

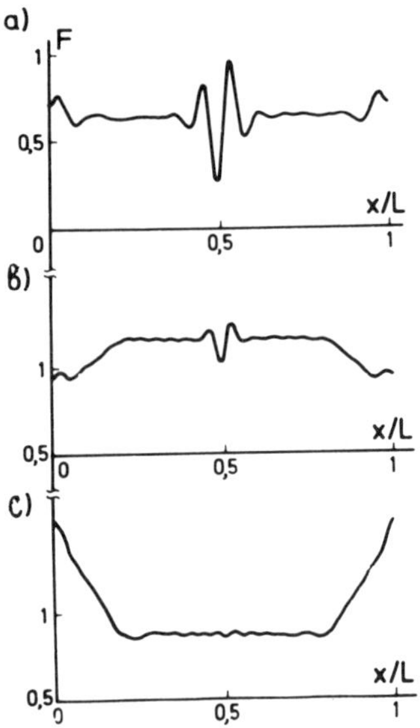

Fig. 6.3. Dependencies of $F(\sigma_{ij})$ function on the x coordinate for inner shell surface at $V_P = 10$ (a), 20 (b), 40 (c); $\tau = 1.0$ (a), 0.6 (b), 0.4 (c).

conclude that the process of shell failure can start in any stage of deformation: momentless, axisymmetric and non-axisymmetric, depending on the loading rate, conditions of end restraint, extent and kind of non-axisymmetric initial imperfections.

Let us consider in more detail the results of calculating the shell failure obtained by the procedure of solving the problem of non-axisymmetric dynamic buckling, presented in paragraph 5.4. The geometric shell parameters are: $R = 1$ m, $L/R = 2$; thickness is indicated for each particular case. The material of a monolayer is prepreg T 300/5208 having the following elastic and strength characteristics:[259]

$$E_1 = 15,4 \cdot 10^{10}\,\text{N/m}^2;\;\; E_2 = 1,08 \cdot 10^{10}\,\text{N/m}^2;\;\; G_{12} = 0,57 \cdot 10^{10}\,\text{N/m}^2;$$
$$\nu_{21} = 0,28;\;\; r_{100} = 1470\,\text{MPa};\;\; r_{\bar{1}00} = 1400\,\text{MPa};\;\; r_{020} = 43\,\text{MPa};$$
$$r_{0\bar{2}0} = 147\,\text{MPa};\;\; r_{006} = 91\,\text{MPa};\;\; \rho = 1,6 \cdot 10^3\,\text{kg/m}^3. \tag{6.9}$$

The designations of strength characteristics are assumed according to A. K. Malmeister.[184]

Let us specify the Fourier coefficients for the initial deflection:

$$W_{mn}{}^0 = \begin{cases} a_0 \dfrac{(-1)^{l+1}}{m^2} & \text{at} \quad n=3; \\ 0 & \text{at} \quad n \neq 3, \end{cases} \tag{6.10}$$

TABLE 6.3

Calculation Results for Initial In-Plane Failure Characteristics under Axial Dynamic Compression

Nos. of variants	R/h	V_p	a_0/h	τ^*	k_d	x^*/L	$\tilde{w}^{max}/h$	σ_{11}^*	σ_{22}^*
1	200	0.5	0.2	6.46	3.23	0.48	0.81	0.07	0.99
2	200	0.5	2.0	3.93	1.96	0.48	0.94	0.11	0.98
3	200	0.5	0.02	8.02	4.01	0.48	0.71	0.06	0.99
4	500	0.5	0.2	7.48	3.74	0.48	1.73	0.15	0.97
5	50	0.5	0.2	4.28	2.14	0.45	0.30	−0.16	1.02
6	50	0.5	2.0	1.76	0.88	0.47	0.25	0.14	0.96
7	200	5.0	0.2	1.80	9.0	0.49	0.62	0.06	1.00
8	50	5.0	2.0	0.52	2.60	0.49	0.12	−1.00	0.04
9	200	0.05	0.2	39.0	1.95	0.47	0.84	0.08	0.99
10	50	0.05	0.2	26.5	1.32	0.44	0.46	−0.06	1.01
11	50	0.05	2.0	7.55	0.38	0.46	0.50	0.10	0.98

where $l = m/2$ at even m, and $l = (m+1)/2$ at odd m. The linear law $P(\tau) = V_P P^* \tau$ is assumed.

The results of calculations of the initial failure for a single-layer shell (the material axis 1 is directed along the x-axis) are presented in Table 6.3. Along with the earlier introduced designations the following ones are used: σ_{11}^* and σ_{22}^* are the stresses in the point of initial failure $\{x^*, y^*, z^*\}$ (at $\tau = \tau^*$) related to respective material strengths (in tension or compression depending on the sign of the stress); $\tilde{w} = w - w^0$ is additional deflection. For all variants it was established that $y^* = \pi R$.

Let us also note that in all the cases treated (except variant 8), the initial failure occurs on the outer shell surface due to tensile circumferential stresses. The variant 8 corresponds to the greatest shell thickness and the highest loading rate, thereby, the membrane longitudinal compressive stresses, generating compressive deformation along fibers of the order of 1%, play the dominant role. The shell fails prior to the onset of pronounced non-axisymmetric buckling (the values of additional deflection at the instant of initial failure is a convincing evidence of this fact).

Further let us consider laminated shells of symmetric through the thickness composition of layers (the axis 1 of the material in the i-th layer forms an angle Θ_i with the shell generatrix). The angles of ply layup over the thickness, membrane C_{jk} and flexural D_{jk} stiffnesses of multi-layer packages as well as the static critical traction P^* for each of them are presented in Table 6.4. The results of calculating the first-ply failure at two loading rates are presented in Tables 6.5 and 6.6.

Having considered the specific features of initial failure of laminated

TABLE 6.4

Elements of Stiffness Matrix and Critical Static loads for Multilayered Shells

Number of variants[a]	*Package Structure*	$C_{ij} \cdot 10^{-8}$, N/m				$D_{ij} \cdot 10^{-3}$, N/m				$P^* \cdot 10^{-6}$, N/m	$q^* \cdot 10^{-5}$, N/m^2
		C_{11}	C_{22}	C_{12}	C_{66}	D_{11}	D_{22}	D_{12}	D_{66}		
						$R/h = 100$					
1.1	0°	15.4	1.1	0.3	0.6	12.9	0.9	0.25	0.47	1.3	0.8
1.2	[0°/90°]s	8.3	8.3	0.3	0.6	11.4	2.4	0.25	0.47	1.6	1.5
1.3	[90°/0°]s	8.3	8.3	0.3	0.6	2.4	11.4	0.25	0.47	1.6	4.2
1.4	[90°/0°/0°]s	10.7	5.9	0.3	0.6	4.5	9.3	0.25	0.47	1.7	3.9
1.5	[0°/90°/0°]s	10.7	5.9	0.3	0.6	9.8	4.0	0.25	0.47	1.7	2.1
1.6	[0°/30°/60°/90°]s	7.0	7.0	1.6	1.8	10.0	2.1	1.12	1.34	2.4	1.5
						$R/h = 50$					
2.1	90°	2.2	31.0	0.6	1.1	7.2	103	2.03	3.80	5.2	19.5
2.2	[90°/0°]s	16.6	16.6	0.6	1.1	19.2	91.2	2.03	3.80	6.4	22.7
2.3	[0°/90°]s	16.6	16.6	0.6	1.1	91.2	19.2	2.03	3.80	6.4	8.1
2.4	[90°/0°/0°]s	21.4	11.8	0.6	1.1	35.7	74.8	2.03	3.80	6.9	20.7
2.5	[0°/90°/0°]s	21.4	11.8	0.6	1.1	78.3	32.1	2.03	3.80	7.1	11.6
2.6	[90°/60°/30°/0°]s	14.0	14.0	3.2	3.7	16.8	79.8	8.97	10.70	9.5	25.4

[a]In the case of double numeration of variants the multi-layer shells, corresponding to the first number 1, are loaded with axial compression, while those corresponding to the first number 2, are loaded with external pressure.

TABLE 6.5

Calculational Results of First-Ply Failure and Transverse Stresses in Multi-Layered Shells under Axial Dynamic Compression ($R/h = 100$, $V_P = 0.5$, $a_0 = 0.2h$)

No. of variants	τ^*	$\frac{x^*}{L}$	$\frac{\tilde{w}^{max}}{h}$	σ_{11}^*	σ_{22}^*	$\lvert\tau_{xz}^{max}\rvert$ *MPa*	$\lvert\tau_{yz}^{max}\rvert$ *MPa*	$\frac{z_{yz}^*}{h}$	$-\sigma_z^{max}$ *MPa*	$\frac{z_z^-}{h}$	σ_z^{max} *MPa*	$\frac{z_z^+}{h}$	m^*	λ_x^*
1.1	5.42	0.47	0.49	−0.03	1.01	9.1	0.22	0	0.07	−0.30	0.05	0.30	6	33
1.2	8.24	0.48	0.74	−0.99	−0.34	11.7	1.22	−0.25	0.46	−0.25	0.42	0.25	7	29
1.3	5.08	0.49	0.38	0.31	0.92	20.4	1.84	0.25	0.25	0.25	0.15	−0.25	14	14
1.4	5.86	0.48	0.56	0.36	0.84	21.7	1.95	0.33	0.32	0.33	0.18	−0.33	11	18
1.5	7.62	0.48	0.81	−0.98	−0.41	16.8	1.30	0.17	0.40	−0.33	0.36	0.33	8	25
1.6	7.75	0.48	0.29	−1.00	0.04	7.0	2.02	0.10	0.25	−0.25	0.20	0.25	9	22

TABLE 6.6

Calculational Results of First-Ply Failure Characteristics and Transverse Stresses in Multi-Layered Shells under Axial Dynamic Compression ($R/h = 100$, $V_P = 5$, $a_0 = 0.2h$)

No. of variants	τ^*	$\frac{z^*}{h}$	$\frac{\tilde{w}^{max}}{h}$	σ_{11}^*	σ_{22}^*	$\lvert\tau_{xz}^{max}\rvert$, MPa	$\lvert\tau_{yz}^{max}\rvert$, MPa	$\frac{z_{yz}^*}{h}$	$-\sigma_z^{max}$ MPa	$\frac{z_z^-}{h}$	σ_z^{max} MPa	$\frac{z_z^+}{h}$	m^*	λ_x^*
1.1	1.41	−0.5	0.21	−0.90	0.24	20.0	0.28	0	0.29	−0.30	0.26	0.30	12	17
1.2	1.12	−0.5	0.07	−1.08	−0.16	6.5	0.26	−0.25	0.17	−0.30	0.16	0.30	8	25
1.3	0.95	−0.25	0.10	−1.04	−0.13	18.3	0.82	0.25	0.52	−0.20	0.37	0.25	18	11
1.4	1.14	−0.33	0.14	−1.03	−0.08	21.1	0.78	0.33	0.54	−0.23	0.46	0.23	14	14
1.5	0.91	−0.5	0.04	−1.01	−0.02	3.6	0.49	−0.25	0.10	−0.26	0.09	0.30	8	25

shells under axial dynamic compression, let us note that at $V_P = 0.5$ (Table 6.5) the greatest τ^*-values correspond to variants 1.2, 1.5, 1.6 with fairly greater stiffnesses C_{22} and D_{22} than for unidirectionally reinforced material. The initial failure of the packages indicated occurs in the layer, reinforced along the generatrix, due to compressive stresses along the fibers, whereas for the variant with axial reinforcement of the entire package — due to tensile stresses perpendicular to fibers. The increase in hoop stiffness leads to the change of initial failure mode and a notable increase in corresponding dynamic load. The D_{11}-value is also of importance, its decrease reduces abruptly τ^* (it is best seen from the comparison of the results for variants 1.2 and 1.3). Comparatively high resistance to axial dynamic compression of the shell, corresponding to variant 1.6, can be accounted for by the highest among all the packages considered value P^*, defining the latest onset of non-axisymmetric dynamic buckling.

At the loading rate $V_P = 5$ (see Table 6.6) the highest τ^*-value corresponds to the shell reinforced along the generatrix. For all the variants treated, one and the same mode of initial failure is realized — due to compression along the fibers. The contribution of the flexural components to stresses is negligible. It is natural that in such a situation the membrane stiffness of a shell in the x-direction, reaching its maximum in the case of reinforcement along the generatrix, is of paramount significance. The values of hoop membrane C_{22} and flexural D_{22} stiffnesses affect slightly the resistance of a shell to high-speed longitudinal compression.

The general conclusion can be made from the set of results: the effect of each of the above treated factors (membrane and flexural stiffnesses, magnitudes of critical static axial traction) on the characteristics of initial shell failure depends greatly on the rate of axial dynamic loading.

6.3 INITIAL FAILURE UNDER DYNAMIC LATERAL PRESSURE

Let us consider the results of calculating the initial failure of shells, made up of a unidirectionally reinforced graphite/epoxy layers, having the characteristics of (6.9) and loaded with linearly increasing in time external pressure $q(\tau) = V_q q^* \tau$. The distribution of Fourier coefficients for initial deflection is assumed as

$$W_{mn}{}^0 = 0{,}01h(-1)^{l+n} \exp\left[-\frac{(n-4)^2}{10} - \frac{(m-1)^2}{10}\right], \quad (6.11)$$

where $l = m/2$ at even m and $l = (m+1)/2$ at odd m.

The results for a single-layer shell (the material axis 1 is directed along the y-axis) are presented in Table 6.7. Let us note that both modes of failure — due to compression along fibers at high loading rates (variants 2 and 5) and due to tension across the fibers at lower loading rates (variants 1, 4 and 6) and for a fairly thinner shell (variant 8) — are possible. As is seen, the dynamicity coefficient depends greatly on the loading rate and R/h relation. It increases with increasing V_q (within the indicated range of thickness, the increase in V_q by two orders brings about the increase in K_d, 15–17 times). In its turn, the decrease in thickness by an order results in the increase of K_d 3.5–10 times depending on the V_q-value.

The calculation results of the initial failure of multi-layer shells, described in 6.4, are presented in Tables 6.8 and 6.9 for two values of V_q. At $V_q = 0.2$, the highest τ^*-values are for variants 2.2 and 2.6, for which the magnitudes of critical static external pressure q^* are maximum (see Table 6.4). The τ^*-values are somewhat lower for variants 2.4 and 2.1 and the smallest — for variants 2.3 and 2.5, for which q^* are minimal. In such a way, at comparatively low loading rate the magnitude of critical static pressure, defining the onset of intensive non-axisymmetric buckling of each concrete shell, plays the main role.

At the loading rate $V_q = 2$ (see Table 6.9) the τ^*-value is maximum for a shell, reinforced in the circumferential direction (i.e. a shell having the highest hoop stiffness). Failure is due to tensile stresses, perpendicular to fibers. For all other variants of ply layup the failure due to compression along the fibers is stated. It can be explained by the dominant role of momentless hoop stresses at the time instant when first-ply failure starts. The smallest value of τ^* is for variants 2.4 and 2.5 corresponding to the lowest stiffness C_{22}. Let us note that a considerable increase in stiffness D_{22}

TABLE 6.7

Calculational Results of Initial In-Plane Failure Characteristics under Dynamic External Pressure

No. of variants	R/h	V_q	τ^*	k_d	x^*/L	z^*/h	$\tilde{w}^{max}/h$	σ^*_{11}	σ^*_{22}
1	200	2.0	8.9	17.7	0.42	0.5	4.0	0.20	0.95
2	200	20	3.8	76.4	0.42	−0.5	3.5	−0.90	0.25
3	200	0.2	24.4	4.9	0.36	0.5	5.7	0.33	0.88
4	50	2.0	3.4	6.9	0.43	0.5	0.5	−0.05	1.01
5	50	20	0.7	14.2	0.41	−0.5	0.03	−0.98	−0.003
6	50	0.2	12.2	2.4	0.39	0.5	0.7	0.20	0.95
7	500	2.0	15.4	30.8	0.41	0.5	16.1	0.21	0.93
8	500	20	6.8	136	0.43	0.5	13.5	−0.05	1.01
9	500	0.2	38.2	7.7	0.35	0.5	21.1	0.31	0.86

TABLE 6.8

Calculational Results of First-Ply Failure Characteristics and Transverse Stresses in Multi-Layered Shells under Dynamic External Pressure ($R/h = 50, V_q = 0{\cdot}2$)

No. of variants	τ^*	$\frac{x^*}{L}$	$\frac{\tilde{w}^{max}}{h}$	σ_{11}^*	σ_{22}^*	$\lvert\tau_{xz}^{max}\rvert$, MPa	$\frac{z_{xz}}{h}$	$\lvert\tau_{yz}^{max}\rvert$, MPa	$-\sigma_z^{max}$ MPa	$\frac{z_z}{h}$	n^*	λ_y^*
2·1	12·2	0·39	0·72	0·20	0·95	0·41	0	13·7	2·66	0	5	32
2·2	13·4	0·40	1·26	0·40	0·82	1·01	−0·1	20·2	3·91	−0·1	5	32
2·3	8·5	0·37	0·62	0·14	0·97	1·73	0·25	8·9	1·28	0	6	26
2·4	12·3	0·40	0·98	−1·08	−0·21	1·22	−0·067	14·6	2·68	−0·067	5	32
2·5	9·7	0·38	0·81	0·17	0·98	1·71	−0·033	12·2	1·96	−0·167	6	26
2·6	13·1	0·43	0·95	−1·04	−0·08	3·61	0·125	16·0	2·87	−0·175	5	32

TABLE 6.9

Calculational Results of First-Ply Failure Characteristics and Transverse Stresses in Multi-Layered Shells under Dynamic External Pressure ($R/h = 50, V_q = 2$)

No. of variants	τ^*	$\frac{z^*}{h}$	$\frac{\tilde{w}^{max}}{h}$	σ_{11}^*	σ_{22}^*	$\frac{\lvert\tau_{xz}^{max}\rvert}{MPa}$,	$\frac{z_{xz}}{h}$	$\frac{\lvert\tau_{yz}^{max}\rvert}{MPa}$,	$\frac{z_{yz}}{h}$	$\frac{-\sigma_z^{max}}{MPa}$	$\frac{z_z}{h}$	n^*	λ_y^*
2·1	3·45	0·5	0·49	–0·05	1·01	0·83	0	19·1	0	2·92	0	6	26
2·2	3·12	–0·5	0·32	–1·07	–0·23	1·10	–0·05	10·1	0·1	1·52	–0·15	6	26
2·3	3·18	–0·25	0·50	–1·04	–0·18	3·23	0·20	8·8	0	1·18	0	7	22
2·4	2·60	–0·5	0·16	–1·03	–0·16	0·80	–0·03	3·1	0·1	0·61	–0·07	6	26
2·5	2·63	–0·33	0·20	–1·08	–0·20	1·20	–0·03	3·6	0·07	0·53	–0·17	6	26
2·6	2·78	–0·5	0·18	–1·01	–0·06	1·45	–0·12	5·5	0	0·79	0	7	22

critical static pressure q^* (see variants 2.2 and 2.3) leads to a slight increase in τ^*. In such a way, at a comparatively high loading rate, the hoop membrane stiffness exerts the greatest influence on the value of critical dynamic load and mode of first-ply failure.

On the whole, a study of the characteristic features of the initial failure of cylindrical shells under dynamic external pressure allows us to conclude that within a certain, limited from below, range of loading rates the shell failure starts already at the momentless stage of deformation, prior to the development of notable non-axisymmetric deformations. At lower loading rates, the process of shell failure starts practically simultaneously with its intensive non-axisymmetric buckling. At sufficiently low loading rates, repeated 'bulgings' of a rather thin shell (described by non-linear solution obtained in paragraph 5.4), which is not accompanied by damages in composite material, become possible.

6.4 THE METHOD FOR CALCULATION OF TRANSVERSE AND INTERLAMINAR STRESSES UNDER DYNAMIC COMPRESSIVE LOADS

The calculation procedure, outlined in Chapter 5, enables us to determine three stress tensor components acting within the plane of shell layers. Correspondingly, the method of analyzing the initial failure, described in 6.1, is based on the strength criterion (6.4), comprising the same three stress components. For laminated shells made of anisotropic composites and subjected to compressive loads, the transverse shear stresses and the stresses, acting on the interfaces between layers, can prove to be dangerous, too.

In order to assess the danger of realization of the modes of failure, corresponding to transverse and interlaminar stresses, first of all, we are in need of the procedures for their calculation. In this paragraph, the simplest approach to calculation of transverse shear and normal stresses in laminated cylindrical shells is considered. The analysis of first-ply failure, taking into account the volumetric stress state, as well as the analysis of initial failure referred to the interfaces can be based on this approach. Obviously, problems of such kind have not been hitherto treated.

In developing the procedures for calculating transverse stresses in a laminated shell, we shall employ the methods for calculating stress–strain state of the layer, outlined in Chapter 5. This is possible, provided the approach, proposed by S. G. Lekhnitskii[179] for laminated plates and later on developed by S. A. Ambartsumian[16] for laminated shallow shells. This approach is rather popular, but it was used only in the simplest linear static problems of transverse bending. It has one serious demerit — the formulas obtained for transverse stresses, do not comprise transverse shear moduli and the modulus of transverse deformation of the layer. Therefore, this approach requires that the applicability of it to concrete classes of laminated structures and loading conditions should be comprehensively substantiated. Special attention will be paid to corresponding substantiation in problems considered here.

Let us treat the cylindrical shell, consisting of $2m + 1$ anisotropic layers. The structure of a shell over the thickness is symmetric in respect to the middle surface of the $(m + 1)$-th layer. This surface, in such a way, is also the middle surface of the shell. Each layer has one plane of elastic symmetry, parallel to the middle surface. The numeration of layers starts from the inner shell surface. The coordinate lines x and y are directed along the generatrix and along the circumference, the coordinate line z is directed along the outer normal to the middle surface.

It is assumed that the compressive end forces, external pressure, as well as any their combinations (either in dynamic or static variant) can be applied to a shell. The shell has a general field of initial non-axisymmetric shape imperfections. As was shown in Chapter 5, using linearized equations of motion in the problem about axial dynamic compression, we can obtain not only a qualitatively reliable characterization of the buckling process, but also sufficiently accurate, for practical purposes, results of stress–strain state calculation right up to the instant of initial failure. The same refers to the problem of dynamic external pressure within a limited from below range of loading rates.

Taking into account the specific features of problems under consideration, the methodology[16] should first be modified. Calculation of transverse stresses in a shell layer should be performed, proceeding from the linearized equations of the theory of elasticity. These equations, written in terms of 'disturbances' of stresses, displacements and surface tractions, follow from the equations of non-linear theory of elasticity[210] by neglecting the non-linear terms, comprising the products of 'disturbances' and, provided the geometry of axisymmetric deformed state differs slightly from the geometry of non-deformed state.[67] The latter assumption, as shown in paragraph 5.1, is acceptable beyond the zones of axisymmetric edge effect. It is assumed, additionally, that the development of non-axisymmetric deformations occurs on the background of homogeneous axisymmetric stress state, characterized by axial $\sigma_x^0(t)$ and hoop $\sigma_y^0(t)$ stresses. The possibility of adopting such an assumption beyond the zones of axisymmetric edge effect was substantiated in paragraph 5.3.

In such a way, the calculation method refers to transverse stresses arising in a shell as a result of initiating of non-axisymmetric flexural deformations. Thereby, the zones of axisymmetric edge effect have been left out of consideration. Taking into consideration the stress component σ_z, having arisen in the process of wave propagation from the impacted outer side, needs a special study.

In accordance with the assumptions indicated, the system of equations of motion for the i-th shell layer is put down as

$$\begin{gathered}\frac{\partial \sigma_x{}^i}{\partial x}+\frac{\partial \tau_{xy}{}^i}{\partial y}+\frac{\partial \tau_{xz}{}^i}{\partial z}+P_x{}^i=0;\\ \frac{\partial \tau_{xy}{}^i}{\partial x}+\frac{\partial \sigma_y{}^i}{\partial y}+\frac{\partial \tau_{yz}{}^i}{\partial z}+P_y{}^i=0;\\ \frac{\partial \sigma_z{}^i}{\partial z}-\frac{1}{R}\sigma_y{}^i+\frac{\partial}{\partial x}\left(\tau_{xz}{}^i+\sigma_x{}^0\frac{\partial w}{\partial x}\right)+\frac{\partial}{\partial y}\times\\ \times\left(\tau_{yz}{}^i+\sigma_y{}^0\frac{\partial w}{\partial y}\right)+P_z{}^i=0,\end{gathered} \tag{6.12}$$

where P_x^i, P_y^i, P_z^i are mass forces; w is shell deflection. Integrating (6.12) in respect to the z coordinate, we obtain the following formulas for transverse stresses:

$$\tau_{xz}{}^i=-\int\left[\frac{\partial \sigma_x{}^i}{\partial x}+\frac{\partial \tau_{xy}{}^i}{\partial y}+P_x{}^i\right]dz+\varphi_i(x, y); \tag{6.13}$$

$$\tau_{yz}{}^i=-\int\left[\frac{\partial \sigma_y{}^i}{\partial y}+\frac{\partial \tau_{xy}{}^i}{\partial x}+P_y{}^i\right]dz+\psi_i(x, y); \tag{6.14}$$

$$\begin{gathered}\sigma_z{}^i=-\int\left[-\frac{1}{R}\sigma_y{}^i+\frac{\partial}{\partial x}\left(\tau_{xz}{}^i+\sigma_x{}^0\frac{\partial w}{\partial x}\right)+\right.\\ \left.+\frac{\partial}{\partial y}\left(\tau_{yz}{}^i+\sigma_y{}^0\frac{\partial w}{\partial y}\right)+P_x{}^i\right]dz+\chi_i(x, y).\end{gathered} \tag{6.15}$$

The functions φ_i, ψ_i, χ_i should be determined from the boundary conditions on the side surfaces of a shell and from the continuity condition for stresses τ_{xz}^i, τ_{yz}^i and σ_z^i on the interfaces.

Henceforth, the procedure of calculation is fully identical to that described in Ref. 16. The expressions for in-plane stresses σ_x^i, σ_y^i and τ_{xy}^i in terms of deformations

$$\begin{gathered}\sigma_x{}^i=B_{11}{}^i\varepsilon_1+B_{12}{}^i\varepsilon_2+B_{16}{}^i\gamma+z(B_{11}{}^i\varkappa_1+B_{12}{}^i\varkappa_2+2B_{16}{}^i\tau);\\ \sigma\ {}^i=B_{12}{}^i\varepsilon_1+B_{22}{}^i\varepsilon_2+B_{26}{}^i\gamma+z(B_{12}{}^i\varkappa_1+B_{22}{}^i\varkappa_2+2B_{26}{}^i\tau);\\ \tau_{xy}{}^i=B_{16}{}^i\varepsilon_1+B_{26}{}^i\varepsilon_2+B_{66}{}^i\gamma+z(B_{16}{}^i\varkappa_1+B_{26}{}^i\varkappa_2+2B_{66}{}^i\tau),\end{gathered} \tag{6.16}$$

are substituted into (6.13) and (6.14). Here

$$\varepsilon_1=\frac{\partial u}{\partial x};\quad \varepsilon_2=\frac{\partial v}{\partial y}+\frac{w-w^0}{R};\quad \gamma=\frac{\partial u}{\partial y}+\frac{\partial v}{\partial x};$$

$$\varkappa_1=-\frac{\partial^2(w-w^0)}{\partial x^2};\quad \varkappa_2=-\frac{\partial^2(w-w^0)}{\partial y^2};\quad \tau=-\frac{\partial^2(w-w^0)}{\partial x\partial y}.$$

Here w^0 is the initial deflection.

After integrating in (6.13) and (6.14) and ascertaining φ_i and ψ_i-functions from the continuity conditions for τ_{xz}, τ_{yz} on the interfaces, we obtain the following formulas for transverse tangential stresses. For τ_{xz}^i in the lower half of the package ($i=1,\ldots,m+1$):

$$\begin{aligned}\tau_{xz}{}^i=&-z[\hat{L}_{11}(B_{jk}{}^i)u+\hat{L}_{12}(B_{jk}{}^i)v+\hat{L}_{13}(B_{jk}{}^i)(w-w^0)]+\\&+\frac{z^2}{2}\hat{E}_1(B_{jk}{}^i)(w-w^0)-h_0[\hat{L}_{11}(R_{jk}{}^i)u+\hat{L}_{12}(R_{jk}{}^i)v+\hat{L}_{13}(R_{jk}{}^i)\times\\&\times(w-w^0)]-\frac{h_0{}^2}{2}\hat{E}_1(P_{jk}{}^i)(w-w^0)+X^-,\end{aligned}\tag{6.17}$$

and in the upper half of the package ($i=m+1,\ldots,2m+1$):

$$\begin{aligned}\tau_{xz}{}^i=&-z[\hat{L}_{11}(B_{jk}{}^{2m+2-i})u+\hat{L}_{12}(B_{jk}{}^{2m+2-i})v+\hat{L}_{13}(B_{jk}{}^{2m+2-i})\times\\&\times(w-w^0)]+\frac{z^2}{2}\hat{E}_1(B_{jk}{}^{2m+2-i})(w-w^0)+h_0[\hat{L}_{11}(R_{jk}{}^{2m+2-i})u+\\&+\hat{L}_{12}(R_{jk}{}^{2m+2-i})v+\hat{L}_{13}(R_{jk}{}^{2m+2-i})(w-w^0)]-\frac{h_0{}^2}{2}\hat{E}_1(P_{jk}{}^{2m+2-i})\times\\&\times(w-w^0)+X^+.\end{aligned}\tag{6.18}$$

For τ_{yz}^i in the lower half of the package:

$$\begin{aligned}\tau_{yz}{}^i=&-z\Big[\hat{L}_{12}(B_{jk}{}^i)u+\hat{L}_{22}(B_{jk}{}^i)v+\hat{L}_{23}(B_{jk}{}^i)(w-w^0)+\\&+\frac{z^2}{2}\hat{E}_2(B_{jk}{}^i)(w-w^0)\Big]-h_0[\hat{L}_{12}(R_{jk}{}^i)u+\hat{L}_{22}(R_{jk}{}^i)v+\hat{L}_{23}(R_{jk}{}^i)\times\\&\times(w-w^0)]-\frac{h_0{}^2}{2}\hat{E}_2(P_{jk}{}^i)(w-w^0)+Y^-,\end{aligned}\tag{6.19}$$

and in the upper half of the package:

$$\tau_{yz}{}^i=-z[\hat{L}_{12}(B_{jk}{}^{2m+2-i})u+\hat{L}_{22}(B_{jk}{}^{2m+2-i})v+\hat{L}_{23}(B_{jk}{}^{2m+2-i})\times$$

$$\times(w-w^0)]+\frac{z^2}{2}\hat{E}_2(B_{jk}{}^{2m+2-i})(w-w^0)+h_0[\hat{L}_{12}(R_{jk}{}^{2m+2-i})u+$$
$$+\hat{L}_{22}(R_{jk}{}^{2m+2-i})v+\hat{L}_{23}(R_{jk}{}^{2m+2-i})(w-w^0)]-\frac{h_0{}^2}{2}\hat{E}_2(P_{jk}{}^{2m+2-i})\times$$
$$\times(w-w^0)+Y^+. \tag{6.20}$$

Substitution of (6.16)–(6.20) into (6.15) after integrating and finding out the χ_i functions from the continuity condition for σ_z on the interfaces lead to the following formulas for the stress σ_z. In the lower half of the package:

$$\sigma_z{}^i=z\left[\hat{L}_{13}(B_{jk}{}^i)u+\hat{L}_{23}(B_{jk}{}^i)v+B_{22}{}^i\frac{w-w^0}{R^2}\right]+\frac{z^2}{2}[\hat{E}_1(B_{jk}{}^i)u+$$
$$+\hat{E}_2(B_{jk}{}^i)v]-\frac{z^3}{6}\hat{L}(B_{jk}{}^i)(w-w^0)+h_0z[\hat{E}_1(R_{jk}{}^i)u+$$
$$+\hat{E}_2(R_{jk}{}^i)v+\hat{S}(R_{jk}{}^i)(w-w^0)]+\frac{h_0{}^2}{2}z\hat{L}(P_{jk}{}^i)(w-w^0)+$$
$$+h_0\left[\hat{L}_{13}(R_{jk}{}^i)u+\hat{L}_{23}(R_{jk}{}^i)v+R_{22}{}^i\frac{w-w^0}{R^2}\right]+\frac{h_0{}^2}{2}[\hat{E}_1(P_{jk}{}^i)u+$$
$$+\hat{E}_2(P_{jk}{}^i)v+2\hat{S}(P_{jk}{}^i)(w-w^0)]+\frac{h_0{}^3}{3}\hat{L}(H_{jk}{}^i)(w-w^0)- \tag{6.21}$$
$$-(z+h_1)\frac{\partial}{\partial x}\left(\sigma_x{}^0\frac{\partial w}{\partial x}\right)-(z+h_1)\frac{\partial}{\partial y}\left(\sigma_y{}^0\frac{\partial w}{\partial y}\right)-$$
$$-(z+h_1)\frac{\partial X^-}{\partial x}-(z+h_1)\frac{\partial Y^-}{\partial y}-(z+h_i)P_z{}^i-\sum_{s=1}^{i-1}(h_s-h_{s+1})P_z{}^s+Z^-,$$

and in the upper half of the package:

$$\sigma_z{}^i=z\left[\hat{L}_{13}(B_{jk}{}^{2m+2-i})u+\hat{L}_{23}(B_{jk}{}^{2m+2-i})v+B_{22}{}^{2m+2-i}\frac{w-w^0}{R^2}\right]+$$
$$+\frac{z^2}{2}[\hat{E}_1(B_{jk}{}^{2m+2-i})u+\hat{E}_2(B_{jk}{}^{2m+2-i})v]-\frac{z^3}{6}\hat{L}(B_{jk}{}^{2m+2-i})\times$$
$$\times(w-w^0)-h_0z[\hat{E}_1(R_{jk}{}^{2m+2-i})u+\hat{E}_2(R_{jk}{}^{2m+2-i})v+$$
$$+\hat{S}(R_{jk}{}^{2m+2-i})(w-w^0)]+\frac{h_0{}^2}{2}z\hat{L}(P_{jk}{}^{2m+2-i})(w-w^0)+$$
$$+h_0\left[\hat{L}_{13}(R_{jk}{}^{2m+2-i})u+\hat{L}_{23}(R_{jk}{}^{2m+2-i})v+R_{22}{}^{2m+2-i}\frac{w-w^0}{R^2}\right]+ \tag{6.22}$$

$$+\frac{h_0^2}{2}[\hat{E}_1(P_{jk}{}^{2m+2-i})u+\hat{E}_2(P_{jk}{}^{2m+2-i})v+2\hat{S}(P_{jk}{}^{2m+2-i})(w-w^0)]-$$
$$-\frac{h_0^3}{3}\hat{L}(H_{jk}{}^{2m+2-i})(w-w^0)-(z-h_{2m+1})\frac{\partial}{\partial x}\left(\sigma_x{}^0\frac{\partial w}{\partial x}\right)-$$
$$-(z-h_{2m+1})\frac{\partial}{\partial y}\left(\sigma_y{}^0\frac{\partial w}{\partial y}\right)-(z-h_{2m+1})\frac{\partial X^+}{\partial x}-(z-h_{2m+1})\frac{\partial Y^+}{\partial y}-$$
$$-(z-h_{2m+2-i})P_z{}^{2m+2-i}+\sum_{s=1}^{2m+2-i-1}(h_s-h_{s+1})P_z{}^s+Z^+.$$

In formulas (6.17)–(6.22) B^i_{jk} are components of stiffness matrix for the i-th layer; h_0 is a half-thickness of a package; h_i is the distance from the middle surface to the outer surface of the i-th layer; X^-, Y^-, Z^- and X^+, Y^+, Z^+ are 'disturbances' of surface load components, acting on the inner and outer shell surfaces, respectively. The differential operators $\hat{L}_{11}$, $\hat{L}_{12}$, $\hat{L}_{13}$, $\hat{L}_{22}$, $\hat{L}_{23}$, $\hat{E}_1$, $\hat{E}_2$ and values R^i_{jk}, P^i_{jk} are presented in Ref. 21; operators $\hat{S}$, $\hat{L}$ and the values H^i_{jk} are determined in Ref. 16.

The formulas for the lower (6.17), (6.19) and (6.21) and upper (6.18), (6.20) and (6.22) halves of the package yield the same magnitudes of stresses on the middle surface ($z=0$), only if the three following equations are satisfied:

$$\hat{L}_{11}(C_{jk})u+\hat{L}_{12}(C_{jk})v+\hat{L}_{13}(C_{jk})(w-w^0)+X^+-X^-=0;$$
$$\hat{L}_{12}(C_{jk})u+\hat{L}_{22}(C_{jk})v+\hat{L}_{23}(C_{jk})(w-w^0)+Y^+-Y^-=0;$$
$$\hat{L}_{13}(C_{jk})u+\hat{L}_{23}(C_{jk})v+C_{22}\frac{w-w^0}{R^2}+\hat{L}(D_{jk})(w-w^0)-$$
$$-2h_0\frac{\partial}{\partial x}\left(\sigma_x{}^0\frac{\partial w}{\partial x}\right)-2h_0\frac{\partial}{\partial y}\left(\sigma_y{}^0\frac{\partial w}{\partial y}\right)-h_0\left(\frac{\partial X^-}{\partial x}+\frac{\partial X^+}{\partial x}\right)-$$
$$-h_0\left(\frac{\partial Y^-}{\partial y}+\frac{\partial Y^+}{\partial y}\right)+Z^--Z^+= \tag{6.23}$$
$$=2[h_{m+1}P_z{}^{m+1}+(h_m-h_{m+1})P_z{}^m+\ldots+(h_1-h_2)P_z{}^1],$$

These conditions represent themselves as linearized equations of motion for the middle shell surface; C_{jk} and D_{jk} are matrices of membrane and flexural stiffnesses of a multilayer package.

In such a way, calculation of transverse stresses falls into two stages: solution of equations of motion (6.23), as a result, the displacements of the middle surface u, v, w are determined, and calculation of τ^i_{xz}, τ^i_{yz}, σ^i_z in the i-th layer according to formulas (6.17), (6.19) and (6.21) or (6.18), (6.20) and (6.22). The stresses on the interface between the i-th and $(i+1)$-th layers are calculated according to the same formulas, but at $z=h_i$.

6.5 TRANSVERSE STRESSES IN LAMINATED SHELLS UNDER DYNAMIC COMPRESSIVE LOADS

Let us treat a shell loaded with uniformly distributed over each end face compressive tractions $P(t)$ and uniformly distributed over the outer side surface pressure $q(t)$. Under the assumption that these loads generate momentless homogeneous stress state, the corresponding stresses can be written in the form $\sigma_x^0 = -P(t)/2h_0$, $\sigma_y^0 = -(q(t)R)/2h_0$. The 'disturbances' of surface tractions are: $X^+ = X^- = Y^+ = Y^- = Z^+ = Z^- = 0$. Taking into account that the only mass forces are radial inertia forces $P_z^i = -\rho^i(\partial^2 w/\partial t^2)$ and assuming that the densities of all layers are identical $(\rho^i = \rho)$, we obtain

$$2[h_{m+1}P_z^{m+1} + (h_m - h_{m+1})P_z^m + \ldots + (h_1 - h_2)P_z^1] = -\mu\frac{\partial^2 w}{\partial t^2},$$

where $\mu = \rho h$ is a mass of unit surface of a shell, $h = 2h_0$.

The procedure of solving the linearized equations of motion (6.23) for a shell under free support conditions at the end faces was described in Chapter 5. Finally, the displacements u, v and w are expressed in the form of double Fourier series. Consequently, the transverse stresses are also written down as Fourier series, obtained through term-by-term differentiation of series for displacements in accordance with the formulas (6.17)–(6.22).

Let us consider the results of calculating transverse stresses (stresses in Figs 6.4–6.7 are given in MPa units) in cylindrical shells with unidirectionally reinforced graphite/epoxy layers having the characteristics of (6.9). The linearly increasing in time loads of axial compression or external pressure are acting on the shell. The initial imperfections of the shell are prescribed through (6.10) and (6.11) for each kind of loading, respectively. In Figs 6.4 and 6.5 the graphs of transverse stresses over the thickness are presented for the cases of axial compression (for a shell, reinforced along the generatrix) and of external pressure (for a shell, reinforced in the circumferential direction). Diagrams of transverse stresses through the thickness of laminated shells under respective types of loading are presented in Figs 6.6 and 6.7. The disposition of layers is symmetric to the surface $z = 0$. The total thickness of layers $h = R/100$ for axial compression and $h = R/50$ for external pressure. Each of the curves in Figs 6.4–6.7 refers to the time moment $\tau = \tau^*$ of the initial failure of a layer (first-ply failure) for a concrete shell and concrete variant of loading and corresponds to that point $\{x, y\}$ of the shell surface for which the transverse stress under consideration reaches its maximum. The diagrams

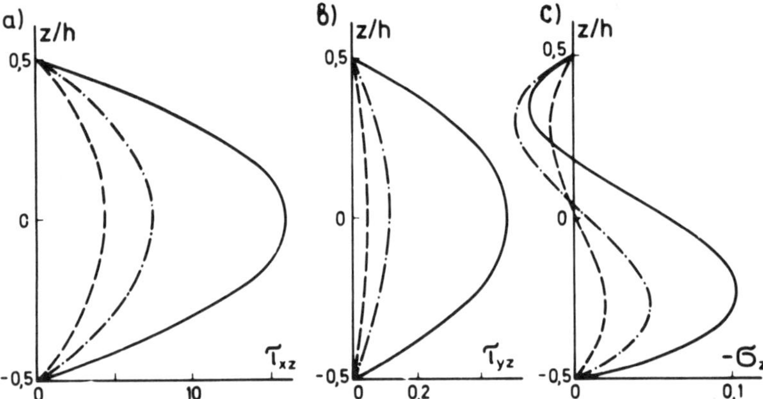

Fig. 6.4. Transverse stresses dependence on the z coordinate for longitudinally reinforced shell loaded with axial dynamic compression ($V_P = 0.5$), at $R/h = 50$ and $a_0 = 0.2h$ (———); $R/h = 200$ and $a_0 = 0.02h$ (–· – ·–); $R/h = 500$ and $a_0 = 0.2h$ (– – –).

obtained allow, in such a way, to find the maximum values of transverse stresses, having arisen in the shell in the instant of first-ply failure.

When analyzing Figs 6.4–6.7, let us pay attention to the coincidence of the magnitudes of transverse stresses 'from above' and 'below' on the middle surface and on the interfaces between adjacent layers. It is a result of using the analytical form (in series) of solution for the system of equations of motion (6.23). At any magnitude of upper summing limits in

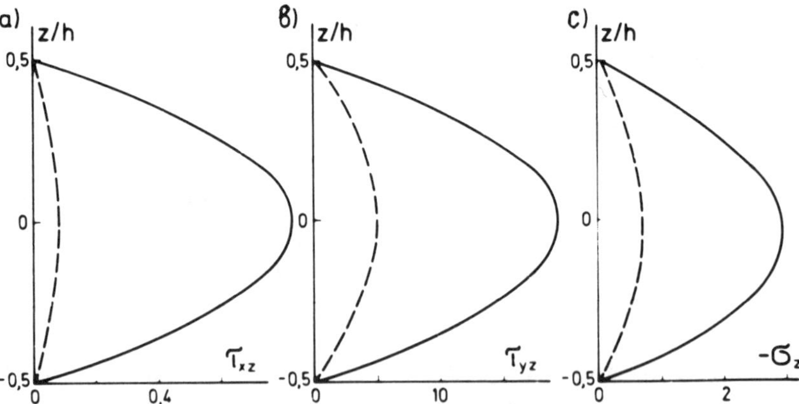

Fig. 6.5. Transverse stresses dependence on the z coordinate for circumferentially reinforced shell loaded with dynamic external pressure at $V_q = 2$ and $R/h = 50$ (———); $V_q = 0.2$ and $R/h = 200$ (– – –).

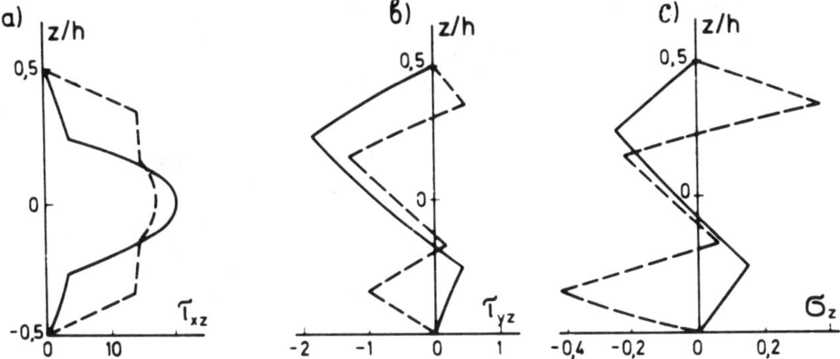

Fig. 6.6. Transverse stresses dependence on the z coordinate under dynamic axial compression ($V_P = 0.5$) for shells having package structures $[90°/0°]_S$ (———) and $[0°/90°/0°]_S$ (– – –); $R/h = 100$, $a_0 = 0.2h$.

Fourier series, this solution for deflection can be obtained, in principle, with any prescribed accuracy by specifying the respective accuracy for numerical integration of differential equations in respect to $W_{mn}(t)$. In the cases, when it is impossible to get an analytical solution for the equations of motion (6.23), the difference between the magnitudes of transverse stresses on the middle surface 'from above' and 'below' can serve as an obvious criterion of evaluation of the accuracy of the solution obtained.

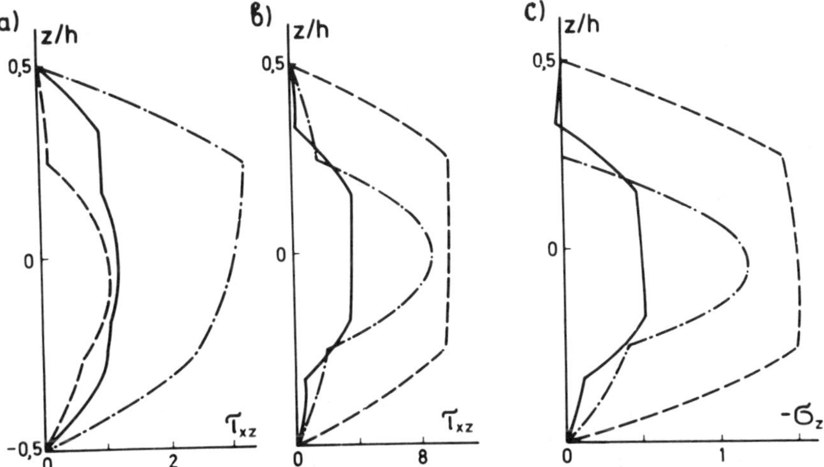

Fig. 6.7. Transverse stresses dependence on the z coordinate under dynamic external pressure ($V_q = 2$) for shells having package structures $[0°/90°/0°]_S$ (———); $[90°/0°]_S$ (– – –); $[0°/90°]_S$ (–· – ·–); $R/h = 50$.

The diagrams of transverse shear stresses, shown in Figs 6.4 and 6.5, have the form of quadratic parabolas; the maximum in all the cases is reached on the middle surface. Let us note that the diagram of transverse normal stress changes qualitatively depending on the kind of loading: under axial compression the stress σ_z has two extremums — positive and negative, while under external pressure σ_z is of the same sign with a maximum on the middle surface. The diagrams of transverse stresses throughout the thickness of laminated shells (Figs 6.6 and 6.7) reflect fairly shallow regions of parabolas within each layer with strong breaks on the interfaces. It is seen from Figs 6.6 and 6.7 that in the case of laminated shells not only σ_z diagrams, but also those for τ_{xz} and τ_{yz} differ qualitatively for the two kinds of loading.

Let us also note that the maximums of transverse stresses in all the cases are reached on the interfaces between layers or on the middle shell surface.

Tables 6.10 and 6.11 reflect the calculations of maximum (in the instant $\tau = \tau^*$) values of transverse shear and normal stresses in homogeneous shells for variants considered in Tables 6.3 and 6.7. The coordinates x_{xz}, x_{yz} and x_z on the generatrix, corresponding to maximums of stresses τ_{xz}, τ_{yz} and σ_z are presented. The maximums of stresses τ_{xz} and σ_z in the circumferential direction for all variants, treated in Tables 6.10 and 6.11, are reached in the $y = \pi R$ section, whereas the maximum of τ_{yz} for all variants, treated in Table 6.10 — in $y = \pi R/2$ section. Let us note that the stresses τ_{xz} and τ_{yz} in all cases have the maximum on the middle surface of the shell.

The magnitudes of transverse stresses at $\tau = \tau^*$ depend on the ratio of geometric parameters h and R, the loading rate and the field of initial

TABLE 6.10

Transverse Stresses at the Instant of Initial Ply Failure under Axial Dynamic Compression

No. of variants	$\frac{\lvert\tau_{xz}^{max}\rvert}{MPa}$,	$\frac{x_{xz}}{L}$	$\frac{\lvert\tau_{yz}^{max}\rvert}{MPa}$,	$\frac{x_{yz}}{L}$	$\frac{-\sigma_z^{max}}{MPa}$,	$\frac{x_z}{L}$	$\frac{z^-}{h}$	$\frac{\sigma_z^{max}}{MPa}$	$\frac{z^{*+}}{h}$	m^*	λ_x^*
1	6·3	0·52	0·11	0·48	0·04	0·48	−0·3	0·03	0·3	9	44
2	6·9	0·52	0·10	0·48	0·04	0·48	−0·3	0·04	0·3	6	67
3	7·2	0·52	0·12	0·48	0·05	0·48	−0·3	0·04	0·3	10	40
4	4·3	0·51	0·04	0·48	0·02	0·49	−0·3	0·01	0·3	15	67
5	15·9	0·51	0·50	0·45	0·10	0·46	−0·2	0·02	0·4	3	33
6	34·9	0·51	0·64	0·47	0·84	0·48	−0·3	0·76	0·3	3	33
7	21·2	0·51	0·20	0·49	0·22	0·49	−0·3	0·20	0·3	15	27
8	83·5	0·51	1·00	0·49	3·30	0·49	−0·3	3·20	0·3	4	25
9	4·2	0·52	0·09	0·47	0·02	0·47	−0·3	0·01	0·3	8	50
10	8·3	0·52	0·42	0·44	0·07	0·45	−0·1	—	—	1	100
11	14·5	0·51	0·44	0·46	0·18	0·47	−0·3	0·10	0·3	1	100

TABLE 6.11

Transverse Stresses at the Instant of Initial Ply Failure under Dynamic External Pressure

No. of variants	$\vert\tau_{xz}^{max}\vert$, *MPa*	$\vert\tau_{yz}^{max}\vert$, *MPa*	$\frac{x_{yz}}{L}$	$\frac{y_{yz}}{2\pi R}$	$-\sigma_z^{max}$, *MPa*	$\frac{x_z}{L}$	n^*	λ_y^*	$-\sigma_z^0$ *MPa* $\left(\frac{h}{2}\right)$
1	0.14	5.0	0.42	0.472	0.67	0.41	8	90	0.10
2	0.13	4.4	0.42	0.472	0.54	0.42	8	78	0.45
3	0.08	4.6	0.36	0.466	0.64	0.37	7	90	0.03
4	0.83	19.1	0.43	0.466	2.92	0.42	6	26	1.32
5	0.06	1.4	0.41	0.466	0.20	0.42	6	26	2.77
6	0.41	13.7	0.39	0.455	2.66	0.39	5	32	0.48
7	0.04	1.9	0.41	0.477	0.20	0.40	9	175	0.02
8	0.04	1.5	0.43	0.477	0.16	0.42	9	175	0.08
9	0.02	2.0	0.35	0.472	0.24	0.37	8	195	0.01

imperfections. In order to interpret the effects, illustrated by the obtained numerical results, it is necessary to introduce into consideration the characteristic lengths of half-waves of buckling produced in the longitudinal $\lambda_x^* = L/(m^*h)$ and circumferential $\lambda_y^* = \pi R/(n^*h)$ directions. The characteristic numbers of axial m^* and circumferential n^* harmonics can be specified from the shape of distribution of additional deflection Fourier coefficients in respect to m and n (see Fig. 5.17) at $\tau = \tau^*$. Henceforth, under m^* and n^* we shall understand numbers of harmonics, corresponding to maximums in the distributions for each calculation variant. The values m^* and λ_x^* are given in Table 6.10, n^* and λ_y^* are given in Table 6.11.

As it is seen, for all variants of loading with axial compression and the variants of loading with external pressure corresponding to $V_q = 0.2$ and 2.0, the transverse stresses increase with increasing shell thickness. This is explained for by the decrease in λ_x^* and λ_y^*. In the case of external pressure with loading rate $V_q = 20$ the decrease of R/h from 500 to 200 also leads to the increase in transverse stresses, however, a further increase in thickness leads to decrease of transverse stresses. The latter result is explained for by the fact that on passing over from $R/h = 200$ to $R/h = 50$, the additional deflection $\tilde{w}$ is reduced by two orders (see Table 6.7), the contribution of flexural components in in-plane stresses falls abruptly and, as a result, the transverse stresses decrease (they are defined really by flexural components).

Considering the effect of loading rate, we can note that under axial compression, the decrease in $\tilde{w}$ with growing V_p exerts a weaker effect than the decrease in λ_x^*. As a result, the transverse stresses increase. Under external pressure, at $R/h = 50$ and $R/h = 200$ the transverse stresses increase with increasing V_q from 0.2 to 2.0 (the decrease in λ_y^* exerts a stronger effect than the decrease in $\tilde{w}$) and decrease during its further increase (λ_y^* does not change practically, while $\tilde{w}$ falls abruptly). At $R/h = 500$, the transverse stresses decrease monotonously with increasing V_q from 2.0 to 20. In this case, the decrease in $\tilde{w}$ turns out to be dominant.

Summarizing the analysis, it can be stated that the magnitudes reached by transverse stresses at $\tau = \tau^*$ are defined by two basic factors. Firstly, they increase with decreasing relative characteristic lengths of half-waves in longitudinal and circumferential directions $\lambda_x^* = L/(m^*h)$ and $\lambda_y^* = \pi R/(n^*h)$. This is observed by the increase of h, V_P and V_q. Secondly, the transverse stresses decrease with decreasing contribution of flexural components in stresses σ_x, σ_y at $\tau = \tau^*$ (this also takes place in the increase of h, V_P, V_q). The sufficiently complicated character of the dependence of transverse stresses on shell thickness and loading rate can be attributed to the presence of two competitive factors.

In accordance with the data of Tables 6.10 and 6.11, under axial compression the stress τ_{xz} is the largest of the three transverse stresses, while under external pressure the largest is τ_{yz}. The stress σ_z is by two orders lower than the highest of transverse shear stresses under axial compression, whereas under external pressure it is only 5–10 times lower. The maximum magnitudes of transverse shear stresses for all calculation variants, except variants 6 and 8 (Table 6.10), are lower than the typical values of transverse normal strength in tension of carbon plastics (in accordance with Ref. 267 equal to 10–17 MPa). In such a way, only for one calculation variant — for a shell having $R/h = 50$ and $a_0 = 2h$ — the failure due to transverse shear stress starts earlier than the in-plane failure. Obviously, it can be supposed with sufficient assurance that for a shell having $R/h > 100$ within a wide range of loading rates (both for axial compression and external pressure), the in-plane stresses are the most dangerous from the viewpoint of initial failure.

The magnitudes reached by transverse stresses in a multi-layer shell in the instant τ^*, on the whole, are defined by the same factors as in the case of homogeneous shells: R/h ratio, loading rates V_p and V_q. The structure of ply layup through the thickness also exerts a certain effect. A sufficiently complete interpretation of the results, presented in Tables 6.5, 6.6, 6.8 and 6.9, can be made using the characteristic lengths of half-waves λ_x^* and λ_y^* and the magnitudes of additional deflection $\tilde{w}$, having been reached in the instant $\tau = \tau^*$. Thus, in accordance with Table 6.5, the highest magnitudes of τ_{xz} have been obtained for variants 1.3 and 1.4 (Table 6.4), to which corresponds to the lowest magnitude λ_x^*; the minimum value of τ_{xz} was obtained for the variant 1.6, to which corresponds the lowest value of $\tilde{w}$. Table 6.6 shows the analogous tendency.

In the case of loading with dynamic external pressure, the values λ_y^* for all packages differ slightly, therefore, the magnitudes of transverse stresses are defined mainly by the $\tilde{w}$-value. Thus, in Table 6.8, it is distinctly seen that the stress τ_{yz} increases with increasing $\tilde{w}$. The results, presented in Table 6.9, in general, also reflect this tendency.

The maximum magnitudes of transverse and interlaminar stresses at $\tau = \tau^*$, as follows from the data presented in Tables 6.5, 6.7–6.9, for all the variants of ply layup are lower than the typical magnitudes of interlaminar shear strength and transverse normal tensile strength of high-quality laminated carbon plastics. In such a way, in the problem treated, just as in the case of unidirectionally reinforced shells, for $R/h > 100$, the in-plane stresses are not dangerous, from the viewpoint of initial failure.

6.6 ON THE SUBSTANTIATION OF THE METHOD FOR CALCULATION OF TRANSVERSE STRESSES

Further let us dwell upon the substantiation of the applicability of the procedure, outlined in paragraph 6.4. Of course, the most convincing way of such substantiation consists in comparison of the magnitudes of transverse stresses, obtained by this procedure, with the data of calculating the shell as a laminated cylinder, described by linearized equations of three-dimensional theory of elasticity.

However, there are neither procedures for calculation nor for the results, permitting such a comparison to be made. Therefore, let us employ certain analogies between the problem of dynamic buckling of an imperfect shell loaded with axial compressive tractions or external pressure and the problem of transverse dynamic bending of laminated beam. The latter problem was solved in two-dimensional approach on the basis of equations of the theory of elasticity for a piecewise-homogeneous orthotropic media.[63] As was shown in Ref. 63, the basic parameter, defining the applicability of the simplified models of non-deformable normal, straight or broken normal, is the relation $\lambda_k = L/hk$, where L and h are length and thickness of a beam; k is the number of half-waves over the beam length, formed in transverse bending under the load of the type $q(x, t) = Q(t)\sin(\pi k/L)x$. Analogous parameters $\lambda_x^* = L/hm^*$ and $\lambda_y^* = \pi R/hn^*$ were introduced in paragraph 6.5 for a cylindrical shell. Consequently, evaluation of the accuracy of calculating the transverse stresses in a beam at fixed λ_k-value allows to estimate the accuracy of calculating the transverse stresses in a shell for geometric characteristics, numbers of axial and circumferential harmonics that would satisfy the conditions $\lambda_x^* = \lambda_k$ and $\lambda_y^* = \lambda_k$. In such a way, it is possible to determine the error in calculations for each of the harmonics, being taken into account in summing up Fourier series for transverse stresses in a shell.

As the numerical calculations have shown, in loading with axial compression at rates $V_P \leq 5$ and variegated functions of initial imperfections, 1% accuracy of summing up series for transverse stresses in respect to m is reached at $m_0 = 1$, $M = 50$, for $L/h = 400$ and $M = 20$, for $L/h = 100$. To these limits for summing correspond: $\lambda_{m_0} = 400$, $\lambda_M = 8$ in the first case and $\lambda_{m_0} = 100$, $\lambda_M = 5$ in the second case. Under external pressure, for loading rates $V_q \leq 20$ such accuracy of summing in respect to n is reached at $n_0 = 1$, $N = 20$ for $R/h = 200$ and $N = 15$ for $R/h = 50$, to which correspond $\lambda_{n_0} = 628$, $\lambda_N = 63$ and $\lambda_{n_0} = 157$, $\lambda_N = 13$.

After evaluating the boundaries of varying λ_m and λ_n, let us calculate the stress τ_{xz} in beams of above-described graphite/epoxy material T 300/5208. Figures 6.8 and 6.9 show the diagrams obtained in accordance with the procedure of A. E. Bogdanovich and E. V. Yarve,[63] based on two-dimensional equations of the theory of elasticity for a medium with piecewise-constant characteristics and in accordance with the procedure based on the hypothesis about non-deformed normal for a laminated beam. In the latter case, the procedure of calculating τ_{xz} is identical to that elaborated in paragraph 6.4 for a laminated shell. The calculations were performed for a single-layer beam, reinforced along its longitudinal axis (Figs 6.8(a, b)) and four-ply beams of symmetric through the thickness structure with ply layup $[0°/90°]_S$ (Fig. 6.8(c, d)) and $[90°/0°]_S$ (Fig. 6.8(e, f)). As is seen, at $\lambda_k = 100$ the relative error of the calculation of τ_{x2} in the beam reinforced along its axis is 5%, and for the laminations [0°/90°]s and [90°/0°]s it is 3% and 1%, respectively. At $\lambda_k = 2$, these errors are equal to 15, 10 and 24%, respectively. A comparative analysis of results, presented in Fig. 6.8 ($\tau = 1$) and 6.9 ($\tau = 2.5$), shows that the difference in results does not practically change in time.

As it was noted above, within a sufficiently wide range of loading rates and geometric parameters of thin-wall shells, the harmonics with $\lambda_m > 10$ and $\lambda_n > 10$ make the main contribution to transverse stresses. The estimate of accuracy in calculating the transverse shear stress, obtained for a beam, allows to assume that the approach proposed in paragraph 6.4 allows to evaluate the danger of initiation of initial failure under dynamic loads of axial compression and external pressure in thin laminated composite cylindrical shells.

The outlined procedure does not allow to take into account momentless stress component σ_z, defined by the process of stress wave propagation from the loaded surface in the z-direction. The study of wave processes in a multi-layer cylinder under dynamic loads applied to its side surface is a

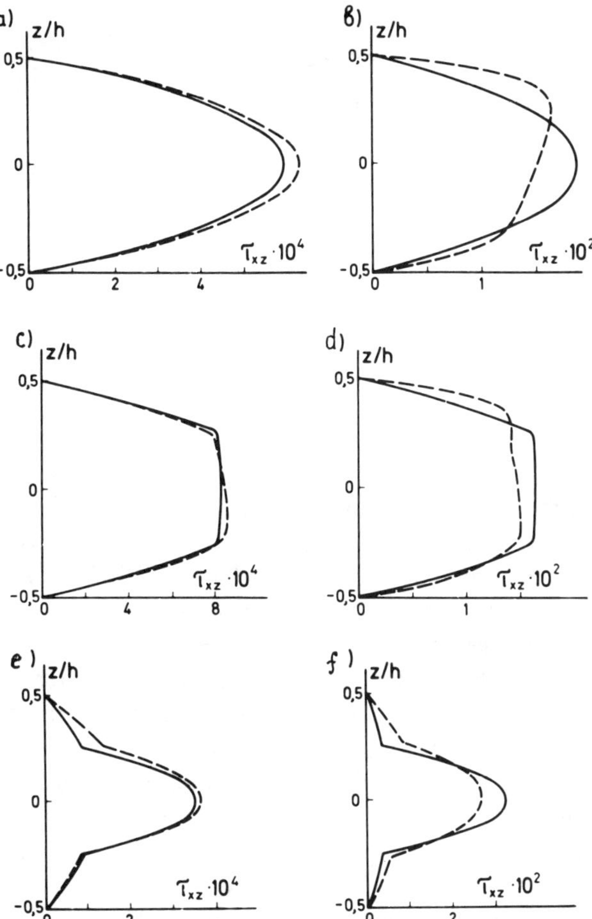

Fig. 6.8. Transverse shear stress dependence on the z coordinate in a unidirectionally reinforced (a, b) and laminated $[0°/90°]_S$ (c, d), $[90°/0°]_S$ (e, f) graphite/epoxy beams at $\tau = 1$; $\lambda_k = 100$ (a, c, e) and 2 (b, d, f); calculations in accordance with the procedure of Ref. 63 (– – –) and on the basis of a model of non-deformable normal (———).

separate complex problem.[112] For thick-wall cylinders, loaded with short-time intensive impulses, at the very first passages of a wave through the thickness, there arise tensile radial stress in the vicinities of the interfaces and free surface, which can lead to local sites of fracture.[111]

Therefore, the employed approach is applicable to comparatively slowly increasing loads, when the stress σ_z in the initial stage of the process is small in comparison to tensile strength in the direction normal to the middle surface of a shell.

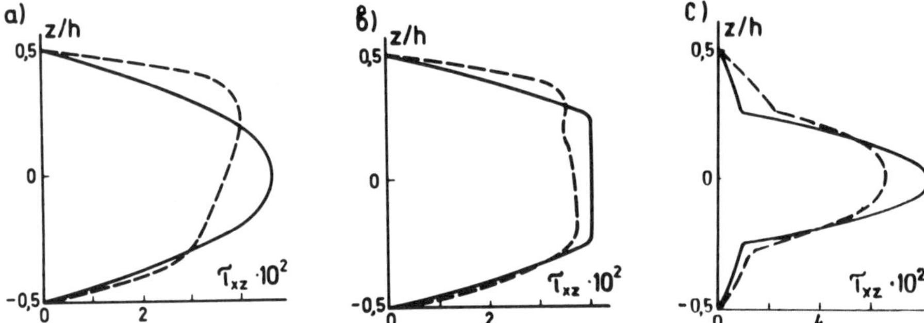

Fig. 6.9. The transverse shear dependencies on the z coordinate in single-ply (a) and four-ply $[0°/90°]_S$ (b), $[90°/0°]_S$ (c) graphite/epoxy beams at $\tau = 2.5$; $\lambda_k = 2$. Designations the same as in Fig. 6.8.

In laminated beams, already by the time of tenfold wave traversing through the thickness, the distribution of σ_z is rather close to the linear one with a maximum on the loaded surface.[63] The tensile stresses, generated in the vicinity of the free surface, are negligible. Obviously, such an effect is also characteristic for distribution of momentless component σ_z^0 of transverse normal stress in a shell. Therefore, we can restrict ourselves here to analysis of the maximum values $\sigma_z^0(h/2)$, which are reached by the instant $\tau = \tau^*$. They are presented in the last column of Table 6.11. As is seen, the maximum of momentless component in all cases, except variant 5, is less than the maximum of flexural component, caused by the process of non-axisymmetric dynamic buckling. The relative contribution of σ_z^0 to the total stress σ_z falls abruptly with decreasing loading rate and shell thickness.

In such a way, the proposed procedure allows to perform the analysis of initial failure of laminated cylindrical shells under comparatively slow dynamic loading with axisymmetric compressive forces at the end faces and uniformly distributed external pressure. The computer programs, evolved on the basis of this procedure, gives one the possibility to calculate the magnitude of dynamic load (or impulse), having caused the initial failure of a shell, to assess the location of initial failure zones in the plane of a layer, and monitor simultaneously the magnitudes of transverse and interlaminar stresses. In the case, when these stresses exceed the corresponding ultimate values, the location of zones of initial delaminations is fixed. For practical use of the algorithms and programs described, it is necessary to know the load dependence on time and field of initial shell deflection, alongside with geometric shell parameters, elastic and strength characteristics of the material.

6.7 NUMERICAL ANALYSIS OF THE PROCESS OF LAYER-BY-LAYER FAILURE

The strength analysis of laminated structural elements, based on the criterion of first-ply failure, allows one to determine the lower bound of ultimate loads. This analysis is expedient when there is a requirement to maintain integrity and hermeticity of walls of the structure during all the time of its exploitation. However, in those situations, when it is required that the load-carrying capacity of the structural element loaded by intensive, comparatively short-time impulses should be ensured, without any attention for its state after unloading, this approach to the strength analysis needs improvement. Matrix cracking and fiber debonding from the matrix in layers as well as 'refusal' of a part of the layers as load-carrying elements due to the loss of fiber stability become admissible. In such a way, taking into consideration that each failure event occurs only in one layer and corresponds to a fairly definite mode of failure in the reinforced material, makes it possible to raise the lower bound of ultimate loads. The corresponding calculation models, the number of which is sufficiently large at present, are based on a certain scheme of reducing stiffnesses of a single layer and, consequently, stiffnesses of the entire multi-layer package. Such a reduction is needed in those time instants, when, in accordance with the strength criterion, one of the possible failure modes is being realized in one layer. The main approaches to constructing such models are discussed in Ref. 255.

One of the simplest schemes of stiffness reduction in a multi-layer composite during the process of its layer-by-layer failure described in Ref. 266, was used in Ref. 48 for calculating imperfect cylindrical shells, loaded with dynamic external pressure. The results obtained in Ref. 48 have shown that the load of 'complete load-carrying capacity exhaustion' can considerably exceed the load of 'initial failure'. This effect is manifested the stronger, the greater is the role played by geometric non-linearity. It is enforced also with decreasing loading rate, since the process of layer-by-layer failure of a shell is prolonged in time.

In this paragraph, the results of calculating the layer-by-layer failure processes of graphite/epoxy cylindrical shells loaded with impulses of axial compression and external pressure are presented. A more perfect model than that described in Ref. 226 for layer stiffness reduction, proposed in Ref. 216 is used. This model presupposes that there are seven variants of the failure conditions in the *i*-th unidirectionally reinforced layer, which are characterized in Table 6.12. The dimensionless stresses $\sigma_{11}^{*(i)}$ and $\sigma_{22}^{*(i)}$ along and across the fibers are determined according to paragraph 6.2;

TABLE 6.12

Schemes for Reduction of Elastic Moduli in a Layer

No. of variants	*Failure mode*	$E_2^{(i)}$	$E_1^{(i)}$	$G_{12}^{(i)}$
1	$\sigma_{11}^{*(i)} \geq 1$	0	0	0
2	$\sigma_{11}^{*(i)} \leq -1; \sigma_{22}^{*(i)} \geq 0$	0	0	0
3	$\sigma_{11}^{*(i)} \leq -1; \sigma_{22}^{*(i)} < 0$	0	$E_2^{0(i)}$	0
4	$\sigma_{22}^{*(i)} \geq 1$	$E_1^{0(i)}$	0	0
5	$\sigma_{22}^{*(i)} \leq -1$	$E_1^{0(i)}$	0	$\bar{G}_{12}^{(i)}$
6	$\sigma_{12}^{*(i)} \geq 1; \sigma_{22}^{*(i)} \geq 0$	$E_1^{0(i)}$	0	0
7	$\sigma_{12}^{*(i)} \geq 1; \sigma_{22}^{*(i)} < 0$	$E_1^{0(i)}$	$E_2^{0(i)}$	$\bar{G}_{12}^{(i)}$

$\sigma_{12}^{*(i)}$ is tangential in-plane stress related to the shear strength. Correspondingly, $E_1^{(i)}$ and $E_2^{(i)}$ are elastic moduli of the layer along and across the fibers, while $G_{12}^{(i)}$ is the in-plane shear modulus; $E_1^{0(i)}$, $E_2^{0(i)}$, $G_{12}^{0(i)}$ are the elastic and shear moduli of a layer in the initial undamaged state. In addition to the model described in Ref. 216 the information about the non-linear shear behavior of a unidirectional composite[365] is used (for variants 5 and 7 from Table 6.12); $\tilde{G}_{12}^{(i)}$ is the current shear modulus, which can be found from the experimental relation $\sigma_{12} = \sigma_{12}(\varepsilon_{12})$ for a monolayer which is to be used at the magnitude of shear strain generated in the i-th layer at the time moment corresponding to realization of variants 5 or 7.

The algorithm for the analysis consists in the following. Let us suppose that at the time instant $\tau = \tau^* = \tau_1^{(i)}$ in a certain point of the i-th shell layer the first-ply failure condition (6.4) is being satisfied. At this instant the values $\sigma_{11}^{*(i)}$, $\sigma_{22}^{*(i)}$, $\sigma_{12}^{*(i)}$ are calculated and all the conditions presented in Table 6.12 are checked. As it follows from the results of Tables 6.3, 6.5, 6.6–6.9, there are situations when none of the conditions is satisfied (for example, for variants 1.3 and 1.4 of Table 6.5, 1.1 of Table 6.6, 2 and 9 of Table 6.7, 2.2 of Table 6.8). And this is natural, since the tensor-polynomial strength criterion (6.4) and maximum stress criterion are inequivalent. There are two possible ways out of this confusing situation. Firstly, not to reduce at $\tau = \tau_1^{(i)}$, the moduli of that layer, in which according to (6.4), the initial failure has been detected, but to continue calculations with initial moduli until one of the conditions, formulated in Table 6.1, has been fulfilled.

Secondly, reducing at $\tau = \tau_1^{(i)}$ the moduli of the layer, in which the initial failure has been detected, proceed from the fact — which of the stresses $\sigma_{11}^{*(i)}$, $\sigma_{22}^{*(i)}$ or $\sigma_{12}^{*(i)}$ is the closest to its ultimate value. Obviously, with consistent use of the second approach (henceforth we shall stick to it), a shorter time interval prior to complete load-carrying capacity exhaustion would be eventually obtained.

Thus, at $\tau = \tau_1^{(i)}$, depending on the mode of failure, the values of certain components of stiffness matrices of a multi-layer package C_{kl}, D_{kl} are reduced stepwise, having after reduction the values $C_{kl}^{(1)}$, $D_{kl}^{(1)}$. They are inserted into equations of motion for a shell and, as a result of their solution at $\tau > \tau_1^{(j)}$, deflection, traction function, shell deformations and stresses in layers are determined.

In the process of integration in time, the condition (6.4) is being checked in each step. Let it be fulfilled in the j-th layer (coincidence of j and i is possible) at $\tau = \tau_2^{(i)} > \tau_1^{(i)}$. At this time instant the mode of failure is established firstly; after that the moduli of the layer are reduced and respective stiffnesses of a multilayer package are modified. Further, at $\tau > \tau_2^{(j)}$, the equations of motion for a shell having stiffnesses $C_{kl}^{(2)} < C_{kl}^{(1)}$, $D_{kl}^{(2)} < D_{kl}^{(1)}$ are being solved. Proceeding with this procedure, we obtain a sequence of failure moments $\tau_1^{(i)}, \tau_2^{(j)}, \ldots, \tau_n^{(q)}$ (the upper indices indicate the number of the layer, lower index $n \geq N$, where N is the number of layers in a package). The latter one, $\tau_n^{(q)}$, is determined in accordance with the condition of tending to zero, at least one of the stiffnesses C_{11}, C_{22} or C_{66} of the package as a whole. The algorithm expounded here allows also to establish the sequence of failure events in layers and to obtain stepwise-

TABLE 6.13

Characteristics of Layer-by-Layer Failure Process under Axial Dynamic Compression for the Loading Program (a)

No. of failure event, k	$\tau_k^{(j)}$	*No. of layer, j*	*Failure mode*	x^*/L	y^*/R
		$[0/90°]_S$			
1	85.50	4	3	0.96	3.14
2	88.01	4	5	0.96	0
3	88.26	1	6	0.96	3.14
4; 5	88.27	3; 2	4	0.96	0
		$[0/90/0/90°]_S$			
1	79.75	8	3	0.96	3.14
2	80.00	6	3	1.10	0
3	80.26	8	5	0.96	0
4	80.27	1	6	0.96	0
5	80.52	3	6	0.96	2.40
6	80.53	6	5	0.96	2.40
7; 8; 9	80.54	2; 4; 5	4	0.96	0

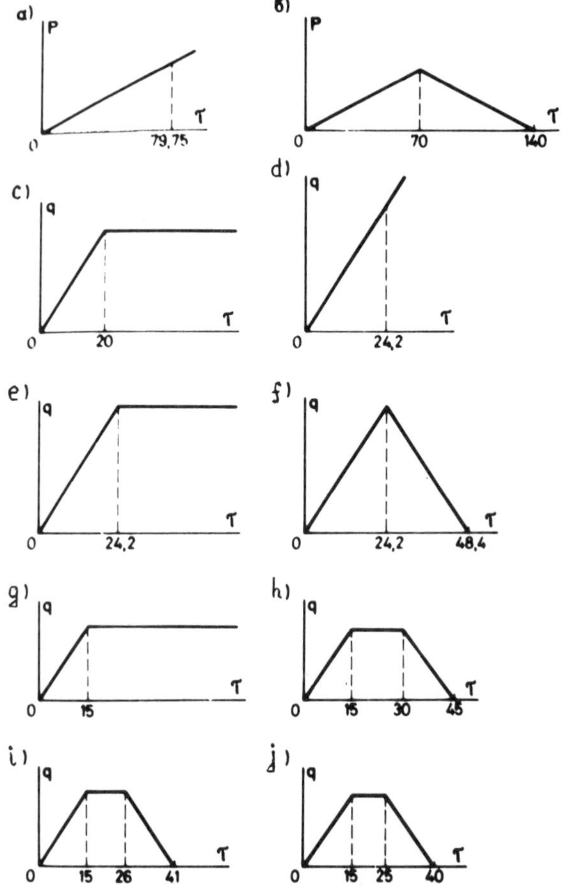

Fig. 6.10. Programs of dynamic loading with axial compression (a, b) and external pressure (c–j).

continuous time dependence of characteristics of the stress–strain state in a shell.

Let us consider some numerical examples. The variance of axial load $P(\tau)$ and external pressure $q(\tau)$ in time is shown in Fig. 6.10. The rates of their growth are $V_P = 0.05$ and $V_q = 0.2$, respectively. We shall determine the initial imperfections for a shell loaded by axial compression in accordance with the formula (6.10) with $a_0 = 0.2h$, whereas for a shell loaded by external pressure — to formula (6.11). The material stiffnesses and strengths of a monolayer we shall specify according to (6.9). The geometric shell parameters are: $R = 1$ m, $L/R = 2$, $R/h = 200$.

The results of calculating the initial and subsequent failure events under

TABLE 6.14

Characteristics of Layer-by-Layer Failure Process under Dynamic External Pressure at Loading Program (c)

No. of failure event, k	$\tau_k^{(j)}$	*No. of layer, j*	*Failure mode*	x^*/L	y^*/R
1	2	3	4	5	6
		$[90/0°]_S$			
1	26.10	3	3	0.92	3.14
2	26.65	2	3	0.96	2.72
3	26.66	4	3	0.84	3.14
4	27.20	4	7	0.90	3.14
5	27.80	1	3	0.92	2.72
		$[90/90/0/0°]_S$			
1	26.10	6	3	0.92	3.14
2	26.65	3	3	0.96	2.72
3	26.66	8	3	0.84	3.14
4	26.81	5	3	0.94	3.14
5	27.01	7	3	0.84	3.14
6	27.21	8	7	0.90	3.14
7	27.56	4	3	0.98	2.72
8	27.71	1	7	0.92	2.72
9	27.96	7	7	0.90	3.14
10	28.61	2	1	0.94	2.72
		$[90/90/90/90/0/0/0/0°]_S$			
1	26.10	12	3	0.92	3.14
2	26.40	11	3	0.92	3.14
3	26.65	5	3	0.96	2.72
4	26.66	16	3	0.82	3.14
5; 6	26.81	10; 15	3	0.94	3.14
7	27.01	6	3	0.96	2.72
8	27.02	14	3	0.84	3.14
9, 10	27.21	13, 16	3	0.84	3.14
11	27.26	9	3	0.96	3.14
12	27.52	7	3	0.98	2.72
13	27.57	15	7	0.90	3.14
14	27.72	1	7	0.92	2.72
15	27.92	14	7	0.90	3.14
16	28.12	2	7	0.92	2.72
17	28.37	13	7	0.90	3.14
18	28.72	8	3	0.98	2.09
19	28.78	3	7	0.94	2.72
20	29.43	4	7	0.96	2.72
		$[90/90/90/90/90/90/90/90/0/0/0/0/0/0/0/0°]_S$			
1	26.10	24	3	0.92	3.14
2	26.25	23	3	0.92	3.14
3	26.40	22	3	0.92	3.14
4; 5	26.60	21; 32	3	0.94	3.14
6	26.66	9	3	0.96	2.72
7	26.71	31	3	0.84	3.14

TABLE 6.14 Continued

No. of failure event, k	$\tau_k^{(j)}$	*No. of layer, j*	*Failure mode*	x^*/L	y^*/R
1	2	3	4	5	6
8; 9; 10	26.81	20; 10; 30	3	0.94	3.14
11; 12	27.01	11; 19	3	0.96	3.14
13; 14	27.02	29; 28	3	0.84	3.14
15	27.22	12	3	0.98	2.72
16; 17	27.24	27; 26	3	0.84	3.14
18	27.27	18	3	0.96	3.14
19; 20	27.37	25; 31	3	0.84	3.14
21	27.48	13	3	0.98	2.72
22	27.53	17	3	0.96	3.14
23	27.58	30	7	0.90	3.14
24	27.68	1	7	0.92	2.72
25	27.73	29	7	0.90	3.14
26; 27	27.88	14; 2	3	0.98	2.72
28	27.98	28	7	0.90	3.14
29	28.13	3	7	0.92	2.72
30	28.19	27	7	0.90	3.14
31	28.38	15	3	1.00	2.82
32	28.39	4	7	0.92	2.72
33	28.44	26	7	0.90	3.14
34	28.54	16	3	0.98	3.14
35	28.64	25	7	0.90	3.14
36	28.84	5	7	0.94	2.72
37	29.14	6	7	0.94	2.72
38	29.50	7	7	0.96	2.72
39	30.15	8	7	0.96	2.72

the loading variant of Fig. 6.10(a) for four-ply and eight-ply shells are presented in Table 6.13; $\tau_k^{(j)}$ is the instant of the k-th failure in the j-th layer; x^*, y^* are coordinates of failure site. For both packages, load-carrying capacity exhaustion in a shell is due to the condition $C_{22} = 0$ being fulfilled. For the loading variant (b) (see Fig. 6.10), there were no failures detected in a shell, whereas for intermediate variants between (a) and (b) with $P(\tau)$ changing within the interval $\tau \in [70; 140]$ both — fulfilment of the condition $C_{22} = 0$ as well as diverse realizations of failure modes in a part of the shell layers could be detected. It should be noted that the time interval between the first and last failure event is short, compared to time interval prior to the first failure. This is associated with an abrupt increase of stress–strain state characteristics in the course of a comparatively short time when longitudinal dynamic compression is applied. This effect was illustrated in detail in Chapter 5.

The results of calculating the initial and subsequent failure events in

TABLE 6.15

Characteristics of Layer-by-Layer Failure Process under Dynamic External Pressure for a Shell with $[90°/0°/90°/0°]_S$ Package

No. of failure event, k	$\tau_k^{(j)}$	*No. of layer, j*	*Failure mode*	x^*/L	y^*/R
1	2	3	4	5	6
		Loading program (d)			
1	24.20	7	3	0.86	3.14
2	24.62	2	3	0.90	2.72
3	24.96	8	3	0.80	3.14
4	25.20	5	3	0.90	3.14
5	25.37	8	7	0.86	3.14
6; 7	25.73	4; 1	3; 7	0.98	2.72
8	25.73	6	3	0.82	3.14
9	26.25	6	7	0.90	3.14
10	26.71	3	7	0.98	2.72
		Loading program (e)			
1	24.20	7	3	0.86	3.14
2	24.65	2	3	0.90	2.72
3	25.00	8	3	0.80	3.14
4	25.20	5	3	0.92	3.14
5	25.40	8	7	0.86	3.14
6; 7	25.76	4; 1	3; 7	0.98	2.72
8	25.77	6	3	0.82	3.14
9	26.31	6	7	0.90	3.14
10	26.76	3	7	0.98	2.72
		Loading program (f)			
1	24.20	7	3	0.86	3.14
2	24.65	2	3	0.90	2.72
3	25.00	8	3	0.80	3.14
4	25.20	5	3	0.92	3.14
5	25.41	8	7	0.86	3.14
6	25.70	4	3	0.98	2.72
7	25.71	6	3	0.82	3.14
8	25.76	1	7	0.90	3.14
9	26.31	6	7	0.90	3.14
10	26.76	3	7	0.96	2.72
		Loading program (g)			
1	26.00	7	3	0.88	3.14
2	26.70	2	3	0.92	2.72
3	27.65	8	3	0.84	3.14
4	27.75	5	3	0.94	3.14
5	28.46	8	7	0.90	3.14
6	29.00	4	3	0.98	2.72
7	29.11	6	3	0.84	3.14
8	29.31	1	7	0.90	2.72
9	30.56	6	7	0.90	3.14
10	31.36	3	7	0.94	2.72

TABLE 6.15 Continued

No. of failure event, k	$\tau_k^{(j)}$	*No. of layer, j*	*Failure mode*	x^*/L	y^*/R
1	2	3	4	5	6
		Loading program (h)			
9	30.51	6	7	0.90	3.14
10	31.31	3	7	0.94	2.72
		Loading program (i)			
1	26.00	7	3	0.88	3.14
2	26.65	2	3	0.92	2.72
3	27.65	8	3	0.84	3.14
4	27.75	5	3	0.94	3.14
5	28.60	8	7	0.90	3.14
6	29.60	4	3	0.98	2.72
7	29.61	6	3	0.82	3.14
8	30.51	1	7	0.88	2.72
9	33.26	6	7	0.76	3.14
10	35.11	3	7	0.70	2.82
		Loading program (j)			
1	26.00	7	3	0.88	3.14
2	26.70	2	3	0.92	2.72
3	27.80	8	3	0.84	3.14
4	27.91	5	3	0.94	3.14
5	29.45	8	7	0.88	3.14

four-, eight-, sixteen- and thirty-two-ply shells loaded with external pressure (see Fig. 6.10(c)) are presented in Table 6.14. Although from the viewpoint of membrane and flexural stiffnesses, all these multi-layer shells are identical, the process of layer-by-layer failure is prolonged in time essentially upon dividing a monolayer through its thickness into several sublayers having identical mechanical characteristics. Let us note that in all cases only two variants of failure (3 and 7) are being realized (see Table 6.12). The load-carrying capacity exhaustion is the result of fulfillment of the condition $C_{22} = 0$.

The effect of external pressure impulse shape on the process of layer-by-layer failure is illustrated in Table 6.15 by an example of an eight-ply shell under loading variants of Fig. 6.10(d–j). For variants (g) and (h), the characteristics of the first eight failures are identical, therefore, they are not presented repeatedly.

It follows from the consideration of the data of Table 6.15 that for all external pressure impulse shapes there occur failures of type 3 and 7 (see Table 6.12). The fact should also be emphasized that both — the sequence of layer failures and the failure mode realization in each layer, as well as coordinates of failure sites — do not practically depend on the shape of applied impulse. In all the cases, load-carrying capacity exhaustion occurs as a result of fulfillment of the condition $C_{22} = 0$. It follows from Table 6.15 that for strongly differing impulses (Fig. 6.10(e, f)) the characteristics of the process of layer-by-layer failure are rather identical, whereas for similar in shape and duration impulses (i, j), this process essentially differs. If in the case of Fig. 6.10(i), complete exhaustion of load-carrying capacity occurs in a shell after ten failure events in layers, then in the case (j) there occur only five failure events concerning only four layers. At $\tau > 29.45$, the condition (6.4) is not fulfilled in any shell point. The latter example illustrates the existence of such dynamic load impulses, the effect of which is inadmissible, from the viewpoint of the initial failure criterion, though they do not lead to damages over the entire wall thickness in a multi-layer structure.

The described algorithm for analyzing layer-by-layer failure does not take into account that the character of damage initiation is local, because of strictly definite sites of intensive non-axisymmetric deformation of an imperfect shell. Since in the algorithm proposed, after each failure in a layer has been detected, the stiffness of a layer as a whole is reduced, the obtained time until complete exhaustion of load-carrying capacity of the shell will eventually be under-rated. More perfect algorithms are certainly needed, if one wants to take into account that in the process of non-axisymmetric dynamic buckling only local damages occur and, after the strength condition for a layer has been fulfilled, a considerable part of a layer still maintains its load-carrying capacity. Development of such algorithms will require that special models for describing the generation and accumulation of local damages in a layer of composite material under highly inhomogeneous stress state should be elaborated. The finite element method may turn out to be effective for solving the problems of such kind. Further treatment of the questions pertaining to the processes of layer-by-layer failure is beyond the scope of this monograph.

References

1. *Abolin'sh D. S.* Compliance tensor for an elastic material reinforced in two directions. — Polym. Mech., 1966, vol. 2, N 3, p. 233–237 (The Faraday Press, Inc., Transl. from Russian).
2. *Abrosimov N. A., Bazhenov V. G.* On the calculation of glass reinforced plastic cylindrical shells under impulse loading. — Mech. Compos. Mater., 1983, N 5, p. 820–823 (in Russian).
3. *Agamirov V. L.* A survey of studies on stability of structures under impulse loading. — In: Design of spatial structures. Moscow, Stroyizdat, 1969, Issue 12, p. 186–200 (in Russian).
4. *Agamirov V. L., Vol'mir A. S.* Behavior of cylindrical shells under dynamic applications of lateral pressure and axial compression. — Trans. USSR Acad. Sci., Department of Technical Sciences. Mechanics and Mechanical Engineering, 1959, N 3, p. 78–83 (in Russian).
5. *Agamirov V. L., Vol'mir A. S.* On stability of a cylindrical shell under longitudinal impact. — Rep. USSR Acad. Sci., 1964, vol. 157, N 2, p. 307–308.
6. *Agamirov V. L., Vol'mir A. S.* Behavior of cylindrical shells under longitudinal impact. — In: Stability problems in structural mechanics. Moscow, 1965, p. 143–152.
7. *Adamovich I. S., Rikards R. B.* Weight optimization of an orthotropic cylindrical shell with varying properties under limitations on natural frequency. — Trans. USSR Acad. Sci., Mechanics of Solids, 1977, N 2, p. 120–125.
8. *Adamovich I. S., Rikards R. B.* Some problems of the optimal design of cylindrical shells from composite materials under dynamic constraints. — Mech. Compos. Mater., 1979, N 4, p. 641–646 (in Russian).
9. *Ainola L. J.* The non-linear Timoshenko type theory for elastic shells. — Transactions of the Estonian SSR Acad. of Sci., Ser. Phys.-Math. Tech. Sci., 1965, Vol. 14, N 3, p. 337–344.
10. *Ainola L. J., Nigul U. K.* Wave deformation processes of elastic plates and shells. — Trans. Estonian SSR Acad. Sci., Ser. of Phys.-Math. Tech. Sci., 1965, vol. 14, N 1, p. 3–63.
11. *Ains E. L.* Ordinary differential equations. Kharkov, Department of Scientific-Technical Information, 1939.
12. *Alpaidze Z. G., Babich Yu. N., Galiev Sh. U.* The Analysis of non-stationary processes caused by nonaxisymmetric loading of multilayer finite cylinders. — Probl. Strength, 1983, N 9, p. 3–7 (in Russian).
13. *Alumyae N. A.* Transient deformation processes of elastic shells and plates. — In:

Proc. VIth All-Union Conf. Theory of Shells and Plates. Moscow, Nauka Publ., 1966, p. 883–889 (in Russian).

14. *Alfutov, N. A.* Fundamentals of stability calculation for elastic systems. Moscow, Mashinostroyeniye Publ., 1978, 312 p.
15. *Ambartsumian S. A.* Some problems in the theory of anisotropic shells. — Trans. Armenian SSR Acad. Sci., 1947, N 9, p. 55–77.
16. *Ambartsumian S. A.* Some basic equations of the theory of a thin laminated shell. — Rep. Armenian SSR Acad. Sci., 1948, vol. 8, N 5, p. 203–210.
17. *Ambartsumian S. A.* Calculation of shallow cylindrical shells composed of anisotropic layers. — Trans. Armenian SSR Acad. Sci., Ser. Phys.-Math., Nat. Tech. Sci., 1951, vol. 4, N 5, p. 373–391.
18. *Ambartsumian S. A.* On the problem of calculation of laminated anisotropic shells. — Trans. Armenian SSR Acad. Sci., 1953, vol. 6, N 3, p. 15–35.
19. *Ambartsumian S. A.* On the design of two-ply orthotropic shells. Trans. USSR Acad. Sci., Department of Technical Sciences, 1957, N 7, p. 57–64.
20. *Ambartsumian S. A.* On the general theory of anisotropic shells. — Appl. Math. Mech., 1958, vol. 22, N 2, p. 226–237.
21. *Ambartsumian S. A.* Theory of anisotropic shells. Moscow, Fizmatgiz Publ., 1961, 384 p.
22. *Ambartsumian S. A.* Some problems of the development of the theory of anisotropic laminated shells. — Trans. Armenian SSR Acad. Sci., Ser. Phys.-Math. Sci., 1964, vol. 17, N 3, p. 29–53.
23. *Ambartsumian S. A.* Specific features of a theory of shells of advanced materials. — Trans. Armenian SSR Acad. Sci., Mechanics, 1968, vol. 21, N 4, p. 3–19.
24. *Ambartsumian S. A.* A general theory of anisotropic shells. Moscow, Nauka Publ., 1974, 448 p.
25. *Ambartsumian S. A., Peshtmaldzyan D. V.* On the non-linear theory of shallow orthotropic shells. — Trans. Armenian SSR Acad. Sci., Ser. Phys.-Math. Sci., 1958, vol. 11, N 1, p. 15–26.
26. *Amiro I. Ya.* On determination of critical magnitudes of compressive forces fastly increasing in time. — Appl. Mech., 1979, vol. 15, N 5, p. 54–60.
27. *Andreev A. N.* Axisymmetric bending and initial "failure" of multilayer reinforced cylindrical shells. — In: Spatial structures in Krasnoyarsk region. Krasnoyarsk, 1978, N 11, p. 47–56.
28. *Andronov A. A., Leontovich M. A.* On vibrations of the system with periodically varying parameters. — J. Russian Phys.-Chem. Soc. Ser. Physics, 1927, vol. 59, p. 429–443.
29. *Anik'ev I. I., Mikhailova M. I., Spisovskii A. S. et al.* Reaction of a five-layer polyethylene fiberglass panel to static and shock-wave loading. — Mech. Compos. Mater., vol. 19, N 1, 1983, p. 125–130 (Consultants Bureau, New York, Transl. from Russian).
30. *Bagdasariyan Zh. E., Gnuni V. Ts.* On the theory of dynamic stability of laminated anisotropic shells of revolution. — Trans. Armenian SSR Acad. Sci., Ser. Phys.-Math. Sci., 1960, vol. 13, N 5, p. 27–36.
31. *Bagdasariyan Zh. E., Gnuni V. Ts.* Resonance in forced non-linear vibrations of laminated anisotropic shells. — Trans. Armenian SSR Acad. Sci., Ser. Phys.-Math. Sci., 1961, vol. 1, N 1.
32. *Bagdasariyan Zh. E., Gnuni V. Ts.* Dynamic stability of an anisotropic cylindrical shell. — Rep. Armenian SSR Acad. Sci., 1965, vol. 41, N 5, p. 278–281.
33. *Bagdasariyan R. A.* On the strength optimization of a cylindrical shell made of composite material under impact. — Rep. Armenian SSR Acad. Sci., 1981, vol. 73, N 5, p. 279–283.
34. *Bazhenov V. G.* Non-linear problems of dynamics for thin-wall structures under impulse loading. — In: Applied problems of strength and plasticity: Statics and

dynamics of deformed systems. Gorkii, 1981, p. 57–66.

35. *Bazhenov V. G., Zhuravl'ov E. A.* Variational-difference method of solving non-linear axisymmetric problems of dynamics for laminated shells. — In: Applied problems of strength and plasticity. Gorkii, 1979, Issue 13, p. 36–45.
36. *Bazhenov V. G., Zhuravl'ov E. A.* Non-linear dynamic deformation of multilayer shells of revolution with irregular structure. — In: Applied problems of strength and plasticity. Gorkii, 1980, Issue 16, p. 50–56.
37. *Bazhenov V. G., Igonicheva E. V.* Dynamic loss of stability and post-critical behavior of a thin cylindrical shell with initial imperfections under the effect of axial impact loading. — Applied problems of strength and plasticity. Gorkii, 1977, Issue 6, p. 98–106.
38. *Bazhenov V. G., Igonicheva E. V.* On the interaction of non-axisymmetric modes of buckling of thin cylindrical shells under longitudinal impact loading. — In: Applied problems of strength and plasticity, Gorkii, 1983, Issue 24, p. 47–54.
39. *Beil'in E. A., Dzhanelidze G. Yu.* A review on dynamic stability of elastic systems. — Appl. Math. Mech., 1952, vol. 16, N 5, p. 635–648.
40. *Belubekyan E. V., Gnuni V. Ts.* Optimal problems of the vibrations of anisotropic laminated cylindrical shells. — Polym. Mech., vol. 12, N 5, 1976, p. 769–772 (Transl. from Russian).
41. *Belyaev N. M.* Stability of prismatic bars under the effect of variable longitudinal forces. — In: Engineering structures and structural mechanics, Leningrad, 1924, p. 149–167.
42. *Belyankin F. P., Yatsenko V. F., Dibenko G. I.* Strength and deformability of laminated plastics. Kiev, Naukova Dumka Publ., 1964, 220 p.
43. *Bivin Yu. K., Naida A. A.* Load carrying capacity of cylindrical shells under the effect of dynamic external pressure. — Appl. Mech., 1970, vol. 6, N 10, p. 28–34.
44. *Blokhina A. I.* Dynamic stability of a cylindrical shell with initial bending at a specified velocity of end faces convergence. — Collect., 1961, vol. 31, p. 196–201.
45. *Bogdanovich A. E.* Dynamic stability of an elastic orthotropic cylindrical shell with allowance for transverse shears. — Polym. Mech., vol. 9, N 2, 1973, p. 268–274 (Trans. from Russian).
46. *Bogdanovich A. E.* Dynamic stability of a viscoelastic orthotropic cylindrical shell. — Polym. Mech., vol. 9, N 4, 1973, p. 626–632 (Transl. from Russian).
47. *Bogdanovich A. E.* A Survey of papers on stability of cylindrical shells under axial dynamic compression. Pt. I. Early papers and basic qualitative results; Pt. 2. Recent papers and state-of-the-art. — In: Electrodynamics and mechanics of continuum. Riga, 1980, p. 68–87; 88–105.
48. *Bogdanovich A. E.* On the calculation of layer-by-layer failure of composite cylindrical shells under dynamic loading. — In: Electrodynamics and mechanics of continuum: The use of numerical methods. Riga, 1981, p. 97–102.
49. *Bogdanovich A. E.* Calculation of transverse and interply stresses in cylindrical shells made of laminated composites in dynamic compressive loading. — Mech. Compos. Mater., vol. 21, N 3, 1985, p. 306–315 (Transl. from Russian).
50. *Bogdanovich A. E., Stol'arova L. A.* Some specific features of dynamic instability regions' spectra for composite cylindrical shells. — In: Problems of electrodynamics and mechanics of continuum. Riga, 1978, Issue 4, p. 80–96.
51. *Bogdanovich A. E., Stol'arova L. A.* Optimization of multilayer orthotropic cylindrical shells under parametric vibrations. — In: Conference of young scientists on the problem of "optimization of structures under dynamic loads". Tartu, 1979, p. 31–34.
52. *Bogdanovich A. E., Stolyarova L. A.* Influence of the boundary conditions on the frequency of natural vibrations of composite cylindrical shells with a filler. — Mech. Compos. Mater., vol. 16, N 1, 1980, p. 57–65 (Transl. from Russian).

53. *Bogdanovich A. E., Feldmane E. G.* On the calculation of non-linear parametric vibrations of cylindrical shells. — Trans. USSR Acad. Sci., Mech. Solids, 1979, N 1, p. 171–177.
54. *Bogdanovich A. E., Feldmane E. G.* Non-linear parametric vibrations of viscoelastic orthotropic cylindrical shells. — Appl. Mech., 1980, vol. 16, N 4, p. 49–55.
55. *Bogdanovich A. E., Feldmane E. G.* Calculation of the load-carrying capacity of composite cylindrical shells under dynamic loading. — Mech. Compos. Mater., vol. 16, N 3, 1980, p. 341–352 (Transl. from Russian).
56. *Bogdanovich A. E., Feldmane E. G.* Deformation of cylindrical composite shells under combined dynamic loading. — Mech. Compos. Mater., vol. 17, N 3, 1981, p. 309–320 (Transl. from Russian).
57. *Bogdanovich A. E., Feldmane E. G.* The analysis of non-axisymmetric buckling of cylindrical shells under axial dynamic compression. — Trans. USSR Acad. Sci., Mechanics of Solids, 1982, N 2, p. 144–154.
58. *Bogdanovich A. E., Feldmane E. G.* Strength and axisymmetric deformation of laminate cylindrical shells under axial impact. — Mech. Compos. Mater., vol. 18, N 4, 1982, p. 449–456 (Transl. from Russian).
59. *Bogdanovich A. E., Feldmane E. G.* Numerical investigation of the bulging process and an analysis of the strength of laminated cylindrical shells under axial impact loads. — Mech. Compos. Mater., vol. 18, N 5, 1982, p. 550–558 (Transl. from Russian).
60. *Bogdanovich A. E., Feldmane E. G.* On the buckling of cylindrical shells under dynamic lateral pressure. — In: Electrodynamics and mechanics of continuum: Mathematical modelling. Riga, 1982, p. 109–122.
61. *Bogdanovich A. E., Yushanov S. P.* Analysis of the buckling of cylindrical shells with a random field of initial imperfections under axial dynamic compression. — Mech. Compos. Mater., vol. 17, N 5, 1981, p. 552–560 (Transl. from Russian).
62. *Bogdanovich A. E., Yushanov S. P.* On the anisotropic shell reliability problem solved by means of stochastic vector field excursion technique. — Mech. Compos. Mater., 1983, N 1, p. 80–89 (in Russian).
63. *Bogdanovich A. E., Yarve E. V.* Analysis of stress in a multilayer beam in lateral dynamic bending. — Mech. Compos. Mater., 1983, N 5, p. 824–837 (in Russian).
64. *Bolotin V. V.* On the lateral vibrations of bars caused by periodical longitudinal forces. — In: Lateral vibrations and critical velocities. Moscow, 1951, N 1, p. 46–77.
65. *Bolotin V. V.* Dynamic stability of elastic systems. Moscow, Gostekhizdat Publ., 1956, 600 p.
66. *Bolotin V. V.* Some new problems in the dynamics of shells. — In: Strength calculations. Moscow, Mashgiz Publ., 1959, Issue 4, p. 331–364.
67. *Bolotin V. V.* Non-conservative problems in the theory of elastic stability. Moscow, Fizmatgiz Publ., 1961, 340 p.
68. *Bolotin V. V.* Modern tendencies in dynamics of plates and shells. — In: Theory of Plates and Shells. Kiev, 1962, p. 16–32.
69. *Bolotin V. V.* Problems in the mechanics of reinforced media. — In: Rep. Scientific-Technical Conf. Moscow Inst. Energetics. Section of Energetic Machine Design. Moscow, 1965, p. 5–42.
70. *Bolotin V. V.* On the equations of stability for thin elastic shells. Eng. J., Mech. Solids, 1967, N 4, p. 117–128.
71. *Bolotin V. V.* Effect of processing factors on mechanical safety of composite structures. — Polym. Mech., 1972, N 3, p. 529–540 (in Russian).
72. *Bolotin V. V.* Dynamic stability of distributed systems. — In: Vibrations in engineering. Moscow, Mashinostroyeniye Publ., 1978, vol. 1, p. 240–256.
73. *Bolotin V. V.* Defects of the delamination type in composite structures. — Mech. Compos. Mater., vol. 20, N 2, 1984, p. 173–188 (Transl. from Russian).

74. *Bolotin V. V., Boichenko G. A.* A study of buckling of thin elastic shells under the effect of dynamic loadings. — In: Strength calculations. Moscow, 1960, Issue 5, p. 259–272.
75. *Bolotin V. V., Boichenko G. A., Makarov B. P. et al.* On the loss of stability of thin elastic shells under the effect of impulse loading. — Struct. Mech. Design Struct., 1959, N 2, p. 9–16.
76. *Bolotin V. V., Novichkov Yu. N.* Mechanics of multilayer structures. Moscow, Mashinostroyeniye Publ., 1980, 376 p.
77. *Bondarenko A. A., Galaka P. I.* Parametric instability of GFRP cylindrical shells. — Appl. Mech., 1977, vol. 13, N 4, p. 124–128.
78. *Bondarenko A. A., Telalov A. I.* Dynamic instability of cylindrical shells under longitudinal kinematic disturbance. — Appl. Mech., 1982, vol. 18, N 1, p. 57–61.
79. *Bondarenko G. V.* Hill's equation and its application to the technical vibrations. Moscow, 1936.
80. *Borisenko V. I.* On the stability of cylindrical shells under longitudinal impact. — Appl. Mech., 1965, vol. 1, N 5, p. 100–104.
81. *Borisenko V. I.* The effect of boundary conditions on stability of a cylindrical shell under longitudinal impact. — Appl. Mech., 1965, vol. 1, N 8, p. 57–62.
82. *Borisenko V. I., Voloshin V. T.* Experimental study of stability of cylindrical shells under longitudinal impact. — Appl. Mech., 1967, vol. 3, N 4, p. 45–52.
83. *Borisenko V. I., Klokova A. I.* Post-critical deformation of a cylindrical shell under impact loading. — Appl. Mech., 1966, vol. 2, N 10, p. 29–35.
84. *Brauns J. A., Rikards R. B.* Investigation of the initial imperfections and buckling modes of glass-reinforced plastic shells under hydrostatic pressure. — Polym. Mech., vol. 7, N 6, 1971, p. 940–945 (Transl. from Russian).
85. *Brauns J. A., Rikards R. B., Teters G. A.* A study of initial imperfections and buckling modes of GFRP shells under long-term loading. — In: Theory of shells and plates. Moscow, Nauka Publ., 1973, p. 99–104.
86. *Brizgalin G. I.* On the creep calculation of GFRP plates. — J. Appl. Mech. Tech. Phys., 1963, N 4, p. 132–136.
87. *Bushtyrkov A. A., Naida A. A.* Behavior of glass-reinforced plastic cylindrical shells subjected to an external pressure pulse. — Polym. Mech., vol. 8, N 2, 1972, p. 288–291 (Transl. from Russian).
88. *Buyakov I. A.* Non-linear equations of a Timoshenko-type theory of laminated anisotropic shells. — Mech. Compos. Mater., vol. 15, N 3, 1979, p. 292–296 (Transl. from Russian).
89. *Buyakov I. A.* On the influence of deformation in the normal direction in the non-linear Timoshenko-type theory of multilayer shells. — Mech. Compos. Mater., 1980, N 2, p. 358–359 (in Russian).
90. *Van Fo Fi G. A.* A theory of reinforced materials. Kiev, Naukova Dumka Publ., 1971, 232 p.
91. *Van Fo Fi G. A.* Structures of reinforced plastics. Kiev, Tekhnika Publ., 1971, 220 p.
92. *Vanin G. A., Semenyuk N. P., Yemel'yanov R. F.* Stability of shells made of reinforced materials. Kiev, Naukova Dumka Publ., 1978, 212 p.
93. *Vicario A., Toland R.* Strength criteria and failure analysis for structures of composite materials. — In: Composite Materials, Moscow, Nauka Publ., 1978, vol. 7, Pt. 1, p. 62–107.
94. *Vlasov V. Z.* A general theory for shells and its application in engineering. Moscow-Leningrad, Gostekhizdat Publ., 1949, 784 p.
95. *Vol'mir A. S.* On the stability of cylindrical shells under dynamic loading. — Rep. USSR Acad. Sci., 1958, vol. 123, N 5, p. 806–808.
96. *Vol'mir A. S.* On stability criteria for shells. — In: Design of spatial structures. Moscow, Stroyizdat Publ., 1964, Issue 9, p. 117–129.

97. *Vol'mir A. S.* Stability of deformable systems. Moscow, Nauka Publ., 1967, 984 p.
98. *Vol'mir A. S.* Actual problems of stability and dynamics of shells. — Struct. Mech. Design Struct., 1970, N 2, p. 32–37.
99. *Vol'mir A. S.* Non-linear problems in dynamics of shells. — In: Strength calculations. Moscow, Mashinostroyeniye Publ., 1971, Issue 15, p. 291–311.
100. *Vol'mir A. S.* Non-linear dynamics of plates and shells. Moscow, Nauka Publ., 1972, 432 p.
101. *Vol'mir A. S., Kil'dibekov I. G.* On the study of behavior of shells and plates under impact loading. — Trans. Higher Educ. Inst. Mech. Eng., 1964, N 7, p. 26–30.
102. *Vol'mir A. S., Mineev V. E.* Experimental investigation of buckling process of a shell under dynamic loading. — Rep. USSR Acad. Sci., 1959, vol. 125, N 5, p. 1002–1003.
103. *Vol'mir A. S., Ponomarev A. T.* Non-linear parametric vibrations of cylindrical shells made from composite materials. — Polym. Mech., vol. 9, N 3, 1973, p. 471–477 (Transl. from Russian).
104. *Vol'mir A. S., Smetanina L. N.* Investigation of dynamic stability of GFRP shells. — Polym. Mech., 1968, N 1, p. 109–115 (in Russian).
105. *Vol'mir A. S., Smetanina L. N.* On stability of a cylindrical orthotropic shell under longitudinal impact. — Rep. USSR Acad. Sci., 1970, vol. 193, N 2, p. 306–308.
106. *Vorovich I. I.* About Bubnov-Galerkin's method in the non-linear theory of vibrations of shallow shells. — Rep. USSR Acad. Sci., 1956, vol. 110, N 5, p. 723–726.
107. *Vorovich I. I., Shlenev M. A.* Plates and Shells. — In: Mechanics. 1963. Scientific Totals, Moscow, 1965, p. 91–177.
108. *Wu E. M.* Phenomenological failure criteria for anisotropic media. — In: Mechanics of composite materials. Moscow 1978, Vol 2, p. 401–91.
109. *Gavrilov Yu. V.* Determination of natural frequencies of elastic circular cylindrical shells. — Trans. USSR Acad. Sci., Department of Technical Sciences, Mechanics and Mechanical Engineering, 1961, N 1, p. 163–166.
110. *Galaka, P. I., Koval'chuk P. S., Mendelutsa V. M. et al.* Experimental investigation of non-linear interaction of flexural forms of vibrations in cylindrical shells under parametric disturbance. — In: Problems of mechanical engineering. Kiev, 1981, Issue 13, p. 33–38.
111. *Galiev Sh. U., Romashchenko V. A.* Unsteady dynamics and strength of hollow viscoelastic anisotropic multilayered cylinders of finite length. — Mech. Compos. Mater. vol. 20, N 4, 1984, p. 480–484 (Transl. from Russian).
112. *Galiev Sh. U., Romashchenko V. A., Alpaidze E. G.* The effect of anisotropy and viscosity on wave propagation in multilayer cylinders. — Probl. Strength, 1983, N 9, p. 40–44.
113. *Galimov K. Z.* Fundamentals of the non-linear theory of thin shells. Kazan', 1975, p. 328.
114. *Galimov K. Z.* Theory of shells with allowance for transverse shears. Kazan', 1977, p. 212.
115. *Galinsh A. K.* Calculation of plates and shells according to refined theories. — In: Investigations in theory of plates and shells. Kazan', 1970, Issue 6–7, p. 23–64.
116. Geisenblasen R. E. The effect of initial deviation on natural frequencies and dynamic stability of closed circular cylindrical shells. — In: Trans. Dnepropetrovsk Inst. Eng. Railway Transp. Dnepropetrovsk, 1968, Issue 73, p. 72–82.
117. *Gershtein M. S.* Geometrically non-linear equations of motion of an elastic sandwich shell. - Polym. Mech., vol. 9, N 5, 1973, p. 785–790 (Transl. from Russian).
118. *Gershtein M. S.* On one variant of the non-linear dynamic theory for thin multilayer shells. — Appl. Math. Mech., 1976, vol. 40, Issue 1, p. 180–185.
119. *Gnuni V. Ts.* On the theory of dynamic stability of laminated anisotropic shallow shells. — Trans. Armenian SSR Acad. Sci., Ser. Phys.-Math. Sci., 1960, vol. 13, N 1,

p. 47–58.

120. *Gnuni V. Ts.* On the theory of non-linear dynamic stability of shells. — Trans. USSR Acad. Sci., Department of Technical Sciences, Mechanics and Mechanical Engineering, 1961, N 4, p. 181–182.
121. *Gnuni V. Ts.* On the boundaries of dynamic instability regions of shells. — In: Proc. Conf. Theory of Plates and Shells, 1960. Kazan', 1961, p. 117–123.
122. *Gnuni V. Ts.* About parametrically induced vibrations in laminated anisotropic flexible shells. — Trans. Armenian SSR Acad. Sci., Ser. Phys.-Math. Sci., 1962, vol. 15, N 3, p. 29–36.
123. *Gol'denveiser A. L.* A theory of elastic thin-wall shells. Moscow, Gostekhizdat Publ., 1953, 544 p.
124. *Gol'denveiser A. L., Lidskii V. B., Tovstik P. E.* Free vibrations of thin elastic shells. Moscow, Nauka Publ., 1979, 384 p.
125. *Gol'dsmith V.* Impact. Moscow, Stroyizdat Publ., 1965, 448 p.
126. *Gontkevich V. S.* Natural vibrations of orthotropic cylindrical shells. — In: Proc. of the Conf. on the Theory of Plates and Shells, 1960. Kazan', 1961, p. 124–129.
127. *Gontkevich V. S.* Free vibrations of plates and shells. Kiev, Naukova Dumka Publ., 1964, 288 p.
128. *Gordienko B. A.* Experimental investigation of the behavior of bars and cylindrical shells under impact. — In: Proc. of the VIIth All-Union Conf. on the Theory of Plates and Shells. Moscow, Nauka Publ., 1970, p. 190–193.
129. *Gordienko B. A.* About computer-assisted solution of impact buckling problems for elastic systems by the use of finite differences technique. — Trans. USSR Acad. Sci., Mech. Solids, 1970, N 3, p. 143–148.
130. *Gordienko B. A.* Dynamics of orthotropic cylindrical shells under an axial impact. — Polym. Mech., vol. 13, N 5, p. 749–752, 1977 (Transl. from Russian).
131. *Gordienko B. A., Nechiporuk G. S., Ten En So.* Response of cylindrical and conical shells to axial impact. — In: Theory of shells and plates. Moscow, Nauka Publ., 1973, p. 431–436.
132. *Grogolyuk E. I.* On the strength and stability of cylindrical bimetallic shells. — Eng. Collect., 1953, N 16, p. 119–148.
133. *Grigolyuk E. I.* On the vibrations of a circular cylindrical panel with finite deflections. — Appl. Math. Mech., 1955, vol. 19, N 3, p. 376–382.
134. *Grigolyuk E. I., Kabanov V. V.* Stability of circular cylindrical shells. — In: Mechanics of rigid deformable bodies, 1967, Scientific Totals. Moscow, 1969, 348 p.
135. *Grigolyuk E. I., Kabanov V. V.* Stability of shells. Moscow, Nauka Publ., 1978, 360 p.
136. *Grigolyuk E. I., Kogan F. A.* The state-of-the-art in a theory of multilayer shells. — Appl. Mech., 1972, vol. 8, N 6, p. 3–17.
137. *Grigolyuk E. I., Mamai V. I.* About one variant of equations in a theory of finite displacements of inshallow shells. — Appl. Mech., 1974, vol. 10, N 2, p. 3–13.
138. *Grigolyuk E. I., Selezov I. T.* Nonclassical theories for vibrations of bars, plates and shells. — In: Mechanics of rigid deformable body, totals of science and technology. Moscow, 1973, Issue 5, 271 p.
139. *Grigolyuk E. I., Srebovskii A. I.* Thin circular cylindrical shells under the effect of external pressure impulse. — Eng. J. Mech. Solids, 1968, N 3, p. 110–118.
140. *Grigolyuk E. I., Chulkov P. P.* On the general theory of three-ply shells with large deflection. — Rep. USSR Acad. Sci., 1963, vol. 150, N 5, p. 1012–1015.
141. *Grigolyuk E. I., Chulkov P. P.* A theory of visco-elastic multilayer shells containing a solid core at finite deflections. — J. Appl. Mech. Tech. Phys., 1964, N 5, p. 109–117.
142. *Grigolyuk E. I., Chulkov P. P.* Non-linear equations for thin elastic laminated anisotropic shallow shells with a solid core. — Trans. USSR Acad. Sci., Mechanics, 1965, N 5, p. 68–80.
143. *Grigorenko Ya. M., Vasilenko A. T.* A theory of shells of varying stiffness. Kiev,

Naukova Dumka Publ., 1981, 544 p.
144. *Guz' A. N., Zarutskii V. A. et al.* Experimental investigations of thin-wall structures. Kiev, Naukova Dumka Publ., 1984, 240 p.
145. *Guz' A. N., Khoroshun L. P., Vanin G. A. et al.* Mechanics of composite materials and structural elements. Kiev, Naukova Dumka Publ., 1982, vol. 1, 368 p.
146. *Guz' A. N., Grigorenko Ya. M., Babich I. Yu. et al.* Mechanics of composite materials and structural elements. Kiev, Naukova Dumka Publ., 1983, vol. 2, 464 p.
147. *Guz' A. N., Ignatov I. V., Girchenko A. G. et al.* Mechanics of composite materials and structural elements. Kiev, Naukova Dumka Publ., 1983, vol. 3, 264 p.
148. *Darevskii V. M.* Non-linear equations of the theory of shells and their linearization in stability problems. — In: Proc. VIth All-Union Conf. on the Theory of Shells and Plates. Moscow, Nauka Publ., 1966, p. 355–368.
149. Darevskii V. M. Stability of a shell under dynamic loading. — In: Problems of mechanics of rigid deformable bodies. Leningrad, Sudostroyeniye Publ., 1970, p. 149–160.
150. *Darevskii V. M.* Stability of a cylindrical shell under axial dynamic loading. — Trans. USSR Acad. Sci., Mech. Solids, 1973, N 2, p. 162–172.
151. *Darevskii V. M.* Stability criterion of a momentary state of an elastic body under dynamic loading. — In: Proc. XIIIth All-Union Conf. Theory of Plates and Shells. Tallin, 1983, Pt. 2, p. 47–52.
152. *Dzhanelidze G. Yu.* Stability of elastic systems under dynamic loadings. — In: Stability problems in structural mechanics. Moscow, Stroyizdat Publ., 1965, p. 68–84.
153. Dudka K. K., Preobrazhenskii I. N., Shestakova A. S. One of the approaches to evaluation of CFRP impact resistance. — Mech. Compos. Mater., 1983, N 4, p. 624–628 (in Russian).
154. *Dudchenko A. A., Obraztsov I. F., Luriye S. A.* Anisotropic multilayer plates and shells. — In: Mechanics of a rigid deformable body. Moscow, 1983, vol. 15, p. 3–68.
155. *Ermolenko A. F., Protasov V. D.* Strength of laminated cylindrical shells upon nonproportional loading. — Mech. Compos. Mater., vol. 17, N 6, 1981, p. 672–677 (Transl. from Russian).
156. *Efimov A. B., Malii V. I.* On the buckling mechanism in a cylindrical shell under longitudinal impact. — In: Theory of shells and plates. Moscow, Nauka Publ., 1973, p. 459–463.
157. *Efimov A. B., Malii V. I., Uteshev S. A.* On the instability of a cylindrical shell under longitudinal impact. — Trans. USSR Acad. Sci., Mech. Solids, 1971, N 1, p. 20–23.
158. *Zinov'ev P. A., Tarakanov A. I.* Fracture conditions of laminated composite materials. — In: Application of plastics in mechanical engineering. Moscow, 1976, Issue 15, p. 63–68.
159. *Zinov'ev P. A., Tarakanov A. I., Fomin B. Ya.* Non-linear deformation of cylindrical shells made of reinforced materials. — In: Proc. XIIth All-Union Conf. Theory of Shells and Plates. Erevan, 1980, vol. 2, p. 157–163.
160. *Kabanov V. V.* Equations of Thin Shells in Highly Inhomogeneous Stress–Strain States. — Trans. USSR Acad. Sci., Mech. Solids, 1979, N 4, p. 155–161.
161. *Kabanov V. V.* Stability of inhomogeneous cylindrical shells. Moscow, Mashinostroyeniye Publ., 1982, 253 p.
162. *Kadashevich Yu. P., Pertsev A. K.* On the loss of stability of a cylindrical shell under dynamic loading. — Trans. USSR Acad. Sci., Department of Technical Sciences. Mechanics and Mechanical Engineering, 1960, N 3, p. 30–33.
163. *Karmishin A. V., Liyuskovets V. A., Myachenkov V. I., Frolov A. N.* Statics and dynamics of thin-wall shell-type structures. Moscow, Mashinostroyeniye Publ., 1975, 376 p.
164. *Karmishin A. V., Skurlatov E. D., Startsev V. T., Fel'dshtein V. A.* Nonstationary aeroelasticity of thin-wall structures. Moscow, Mashinostroyeniye Publ., 1982, 240 p.

165. *Kiiko I. A.* A cylindrical shell subjected to axial impact load. — Trans. USSR Acad. Sci., Mech. Solids, 1969, N 2, p. 135–138.
166. *Kiiko I. A.* Longitudinal impact effect on a thin cylindrical shell. – Moscow State Univ. Bull., Math., Mech., 1972, N 3, p. 118–121.
167. *Kiladze B. A., Preobrazhenskii I. N., Ckhvediani A. Sh.* Vibrations of a multilayer cylindrical panel with anisotropic layers at large deflections. — Mech. Compos. Mater., 1982, N 6, p. 1014–1020 (in Russian).
168. *Kovalchuk P. S., Krasnopol'skaya T. S., Podchasov N. P.* On dynamic instability of circular cylindrical shells with an initial deflection. — Appl. Mech., 1982, vol. 18, N 3, p. 28–33.
169. *Kononenko V. O., Telalov A. I.* Investigation of GFRP shell vibrations. — In: Reinforced materials and their structures. Kiev, Naukova Dumka Publ., 1970, p. 51–66.
170. *Coppa A. O.* On the mechanism of buckling of a circular cylindrical shell under longitudinal impact. — Mechanics, 1961, N 6, p. 145–164.
171. *Kornev V. M.* On the modes of instability in elastic shells under intensive loading. — Trans. USSR Acad. Sci., Mech. Solids, 1969, N 2, p. 129–135.
172. *Kornev V. M., Solodovnikov V. N.* Axisymmetric modes of instability in an elastic cylindrical shell under impact. — J. Appl. Mech. Tech. Phys., 1972, N 2, p. 95–100.
173. *Korolev V. I.* Laminated anisotropic plates and shells made of reinforced plastics. Moscow, Mashinostroyeniye Publ., 1965, 272 p.
174. *Krilov N. M., Bogol'ubov N. N.* A study of resonance phenomena at transverse vibrations under the effect of periodic normal forces applied at one end of a bar. — In: Studies of vibrations in structures. Kharkov-Kiev, Department of Scientific-Technical Information, 1935, p. 25–42.
175. *Kulikov G. M.* On the theory of multilayer shallow shells at finite deflection. — Trans. USSR Acad. Sci., Mech. Solids, 1979, N 3, p. 188–192.
176. *Kuptsov V. I.* On natural transverse vibrations of cantilever orthotropic cylindrical shells. — Appl. Mech., 1977, vol. 13, N 4, p. 38–44.
177. *Kuptsov V. I.* Assessment of the effect of some properties of a composite material on the dynamic characteristics of cylindrical shells. — Appl. Mech., 1979, vol. 15, N 6, p. 123–126.
178. *Lavrent'ev M. A., Ishlinskii A. Yu.* Dynamic modes of instability in elastic systems. — Rep. USSR Acad. Sci., 1949, vol. 64, N 6, p. 779–782.
179. *Lekhnitskii S. G.* Bending of inhomogeneous anisotropic thin plates of symmetric structure. — Appl. Math. Mech., 1941, vol. 5, Issue 1, p. 71–92.
180. *Lekhnitskii S. G.* Anisotropic plates. 2nd edn. Moscow, Gostekhizdat Publ., 1957, 463 p.
181. *Luriye A. I.* A general theory of elastic thin shells. — Appl. Math. Mech., 1940, vol. 4, N 2, p. 7–34.
182. *Love A.* A mathematical theory of elasticity. (Transl. from English), Moscow-Leningrad, Department of Scientific-Technical Information, 1935, 676 p.
183. *McLachlan N. V.* A theory and applications of Mathieu's functions. Moscow, Foreign Literature Publ. House, 1953, 475 p.
184. *Malmeister A. K.* Geometry of theories of strength. — Polym. Mech., vol. 2, N 4, 1966, p. 324–331 (Transl. from Russian).
185. *Malmeister A. K., Tamuzh V. P., Teters G. A.* Strength of polymer and composite materials. Riga, Zinatne Publ., 1980, 572 p.
186. *Markov A. N.* Dynamic stability of anisotropic cylindrical shells. — Appl. Math. Mech., 1949, vol. 13, N 2, p. 145–150.
187. *Matyash V. I.* On the dynamic stability of simply supported viscoelastic column. — Polym. Mech., 1971, N 2, p. 293–300 (in Russian).
188. *Mikaelyan G. Z.* Dynamic stability of a multilayer orthotropic circular cylindrical

shell. — Trans. Armenian SSR Acad. Sci., Mechanics, 1968, vol. 21, N 2, p. 42–52.

189. *Mineev V. E.* Stability study of closed cylindrical shells under the effect of all-round compression. — In: Investigations in the theory of plates and shells. Kazan', 1970, Issue 6–7, p. 596–623.

190. *Mishenkov G. V.* On dynamic stability of shallow elastic shells. – Eng. J., 1961, vol. 1, N 2, p. 112–118.

191. *Mishenkov G. V.* On dynamic stability of a shallow cylindrical shell. — In: Proc. Conf. Theory of Plates and Shells, 1960. Kazan', 1961, p. 239–245.

192. *Movsisyan L. A.* On the instability of cylindrical shells under longitudinal impact. — Trans. Armenian SSR Acad. Sci., Ser. Phys.-Math. Sci., 1964, vol. 17, N 6, p. 57–64.

193. *Mushtari H. M.* Some generalizations of the theory of thin shells with the applications to the problems of stability in elastic equilibrium state. — Trans. Phys.-Math. Soc. Kazan' University, 1938, Ser. 3, No. 11, p. 71–150.

194. *Mushtari H. M.* A theory of elastic equilibrium for plates and shells taking into account initial stresses. — Trans. Kazan' Phys.-Tech. Inst. USSR Acad. Sci., Ser. Phys.-Math. Tech. Sci., 1950, N 2, p. 39–52.

195. *Mushtari H. M.* On the elastic equilibrium of a thin shell with initial imperfections in mid-surface. — Appl. Math. Mech., 1951, vol. 15, N 6, p. 743–750.

196. *Mushtari H. M., Galimov K. Z.* A non-linear theory of elastic shells. Kazan', Tatknigoizdat Publ., 1957, 432 p.

197. *Mushtari H. M., Teregulov I. G.* A theory of shallow orthotropic shells of medium thickness. — Trans. USSR Acad. Sci., Department of Technical Sciences. Mechanics and Mechanical Engineering, 1959, N 6, p. 60–67.

198. *Naida A. A.* Behavior of GFRP and metallic cylindrical shells under the same dynamic loading. — Appl. Mech., 1973, vol. 9, N 12, p. 108–110.

199. *Nemirovskii Yu. V.* Stability and buckling of structurally anisotropic and inhomogeneous shells and plates. — In: Mechanics of rigid deformable bodies. Moscow, 1976, vol. 9, p. 5–154.

200. *Nemirovskii Yu. V., Samsonov V. I.* The method of optimization of reinforced laminated and stiffened cylindrical shells at vibrations and stability. — In: Numerical methods of solving problems of the theory of elasticity and plasticity. — Proc. VIIth All-Union Conf., Novosibirsk, 1982, p. 89–96.

201. *Nechiporuk G. S., Ten En So.* An experimental study of buckling of cylindrical and conical shells under impact. — Trans. USSR Acad. Sci., Mech. Solids, 1974, N 3, p. 175–182.

202. *Nechiporuk G. S., Ten En So.* On several models of equations of motion in solving the problems of shell buckling under impact. — In: Dynamics and strength of structures. Novosibirsk, 1976, p. 127–140.

203. *Nigul U. K.* Comparison of the results of an analysis of transient wave processes in shells and plates in calculated accordance with the theory of elasticity and approximate theories. — Appl. Math. Mech., 1969, vol. 33, N 2, p. 308–322.

204. *Nigul U. K.* Wave deformation processes of shells and plates. — In: Proc. VIIth All-Union Conf. Theory of Shells and Plates. Moscow, Nauka Publ., 1970, p. 846–883.

205. *Nikulin M. V.* Natural vibrations of unstiffened and structurally anisotropic cylindrical shells under static loads. — In: Strength and dynamics of aircraft engines. Moscow, Mashinostroyeniye Publ., 1965, Issue 2, p. 52–128.

206. *Novichkov Yu. N.* Wave propagation in laminated cylindrical shells. — Trans. USSR Acad. Sci., Mech. Solids, 1973, N 2, p. 51–60.

207. *Novichkov Yu. N.* The non-linear theory and stability of thick multilayer shells. — Appl. Math. Mech., 1973, vol. 37, N 3, p. 532–543.

208. *Novozhilov V. V.* Fundamentals of the non-linear theory of elasticity. Moscow-Leningrad, Gostekhizdat Publ., 1948, 212 p.

209. *Novozhilov V. V.* A theory of thin shells. Leningrad, Sudpromgiz Publ., 1951, 344 p.

210. *Novozhilov V. V.* A theory of elasticity. Leningrad, Sudpromgiz Publ., 1958, 370 p.
211. *Novozhilov V. V.* A short essay of the development of a shell theory in the USSR. — In: Studies in the theory of plates and shells. Kazan', 1970, Issue 6–7, p. 3–22.
212. *Obraztsov I. F.* About some promising applied problems in mechanics having significance for national economy. — Trans. USSR Acad. Sci., Mech. Solids, 1982, N 4, p. 3–9.
213. *Obraztsov I. F., Vasil'ev V. V., Bunakov V. A.* Optimal reinforcement of shells of revolution made of composite materials. Moscow, Mashinostroyeniye Publ., 1977, 144 p.
214. *Pisarenko G. S., Tchemeris A. N.* On the problem of dynamic stability of a cylindrical shell. — In: Energy dissipation upon vibrations of mechanical systems. Kiev, Naukova Dumka Publ., 1968, p. 107–114.
215. *Protasov V. D., Ermolenko A. F.* Strength problems of wound composite shell structures. — Mech. Compos. Mater., 1983, N 6, p. 1034–1043 (in Russian).
216. *Protasov V. D., Ermolenko A. F., Filipenko A. A., Dimitrienko I. P.* A study of load carrying capacity of laminated cylindrical shells by computer assisted modelling of the failure process. — Mech. Compos. Mater., 1980, N 2, p. 254–261 (in Russian).
217. *Protsenko O. P.* On stability of a cylindrical shell with initial deflection under the effect of aperiodic axial compressive forces. — Appl. Mech., 1965, vol. 1, N 3, p. 27–34.
218. *Protsenko O. P. Pavlovskii V. S.* Dynamic stability of a closed cylindrical shell under combined aperiodic in time loading. — In: Stability problems in structural mechanics, Moscow, Stroyizdat Publ., 1965, p. 296–305.
219. *Prusakov A. P.* Finite deflections in multilayer shallow shells. — Trans. USSR Acad. Sci., Mech. Solids, 1971, N 3, p. 119–125.
220. *Rabotnov Yu. N.* Mechanics of composites. — Bull. USSR Acad. Sci., 1979, N 5, p. 50–58.
221. *Reznikov B. S.* A study of shear failure in reinforced cylindrical shells. — Struct. Mech. Design Struct., 1979, N 3, p. 9–14.
222. *Reznikov E. S.* Initial failure of cylindrical ribbed shells made of reinforced materials. — Mech. Compos. Mater., 1979, N 6, p. 1031–1035 (in Russian).
223. *Rikards R. B., Teters G. A.* Buckling modes of composite cylindrical shells under long-term loading. — Polym. Mech., 1971, N 4, p. 697–703 (in Russian).
224. *Rikards R. B., Teters G. A.* Stability of composite shells. Riga, Zinatne Publ., 1974, 310 p.
225. *Rosato D. V., Grove C. S.* Filament winding. Moscow, Mashinostroyeniye Publ., 1969, 309 p. (Transl. from English).
226. *Rowlands R.* Flow and loss in load carrying capacity of composites in the state of biaxial stress: comparison of theory and experimental date. — In: Inelastic properties of composite materials. Moscow, Mir Publ., 1978, p. 140–179.
227. *Ryzhanskii V. A., Mineev V. N., Ivanov A. G. et al.* Failure of water-filled cylindrical glass-reinforced epoxy shells under internal impulsive loading. — Polym. Mech., vol. 14, N 2, 1978, p. 227–233.
228. *Sabirova R. S.* Natural vibrations of an anisotropic cylindrical shell. — In: Investigations in the theory of plates and shells. Kazan', 1967, Issue 5, p. 424–432.
229. *Samsonov V. I.* Optimization of natural frequencies of a reinforced cylindrical shell. — In: Dynamics of continuum. Novosibirsk, 1979, N 43, p. 173–177.
230. *Slepov B. I.* Dynamic stability of a circular cylindrical shell under the effect of a shock-wave loading. — In: Proc. Conf. Theory of Plates and Shells, 1960. Kazan', 1961, p. 353–357.
231. *Smith D. G., Huang Ju-Chin.* Post-crazing analysis of glass-epoxy laminates. — In: Strength and fracture of composite materials. Riga, Zinatne Publ., 1983, p. 168–174 (Transl. from English).

232. *Spiro V. E.* Variant of the geometrically non-linear theory of anisotropic shells with allowance for transverse shear. — Polym. Mech., vol. 5, N 5, 1969, p. 766–772 (Transl. from Russian).
233. *Spiro V. E.* The simplest version of the geometrically non-linear theory of orthotropic shells of revolution with allowance for transverse shear and the area of its application. — In: Problems in structural mechanics for a ship. Leningrad, 1973, p. 223–242.
234. *Starzhinskii V. M.* A survey of papers on the stability conditions for a trivial solution of the system of linear differential equations with periodic coefficients. — Appl. Math. Mech., 1954, vol. 18, N 4, p. 469–510.
235. *Strett M. D. O.* Lame, Mathieu and their related functions in physics and engineering. Kharkov-Kiev, Department of Scientific-Technical Information, 1935, 238 p.
236. *Sukhova L. N.* Specific behavior of multilayer anisotropic cylindrical shells under axisymmetric dynamic loading. — In: Trans. N. E. Bauman Moscow Higher Technical School. Moscow, 1980, N 342, p. 31–39.
237. *Tamuzs V. P., Teters G. A.* Problems in mechanics of composite materials. — Mech. Compos. Mater., 1979, N 1, p. 34–45 (in Russian).
238. *Tarnopol'skii Yu. M.* Thick-walled wound fibrous composite structures. — Polym. Mech., 1975, N 1, p. 134–144 (in Russian).
239. *Tarnopol'skii Yu. M., Roze A. V.* Characteristics of calculation of reinforced plastic parts. Riga, Zinatne Publ., 1969, 260 p.
240. *Tarnopol'skii Yu. M., Skudra A. M.* Structural strength and deformability of GFRP. Riga, Zinatne Publ., 1966, 276 p.
241. *Tennyson R. C., Wharram G. E., Elliot G.* Application of cubic strength criterion to the failure analysis of composite materials. Riga, Zinatne Publ., 1983, p. 127–135.
242. *Tennyson R. C., MacDonald D., Nanyaro A. P.* Evaluation of the tensor polynomial failure criterion for composite materials. — Mech. Compos. Mater., 1980, N 3, p. 418–423 (in Russian).
243. *Terebushko O. I.* Stability of a cylindrical shell under high-velocity loading with axial force. — Struct. Mech. Design Struct., 1960, N 1, p. 10–12.
244. *Teters G. A., Pelekh B. L.* Creep Stability of orthotropic shells with allowance for transverse shear strains. — Polym. Mech., vol. 2, N 1, 1966, p. 65–68 (Transl. from Russian).
245. *Teters G. A., Rikards R. B., Narusberg V. L.* Optimization of shells made of laminated composites. Riga, Zinatne Publ., 1978, 240 p.
246. *Wilkinson J.* Algebraic eigenvalue problem. Moscow, Nauka Publ., 1970, 564 p.
247. *Wilkinson J., Reinsh G. A.* A handbook in algorithms in ALGOL. Linear algebra. Moscow, Mashinostroyeniye Publ., 1976, 391 p.
248. *Uteshev S. A.* Buckling of a thin cylindrical shell under longitudinal impact load. — In: Proc. 1st All-Union Symp. Impulse Pressure. Moscow, 1974, vol. 1, p. 152–159.
249. *Uteshev S. A.* Buckling of conical and cylindrical polymer shells under impact on the base. — Polym. Mech., vol. 13, N 1, 1977, p. 70–75 (Transl. from Russian).
250. *Fedorenko A. G., Tsipkin V. I., Ivanov A. G., Rusak V. N., Zaikin S. N.* Features of dynamic deformation and fracture of cylindrical shells from GFRP under internal impulsive loading. — Mech. Compos. Mater., 1983, N 1, p. 90–94 (in Russian).
251. *Feodos'ev V. I.* Geometrically non-linear problems in the theory of shells and plates. — In: Proc. VIth All-Union Conf. Theory of Shells and Plates. Moscow, Nauka Publ., 1966, p. 971–976.
252. *Filippov A. P.* Vibrations of cylindrical shells. — Appl. Math. Mech., 1937, vol. 1, Issue 2.
253. *Filippov A. P., Kokhmanyuk S. S., Yanyutin E. G.* Deformation of structural elements under the effect of impact and impulse loadings. Kiev, Naukova Dumka Publ., 1978,

184 p.

254. *Flügge V.* Statics and dynamics of shells. Moscow, Gosstroyizdat Publ., 1961, 306 p. (Transl. from German).
255. *Tsai S. W., Hahn H. T.* An analysis of composites failure. — In: Inelastic properties of composite materials, Moscow, Mir Publ., 1978, p. 104–139 (Transl. from English).
256. *Rusak V. N., Shitov A. T., Ivanov A. G., Tsipkin V. I.* Deformation and failure of glass/epoxy cylindrical shells under internal impulse loading. — Mech. Compos. Mater., 1981, N 2, p. 249–255 (in Russian).
257. *Tchemeris A. N.* A method of study the dynamic stability of cylindrical shells with allowance for damping. — In: Energy dissipation upon vibration of mechanical systems. Kiev, Naukova Dumka Publ., 1968, p. 203–205.
258. *Tchemeris A. N.* Experimental study of dynamic stability of a cylindrical shell. — Probl. Strength, 1969, N 3, p. 99–100.
259. *Chou S. C.* Delamination of T 300/5208 graphite/epoxy laminates. — In: Strength and fracture of composite materials. Riga, Zinatne Publ., 1983, p. 136–145.
260. *Shapovalov L. A.* On the simple variant of equations in geometrically non-linear theory of thin shells. — Trans. USSR Acad. Sci., Mech. Solids, 1968, N 1, p. 56–62.
261. *Shapovalov L. A.* Elasticity equations for a thin shell at nonaxisymmetric deformation. — Trans. USSR Acad. Sci., Mech. Solids, 1976, N 3, p. 62–72.
262. *Shapovalov L. A.* On the formulation of the Kirchhoff-Love's hypothesis in the non-linear theory of thin shells. — Trans. N. E. Bauman Moscow Higher Techn. School, 1980, N 342, p. 4–19.
263. *Shveiko Yu. Yu., Gavrilov Yu. V., Brusilovskii A. D.* On the effect of boundary conditions on the spectrum of natural frequencies of cylindrical shells. — In: Rep. Scientific-Technical Conf. Moscow Inst. Energetics. Section of energetic machine design. Moscow, 1965, p. 131–148.
264. *Shelud'ko G. A., Shupikov A. N.* A synthesis of optimal multilayer plates and shells under transient loading. — The USSR Acad. Sci., Inst. Probl. Mech. Eng., Kharkov, 1981, N 5428-81, 23 p.
265. *Shkutin L. I.* On the representation of equations in the non-linear theory of elastic shells. — In: Proc. VIth All-Union Conf. Theory of Shells and Plates. Moscow, Nauka Publ., 1966, p. 841–848.
266. *Shkutin L. I.* The non-linear model of a shell with deformable transverse fibers. — J. Appl. Mech. Techn. Phys., 1984, N 1, p. 168–174.
267. *Shlitsa R. P., Spridzans Yu. B.* Experimental estimation of the transverse tensile strength of carbon-reinforced plastics. — Polym. Mech., vol. 10, N 2, 1974, p. 206–210 (Transl. from Russian).
268. *Schmidt G.* Parametric vibrations. Moscow, Mir Publ., 1978, 336 p. (Transl. from German).
269. *Shtaerman I. Ya.* On the theory of symmetric deformations of anisotropic elastic shells. — Trans. Kiev Polytech. Agric. Inst., 1924, Pt. 1, Issue 1, p. 54–72.
270. *Schumik M. A.* Stability of glass fiber reinforced plastic shell under dynamic axial loads. — Polym. Mech., 1972, N 4, p. 740–743 (in Russian).
271. *Shupikov A. N.* Mass minimization in multilayer plates and shells under impulse loading conditions. — In: Problems in mechanical engineering, Kiev, 1981, Issue 13, p. 54–59.
272. *Yakubovich V. A., Starzhinskii V. M.* Linear differential equations with periodic coefficients and their applications. Moscow, Nauka Publ., 1972, 718 p.
273. *Abrahamson G. R., Lindberg H. E.* Peak load-impulse characterization of critical pulse loads in structural dynamics. — In: Dynamic response structures. Proc. symp., Stanford, Calif., 1971. New York etc., 1972, p. 31–53.
274. *Ambartsumian S. A., Bagdasarian G. E., Durgarian S. M., Gnuny V. Ts.* Some problems of vibration and stability of shells and plates. — Int. J. Solids Struct., 1966,

vol. 2, p. 59–81.

275. *Amijima S., Fujii T.* On the fracture behavior of F. R. P. subjected to the impact compression. III. The effect of anisotropic properties of F. R. P. on the characteristics of the impact compressive fracture. — Sci. Eng. Rev. Doshisha Univ., 1980, vol. 21, N 1, p. 74–94.
276. *Amijima S., Fujii T.* Compressive strength and fracture characteristics of fiber composites under impact loading. — In: Adv. in composite materials. Proc. 3rd Intern. conf. Oxford etc., 1980, vol. 1, p. 399–413.
277. *Anderson D. L., Lindberg H. E.* Dynamic pulse buckling of cylindrical shells under transient lateral pressures. — AIAA J., 1968, vol. 6, N 4, p. 589–598.
278. *Ando Y., Yagawa G., Kawai T.* Three-dimensional theory on the natural vibration of circular cylindrical shells. — Nuclear Eng. Design, 1971, vol. 15, N 2, p. 135–148.
279. *Arbocz J.* The imperfection data bank, a mean to obtain realistic buckling loads. — In: Buckling shells. Proc. state-of-the art colloq., Univ. Stuttgart, 1982. Berlin etc., 1982, p. 535–567.
280. *Arbocz J.* Shell stability analysis: theory and practice. — In: Collapse: Buckl. structure: Theory and practice symp., London, 1982. Cambridge etc., Univ. Press, 1983, p. 43–74.
281. *Arbocz J., Abramovich H.* The initial imperfection data bank at the Delft University of Technology, 1. Rep. LP-290, Dept. Aerospace Eng. Delft, 1979, Dec.
282. *Arbocz, J., Babcock C. D.* The effect of general imperfection on the buckling of cylindrical shells. — Trans. ASME. Ser. E, 1969, vol. 36, N 1, p. 28–38.
283. *Arbocz J., Babcock C. D.* Computerized stability analysis using measured initial imperfections. — In: ISAC Proc., 1980, 12th Congr. Intern. Counc. Aeron. Sci. Munich, 1980, s. l., s. a., p. 688–701.
284. *Armenakas A. E., Sciamarella C. A.* Response of glassfiber-reinforced epoxy specimens to high rates of tensile loading. — Experimental Mechanics, 1973, vol. 13, N 10, p. 433–440.
285. *Arnold R. N., Warburton G. B.* Flexural vibrations of the walls of thin cylindrical shells having freely supported ends. — Proc. R. Soc. London. Math. Phys. Sci., 1949, vol. A197, N 1049, p. 238–256.
286. *Arnold R. N., Warburton G. B.* The flexural vibrations of thin cylinders. — Proc. Inst. Mech. Eng. Ser. A, 1953, vol. 167, N 1, p. 62–74.
287. *Atluri S.* A perturbation analysis of non-linear free flexural vibrations of circular cylindrical shell. — Int. J. Solids Structures, 1972, vol. 8, N 4, p. 549–569.
288. *Baron M. L., Bleich H. H.* Tables for frequencies and modes of free vibration of infinitely long thin cylindrical shells. — Trans. ASME. J. Appl. Mech., 1954, vol. 21, N 2, p. 178–184.
289. *Bert C. W.* Vibration of composite structures. — In: Recent advances in structure dynamics. Papers Int. Conf. Southampton, 1980, vol. 2, p. 693–712.
290. *Bert C. W., Baker J. L., Egle D. M.* Free vibrations of multilayer anisotropic cylindrical shells. — J. Compos. Mater., 1969, vol. 3, N 3, p. 480–499.
291. *Bert C. W., Egle D. M.* Dynamics of composite, sandwich and stiffened shell-type structures. — J. Spacecraft Rockets, 1969, vol. 6, N 12, p. 1345–1361.
292. *Bird Y. E.* Vibrations of thick walled hollow cylinders: approximate theory. — J. Acoustical Soc. Amer., 1960, vol. 32, N 11, p. 1413–1419.
293. *Booton M., Tennyson R. C.* Buckling of imperfect anisotropic circular cylinders under combined loading. — In: AIAA/ASME 19th structures, structural dynamics and materials conference, Bethesda, 1978. New York, 1978, p. 351–358.
294. *Bradford L. G., Dong S. B.* Natural vibrations of orthotropic cylinders under initial stress. — J. Sound Vibration, 1978, vol. 60, N 2, p. 157–175.
295. *Budiansky B.* Dynamic buckling of elastic structures: criteria and estimates. — In: Dynamic stability of structures. Proc. intern. conf. Oxford etc., 1967, p. 83–106.

296. *Budiansky B., Hutchinson J. W.* Dynamic buckling of imperfect-sensitive structuers. — In: Proc. 11th Intern. congr. applied mechanics. Berlin, 1964, p. 636–651.
297. *Budiansky B., Sanders Y. L., jun.* On the "best" first order linear shell theory. — In: Progress in applied mechanics. The Prager anniversary vol. New York, Macmillan, 1963, p. 129–140.
298. *Bushnell D.* Dynamic response of two-layered cylindrical shells to time-dependent loads. — AIAA J., 1965, vol. 3, N 9, p. 1698–1703.
299. *Butcher B. R.* The impact resistance of unidirectional CFRP under tensile stress. Fibre Sci. Technol., 1979, vol. 12, N 4, p. 294–326.
300. *Cantwell W. J., Curtis P. T., Morton J.* Impact and subsequent fatigue damage growth in carbon fibre laminates. — Int. J. Fatigue, 1984, vol. 6, N 2, p. 113–118.
301. *Chen J. C., Babcock C. D.* Non-linear vibration of cylindrical shells. — AIAA J., 1975, vol. 13, N 7, p. 868–876.
302. *Chien W. Z.* The intrinsic theory of thin shells and plates. 1. General theory. 2. Application to thin plates. 3. Application to thin shells. — Q. Appl. Math., 1943, vol. 1, N 4, p. 297–327; 1944, vol. 2, N 1, p. 43–59; N 2, p. 120–135.
303. *Chu H. N.* Influence of large amplitudes on flexural vibrations of a thin cylindrical shell. — J. Aerospace Sci., 1961, vol. 28, N 8, p. 602–609.
304. *Cooper R. M., Naghdi P. M.* Propagation of nonaxially symmetric waves in elastic cylindrical shells. — J. Acoustical Soc. Amer., 1957, vol. 29, N 12, p. 1365–1373.
305. *Coppa A. P.* Measurement of initial geometrical imperfections of cylindrical shells. — AIAA J., 1966, vol. 4, N 1, p. 172–175.
306. *Daniel I. M., Labedz R. H., Liber T.* New method for testing composites at very high strain rates. — Experimental Mech., 1981, vol. 21, N 2, p. 71–77.
307. *Das Y. C.* Vibrations of orthotropic cylindrical shells. — Appl. Sci. Res., 1964, vol. A 12, p. 317–326.
308. *Dong S. B.* Free vibration of laminated orthotropic cylindrical shells. — J. Acoustical Soc. Amer., 1968, vol. 44, N 12, p. 1628–1635.
309. *Dong S. B., Selna L. G.* Natural vibrations of laminated orthotropic shells of revolution. — J. Compos. Mater., 1970, vol. 4, N 1, p. 2–19.
310. *Dong S. B., Tso F. K. W.* On a laminated orthotropic shell theory, including transverse shear deformation. — J. Appl. Mech. Trans. ASME. Ser, E, 1972, vol. 94, N 4, p. 1091–1097.
311. *Donnell L. H.* Stability of thin-walled tubes under torsion. — NASA Reports, 1933, N 479.
312. *Donnell L. H.* A new theory for the buckling of thin cylinders under axial compression and bending. — Trans. ASME, 1934, vol. 56, p. 795–806.
313. *Dowell E. H., Ventres C. S.* Modal equations for the non-linear flexural vibrations of a cylindrical shell. — Int. J. Solids Struct., 1968, vol. 4, N 10, p. 975–991.
314. *Dym C. L.* Some new results for the vibrations of circular cylinders. — J. Sound Vibration, 1973, vol. 29, N 2, p. 189–205.
315. *El-Raheb M., Babcock C. D., jun.* Some approximations in the linear dynamic equations of thin cylinders. — J. Sound Vibration, 1981, vol. 76, N 4, p. 543–559.
316. *Ertepinar A.* Large amplitude radial oscillations of layered thick walled cylindrical shells. — Int. J. Solids Struct., 1977, vol. 13, N 8, p. 717–723.
317. *Evan-Ivanowski R. M.* On the parametric response of structures. — Appl. Mech. Rev., 1965, vol. 18, N 9, p. 699–702.
318. *Evensen D. A.* Some observations on the non-linear vibration of thin cylindrical shells. — AIAA J., 1963, vol. 1, N 12, p. 2857–2858.
319. *Evensen D. A.* Non-linear vibrations of an infinitely long cylindrical shell. — AIAA J., 1968, vol. 6, N 7, p. 1401–1403.
320. *Evensen D. A., Fulton R. E.* Some studies of the non-linear dynamic response of shell-type structures. — In: Dynamic stability structures. Proc. Intern. conf./Ed. by G.

Herrman. Oxford, 1967.
321. *Flügge W.* Schwingungen zylindrischer Schalen. — Ztschr. Angew. Math. Mech., 1933, Bd 13, S. 425.
322. *Flügge W.* Stresses in shells. Berlin, Springer-Verl., 1973. — 526 p.
323. *Forsberg K.* Influence of boundary conditions on the modal characteristics of thin cylindrical shells. AIAA J., 1964, vol. 2, N 12, p. 2150–2157.
324. *Forsberg K.* Axisymmetric and beam-type vibrations of thin cylindrical shells. — AIAA J., 1969, vol. 7, N 2, p. 221–227.
325. *Fujii T., Miki M.* The studies on impact behavior of unidirectional fiber reinforced plastics. — Mem. Fac. Eng., Osaka City Univ., 1973, N 14, p. 25–35.
326. *Gazis D. C.* Three-dimensional investigation of the propagation of waves in hollow circular cylinders. 1. Analytical foundation. 2. Numerical results. — J. Acoustical Soc. Amer., 1959, vol. 31, N 5, p. 568–573; 573–578.
327. *Ginsberg J. H.* Non-linear resonant vibrations of infinitely long cylindrical shells. — AIAA J., 1972, vol. 10, p. 979–980.
328. *Ginsberg J. H.* Large amplitude forced vibrations of simply supported thin cylindrical shells. — Trans. ASME. Ser. E, 1973, vol. 40, N 2, p. 471–477.
329. *Gottenberg W. G.* Experimental study of the vibrations of a circular cylindrical shell. — J. Acoustical Soc. Amer., 1960, vol. 32, N 8, p. 1002–1006.
330. *Greenberg J. B., Stavsky Y.* Buckling and vibration of orthotropic composite cylindrical shells. — Acta Mech., 1980, vol. 36, N 1/2–3/4, p. 15–29.
331. *Greenberg J. B., Stavsky Y.* Vibrations of axially compressed laminated orthotropic cylindrical shells, including transverse shear deformation. — Acta Mech., 1980, vol. 37, N 1–2, p. 13–28.
332. *Greenberg J. B., Stavsky Y.* Vibrations of laminated filament-wound cylindrical shells. — AIAA J., 1981, vol. 19, N 8, p. 1055–1062.
333. *Greenberg J. B., Stavsky Y.* Stability and vibrations of compressed, aelotropic, composite cylindrical shells. — Trans. ASME. J. Appl. Mech., 1982, vol. 49, N 4, p. 843–848.
334. *Greenspon J.* Flexural vibrations of thick walled circular cylinders. — In: Proc. 3rd U. S. Nat. Congr. Appl. Mech. New York, 1958, p. 163–173.
335. *Greenspon J.* Vibrations of thick cylindrical shells. — J. Acoustical Soc. Amer., 1959, vol. 31, N 12, p. 1682–1683.
336. *Greenspon J.* Flexural vibrations of a thick walled circular cylinder according to the exact theory of elasticity. — J. Aero-Space Sci., 1960, vol. 27, N 1, p. 37–40.
337. *Greenspon J.* Vibrations of a thick-walled cylindrical shell — comparison of the exact theory with approximate theories. — J. Acoustical Soc. Amer., 1960, vol. 32, N 5, p. 571–578.
338. *Greif R.* Inertia effects in the dynamic response of a cylindrical shell. — AIAA J., 1966, vol. 4, N 6, p. 1105–1106.
339. *Gryboś Ryszard.* Statecznošč konstrukcji pod obciazeniem uderzoniowym. Warszawa; Poznań, Pol. Akad. Nauk, 1980. 424 s.
340. *Guess T. R.* Biaxial testing of composite cylinders: experimental — theoretical comparison. — Composites, 1980, vol. 11, N 3, p. 139–148.
341. *Hahn H. T., Tsai S. W.* On the behavior of composite laminates after initial failures. — J. Compos. Mater., 1974, vol. 8, N 3, p. 288–305.
342. *Harari A., Sandman B. E.* Experimental and theoretical dynamic analysis of carbon-graphite composite shells. — Shock Vibration Bull., 1978, vol. 48, N 3, p. 33–37.
343. *Harding J., Welsh L. M.* A tensile testing technique for fiber-reinforced composites at impact rates of strain. — J. Mater. Sci., 1983, vol. 18, N 6, p. 1810–1826.
344. *Herrmann G., Mirsky I.* Three-dimensional and shell-theory analysis of axially symmetric motions of cylinders. — Trans. ASME. Ser. E. J. Appl. Mech., 1956, vol. 23, N 4, p. 563–568.

345. *Hoff N. J.* The accuracy of Donnel's equations. — Trans. ASME. Ser. E. J. Appl. Mech., 1955, vol. 22, N 3, p. 329–334.
346. *Hoff N.* Dynamic stability of structures. — In: Dynamic stability of structures. Oxford etc., Pergamon Press, 1967, p. 7–41.
347. *Hsu T.-M., Wang J.* Rotationally symmetric vibrations of orthotropic layered cylindrical shells. — J. Sound Vibration, 1971, vol. 16, N 4, p. 473–487.
348. *Hughes T. J. R., Liu W. K., Levit I.* Non-linear dynamic finite element analysis of shells. — In: Non-linear finite element analysis. Structural mechanics. Proc. Europ. — U.S. Workshop, Bochum, 1980. Berlin etc., 1981, p. 151–168.
349. *Humpreys J. S., Sve C.* Dynamic buckling of cylinders under axial shock-tube loading. — AIAA J., 1966, vol. 4, N 8, p. 1477–1480.
350. *Hutchinson J. W., Budiansky B.* Dynamic buckling estimates. — AIAA J., 1966, vol. 4, N 3, p. 525–530.
351. *Iyer S. H., Simmonds S. H.* The accuracy of Donnel's theory for very high harmonic loading of closed cylinders. — Trans. ASME. Ser. E, 1972, vol. 39, N 3, p. 200–202.
352. *Jones J. P., Whittier J. S.* Axially symmetric motions of a two-layered Timoshenko-type cylindrical shell. — Trans. ASME. Ser. E, J. Appl. Mech., 1966, vol. 33, N 4, p. 838–844.
353. *Jones J. P., Whittier J. S.* Dynamics of a flexibly bonded two-layered Timoshenko-type cylindrical shell. — AIAA J., 1969, vol. 7, N 2, p. 244–250.
354. *Jones R. M., Morgan H. S.* Buckling and vibration of cross-ply laminated circular cylindrical shells. — AIAA J., 1975, vol. 13, N 5, p. 664–671.
355. *Kagawa Y.* Non-axially symmetrical vibrations of sandwich cylindrical shells. — J. Sound Vibration, 1967, vol. 7, N 1, p. 41–50.
356. *Kawata K., Hashimoto S., Takeda N.* Mechanical behaviours in high velocity tension of composites. — In: Progr. Sci. and Eng. Compos. Proc. 4th Int. Conf., ICCM-IV. Tokyo, 1982, vol. 1, p. 829–836.
357. *Kawata K., Hondo A., Hashimoto S. et al.* Dynamic behaviour analysis of composite materials. — In: Composite materials. Mechanics, mechanical properties and fabrics. Jap.-US Conf., Tokyo, 1981. Barking, 1981, p. 2–11.
358. *Kim H. C., Park Y. H.* Impact behaviour of quasi-isotropic CFRP laminate. — In: Prog. Sci. Eng. Compos. Proc. 4th Int. Conf., ICCM-IV. Tokyo, 1982, vol. 1, p. 895–899.
359. *Klein S.* Vibration of multilayer shells of revolution under dynamic and impulsive loading. — Shock Vibration Bull., Washington, 1966, vol. 35, N 3, p. 27–44.
360. *Koiter W. T.* On the non-linear theory of thin elastic shells, 1, 2, 3. — Proc. K. Nederl. Akad. Wet., 1966, vol. B69, N 1, p. 1–17; 18–32; 33–54.
361. *Kousiounelos P. N., Williams J. H., jun.* Dynamic fracture of unidirectional graphite fiber composite strips. — Int. J. Fracture, 1982, vol. 20, N 1, p. 47–63.
362. *Libai A.* Non-linear shell dynamics–intrinsic and semi-intrinsic approaches. — Trans. ASME. Ser. E, 1983, vol. 50, N 3, p. 531–536.
363. *Libai A., Simmonds J. G.* Non-linear elastic shell theory. — Adv. Appl. Mech., 1983, vol. 23, p. 271–371.
364. *Lifshitz J. M.* Impact strength of angle ply fiber reinforced materials. — J. Compos. Mater., 1976, vol. 10, N 1, p. 92–101.
365. *Lifshitz J. M., Gilat A.* Experimental determination of the non-linear shear behavior of fiber-reinforced laminae under impact loading. — Exp. Mech., 1979, vol. 19, N 12, p. 444–449.
366. *Lin T., Morgan C.* A study of axisymmetric vibrations of a cylindrical shell as affected by rotatory inertia and transverse shear. — Trans. ASME. Ser. E. J. Appl. Mech., 1956, vol. 78, N 2, p. 255–261.
367. *Lindberg H. E.* Buckling of a very thin cylindrical shell due to an impulsive pressure. — Trans. ASME. Ser. E. J. Appl. Mech., 1964, vol. 31, N 2, p. 267–272.

368. *Lindberg H. E., Herbert R. E.* Dynamic buckling of a thin cylindrical shell under axial impact. — Trans. ASME. Ser. E. J. Appl. Mech., 1966, vol. 33, N 1, p. 105–113.
369. *Loo T. T.* An extension of Donnell's equations for circular cylindrical shell. — J. Aeronautical Sci., 1957, vol. 24, N 5, p. 390–391.
370. *Love A.* On the small free vibrations and deformation of thin elastic shell. — Phil. Trans. R. Soc., 1888, vol. 179 (A).
371. *Marguerre K.* Zur Theorie der gekrümmten Platte grober Formänderung. — In: Proc. 5th Int. Congr. Appl. Mech., Cambridge (Mass.), 1938. New York, J. Wiley a. Son, 1939, p. 93–101.
372. *Matsuzaki T., Kobayashi S.* A theoretical and experimental study of the non-linear flexural vibration of thin circular cylindrical shells with clamped ends. — J. Jap. Soc. Aeronautical Space Sci., 1970, vol. 12, N 21, p. 55–62.
373. *Maymon G., Libai A.* Dynamics and failure of cylindrical shells subjected to axial impact. — AIAA J., 1977, vol. 15, N 11, p. 1624–1630.
374. *Mc Ivor I. K.* The elastic cylindrical shell under radial impulse. — Trans. ASME. Ser. E. J. Appl. Mech., 1966, vol. 33, N 4, p. 831–837.
375. *Mc Ivor I. K., Lovell E. G.* Dynamic response of finite-length cylindrical shells to nearly uniform radial impulse. — AIAA J., 1968, vol. 6, N 12, p. 2346–2351.
376. *Mente L. J.* Dynamic non-linear response of cylindrical shells to asymmetric pressure loading. — AIAA J., 1973, vol. 11, N 6, p. 793–800.
377. *Mettler E.* Stability and vibration problems of mechanical systems under harmonic excitation. — In: Dynamic stability of structures. Oxford; New York, Pergamon Press, 1966.
378. *Meyer A., Döhler B., Skurt L.* Simultane Algorithmen für großdimensionierte Eigenwertproblem und ihre Anwendung auf das Schwingungsproblem. — Wiss. Schriftenr. Techn. Hochsch. Karl-Marx-Stadt, 1983, N 8, S. 90.
379. *Mindlin R. D.* Influence of rotatory inertia and shear on flexural motions of isotropic, elastic plates. — Trans. ASME. Ser. E. J. Appl. Mech., 1951, vol. 73, N 1, p. 31–38.
380. *Mindlin R. D., Bleich H. H.* Response of an elastic cylindrical shell to a transverse step shock wave. — Trans. ASME. Ser. E. J. Appl. Mech., 1953, vol. 20, N 2, p. 189–195.
381. *Mirsky I.* Vibrations of orthotropic, thick, cylindrical shells. — J. Acoustical Soc. Amer., 1964, vol. 36, N 1, p. 41–51.
382. *Mirsky I., Herrmann G.* Nonaxially symmetric motions of cylindrical shells. — J. Acoustical Soc. Amer., 1957, vol. 29, N 10, p. 1116–1123.
383. *Mirsky I., Herrmann G.* Axially symmetric motions of thick cylindrical shells. — Trans. ASME. Ser. E. J. Appl. Mech., 1958, vol. 80, N 1, p. 97–103.
384. *Morley L. S. B.* An improvement on Donnel's approximation for thin-walled circular cylinders. — Qt. J. Mech. Appl. Math., 1959, vol. 12, pt 1, p. 89–99.
385. *Mortimer R. W., Blum A.* The effect of pulse duration on the transient response of cylindrical shells subjected to axial impact. — Trans. ASME. Ser. E, 1974, vol. 41, N 1, p. 312–313.
386. *Mortimer R. W., Rose J. L., Chou P. C.* Longitudinal impact of cylindrical shells. — Experimental Mech., 1972, vol. 12, N 1, p. 25–31.
387. *Naghdi P. M.* Some aspects of the non-linear theory of elastic shells. — Bull. Acad. Pol. Sci. Ser. Sci. Techn., 1964, vol. 12, N 11a, suppl., p. 26–27.
388. *Naghdi P. M., Cooper R. M.* Propagation of elastic waves in cylindrical shells, including the effects of transverse shear and rotatory inertia. — J. Acoustical Soc. Amer., 1956, vol. 28, N 1, p. 55–63.
389. *Naghdi P. M., Nordgren R. P.* On the non-linear theory of elastic shells under the Kirchhoff hypothesis. — Q. Appl. Math., 1963, vol. 21, N 1, p. 49–59.
390. *Nelson R. B., Dong S. B., Kalra R. D.* Vibrations and waves in laminated orthotropic circular cylinders. — J. Sound Vibration, 1971, vol. 18, N 3, p. 429–444.

391. *Nolte L.-P., Stumpf H.* Energy-consistent large rotation shell theories in Langrangean description. — Mech. Res. Commun., 1983, vol. 10, N 4, p. 213–221.
392. *Nowinski J. L.* Non-linear transverse vibrations of orthotropic cylindrical shells. — AIAA J., 1963, vol. 1, N 3, p. 617–620.
393. *Nshanian Y. S., Pappas M.* Optimal laminated composite shells for buckling and vibration. — AIAA J., 1983, vol. 21, N 3, p. 430–437.
394. *Ohira H., Uda N.* On the knee-point of cross-ply composite. — In: Progr. Sci. and Eng. Composites. Proc. 4th Int. Conf. ICCS-IV. Tokyo, 1982, vol. 1, p. 473–480.
395. *Okabe N., Yano T., Kamata I., Mori T.* Impact fatigue strength and reliability for fiber reinforced epoxy resin laminates subjected to repeated impact loads. — J. Soc. Mater. Sci., Jap., 1982, vol. 31, N 351, p. 1210–1216.
396. *Olson M. D.* Some experimental observations on the non-linear vibration of cylindrical shells. — AIAA J., 1965, vol. 3, N 9, p. 1175–1177.
397. *Pietraszkiewicz W.* On consistent approximations in the geometrically non-linear theory of shells. — Mitt. Inst. Mech. Ruhr-Univ. Bochum, 1981, N 26, S. 39.
398. *Pietraszkievicz W.* A simplest consistent version of the geometrically non-linear theory of elastic shells undergoing large/small rotations. — Ztschr. Angew. Math. Mech., 1983, Bd 63, N 4, S. 200–202.
399. *Prathap G., Pandalai K. A.* The role of median surface curvature in large amplitude flexural vibrations of thin shells. — J. Sound Vibration, 1978, vol. 60, N 1, p. 119–131.
400. *Puppo A. H., Evensen H. A.* Strength of anisotropic materials under combined stresses. — AIAA J., 1972, vol. 10, N 4, p. 468–474.
401. *Radwan H. R., Gemin J.* Nonlinear vibrations of thin cylinders. — Trans. ASME, Ser. E, 1976, Vol. 43, N2, p. 370–2.
402. *Radwan H. R., Genin J.* Dynamic instability in cylindrical shells. — J. Sound Vibration, 1978, vol. 56, N 3, p. 373–382.
403. *Raj R. G., Manocha L. M., Bahl O. P., Verma D. S.* Impact strength and fracture of glass fibre reinforced epoxy composites. — Fibre Sci. Technol., 1982, vol. 17, N 2, p. 141–148.
404. *Rand R. A., Shen C. N.* Optimum design of composite shells subject to natural frequency constraints. — Comput. Struct., 1973, vol. 3, N 2, p. 247–263.
405. *Rao S. Y. V. K.* Vibrations of layered shells with transverse shear and rotatory inertia effects. — J. Sound Vibration, 1983, vol. 86, N 1, p. 147–150.
406. *Rath B. K., Das Y. C.* Vibration of layered shells. — J. Sound Vibration, 1973, vol. 28, p. 737–757.
407. *Rayleigh J. W. S.* The theory of the sound. 1929, vol. 1. London, Macmillan [Russian Translation, 1940. — 499 c.]
408. *Rhodes M. D., Williams J. G., Starnes J. H., jun.* Low-velocity impact damage in graphite-fiber reinforced epoxy laminates. — In: Techn. Proc. 34th Annu. conf. reinforced plastics/composites Inst. Reinf. Future, New Orleans, 1979. New York, s. a., p. 20D/1–20D/10.
409. *Rotem A., Hashin Z.* Failure modes of angle-ply laminates. — J. Compos. Mater., 1975, vol. 9, p. 191–206.
410. *Roth R. S., Klosner J. M.* Non-linear response of cylindrical shells subjected to dynamic axial loads. — AIAA J., 1964, vol. 2, N 10, p. 1788–1794.
411. *Rowlands R. E.* Analytical — experimental correlation of polyaxial states of stress in thornel-epoxy laminates. — Exp. Mech., 1978, vol. 18, N 7, p. 253–260.
412. *Roylance D.* Stress wave damage in graphite/epoxy laminates. — J. Compos. Mater., 1980, vol. 14, p. 111–119.
413. *Sanders J. L.* Non-linear theories for thin shells. — Qt. Appl. Math., 1963, vol. 21, N 1, p. 21–36.
414. *Sayers K. H., Harris B.* Interlaminar shear strength of a carbon fibre reinforced

composite material under impact conditions. — J. Compos. Mater., 1973, vol. 7, N 2, p. 129–136.
415. *Schmitt A. F.* Dynamic buckling tests of aluminum shells. — Aeronautical Eng. Rev., 1956, vol. 15, N 9, p. 54–56.
416. *Shafer B. P.* Two-dimensional stress wave propagation in thick multilayered cylindrical shells. — In: AIAA structure dynamics and aeroelasticity specialist conf., AIAA. New York, 1969, p. 162–173.
417. *Sharma C. B.* An analytical vibration study of thin circular cylinders. — In: Recent advances in structural dynamics. Papers Int. Conf. Southampton, 1980, vol. 1, p. 61–72.
418. *Shirakawa K.* Dynamic stability of cylindrical shells taking into account in-plane inertia and in-plane disturbance. — Bull. JSME, 1980, vol. 23, N 176, p. 163–169.
419. *Shirakawa K.* Effects of shear deformation and rotatory inertia on vibration and buckling of cylindrical shells. — J. Sound Vibration, 1983, vol. 91, N 3, p. 425–437.
420. *Shivakumar K. N., Krishna M. A. V.* Vibrations of multifiber composite shells — some numerical results. — J. Struct. Mech., 1976, vol. 4, N 4, p. 379–393.
421. *Smith P. W.* Phase velocities and displacement characteristics of free waves in a thin cylindrical shell. — J. Acoustical Soc. Amer., 1955, vol. 27, N 6, p. 1065–1072.
422. *Smith P. W.* Vibrations of cylindrical shells. — J. Acoustical Soc. Amer., 1958, vol. 30, N 1, p. 83–84.
423. *Soedel W.* On the vibration of shells with Timoshenko-Mindlin type shear deflections and rotatory inertia. — J. Sound Vibration, 1982, vol. 83, N 1, p. 67–79.
424. *Soldatos K. P.* On the buckling and vibration of antisymmetric angle-ply laminated circular cylindrical shells. — Int. J. Eng. Sci., 1983, vol. 21, N 3, p. 217–222.
425. *Stavsky Y., Loewy R.* On vibrations of heterogeneous orthotropic cylindrical shells. — J. Sound Vibrations, 1971, vol. 15, N 2, p. 235–256.
426. *Stellbrink K.* On the behaviour of impact damaged CFRP laminates. — Fibre Sci. Technol., 1983, vol. 18, N 2, p. 81–94.
427. *Stevens K. K.* On the parametric excitation of a viscoelastic column. — AIAA J., 1966, vol. 4, N 12, p. 2111–2116.
428. *Stevens K. K.* Transverse vibration of a viscoelastic column with initial curvature under periodic axial load. — Trans. ASME. Ser. E. J. Appl. Mech., 1969, vol. 36, N 4, p. 814–818.
429. *Stevens K. K.* On the experimental determination of instability regions in rheolinear vibration problems. — Trans. ASME. Ser. E, 1972, vol. 39, N 3, p. 831–832.
430. *Stevens K. K., Evan-Iwanowski R. M.* Parametric resonance of viscoelastic columns. — Int. J. Solids a. Struct., 1969, vol. 5, N 7, p. 755–765.
431. *Sun C. T., Sun P. W.* Laminated composite shells under axially symmetric dynamic loadings. — J. Sound Vibration, 1974, vol. 35, N 3, p. 395–415.
432. *Sun C. T., Sun P. W.* Forced vibration of laminated composite cylindrical shells. — J. Math. Phys. Sci., 1975, vol. 9, N 1, p. 111–133.
433. *Sun C. T., Whitney J. M.* Axisymmetric vibrations of laminated composite cylindrical shells. — J. Acoustical Soc. Amer., 1974, vol. 55, p. 1238–1246.
434. *Tamura Y. S., Babcock C. D.* Dynamic stability of cylindrical shells under step loading. — Trans. ASME. Ser. E, 1975, vol. 42, N 1, p. 190–194.
435. *Tennyson R. C.* Buckling of laminated composite cylinders: a review. — Composites, 1975, vol. 6, N 1, p. 17–24.
436. *Tennyson R. C.* Interaction of cylindrical shell buckling experiments with theory. — In: Theory of shells. North-Holland Publ. Co., 1980, p. 65–116.
437. *Tennyson R. C., Tulk J. D.* Dynamic stability of circular cylindrical shells. — In: Proc. 3rd Can. Congr. Appl. Mechanics. Calgary, 1973, p. 355–356.
438. *Tennyson R. C., Tulk J. D., Ricciatti R.* Analysis of the collapse of cylindrical shells using high-speed photography. — J. Soc. Motion Pictures a. Television Eng., 1971,

vol. 80, N 6, p. 477–481.

439. *Timoshenko S. P.* On the correction for shear of the differential equations for transverse vibrations of prismatic bars. — Philos. Mag., 1921, Ser. 6, vol. 41, p. 744–746.
440. *Timoshenko S. P.* On the transverse vibrations of bars of uniform cross section. — Philos. Mag., 1922, Ser. 6, vol. 43, p. 125–131.
441. *Timoshenko S.* Theory of plates and shells. New York. McGraw-Hill, 1940. 440 p.
442. *Tobias S. A.* A theory of imperfection for the vibrations of elastic bodies of revolution. — Engineering, 1951, vol. 172, N 4470, p. 409–410.
443. *Tsai S. W., Wu E. M.* A general theory of strength for anisotropic materials. — J. Compos. Mater., 1971, vol. 5, p. 58–80.
444. *Tsao C. H.* Strain-displacement relations in large displacement theory of shells. — AIAA J., 1964, vol. 2, N 11, p. 2060–2062.
445. *Tsuboi Y., Tosaka N.* Non-linear theory of thin elastic shells. Technol. Rep. Tohoku Univ., 1970, vol. 35, N 2, p. 87–111.
446. *Utida J., Sezawa K.* Dynamic stability of a column under periodic longitudinal forces. — Rep. Aeronautical Res. Inst. (Tokyo), 1940, vol. 15, p. 139–183.
447. *Valid R.* An intristic formulation for the non-linear theory of shells and some approximations. — Comput. Struct., 1979, vol. 10, N 1/2, p. 183–194.
448. *Vanderpool M. E., Bert C. W.* Vibration of a materially monoclinic, thick-wall circular cylindrical shell. — AIAA J., 1981, vol. 19, N 5, p. 634–641.
449. *Vijayaraghavan A., Evan-Iwanowski R. M.* Parametric instability of circular cylindrical shells. — Trans. ASME, Ser. E, 1967, vol. 34, N 4, p. 985–990.
450. *Warburton G. B.* Vibration of thin cylindrical shells. — J. Mech. Eng. Sci., 1965, vol. 7, N 4, p. 399–407.
451. *Warburton G. B., Soni S. R.* Resonant response of orthotropic cylindrical shells. — J. Sound Vibration, 1977, vol. 53, N 1, p. 1–23.
452. *Weidenhammer F.* Nichtlineare Biegenschwingungen des axialpulsierend belasteten Stabes. — Ing. Arch., 1952, Bd 20, S. 315–330.
453. *Weingarten V. I.* Free vibrations of multilayered cylindrical shells. — Exp. Mech., 1964, vol. 4, N 7, p. 200–205.
454. *Weingarten V. I.* Free vibration of thin cylindrical shells. — AIAA J., 1964, vol. 2, N 4, p. 717–722.
455. *White J. C.* The flexural vibrations of thin laminated cylinders. — Trans. ASME. Ser. B, 1961, vol. 83, N 4, p. 397–402.
456. *Whittier J. S., Jones J. P.* Axially symmetric wave propagation in a two-layered cylinder. — Int. J. Solids Structures, 1967, vol. 3, N 4, p. 657–675.
457. *Wilcox M. W., Abhat O. B.* Dynamic response of laminated composite shells under radial or hydrostatic pressure. — Trans. ASME. Ser. J. J. Pressure Vessel Technol., 1974, vol. 96, p. 299–304.
458. *Wood J. D., Koval L. R.* Buckling of cylindrical shells under dynamic loads. — AIAA J., 1963, vol. 1, N 11, p. 2576–2582.
459. *Wu E. M.* Optimal experimental measurements of anisotropic failure tensors. — J. Compos. Mater., 1972, vol. 6, p. 472–489.
460. *Yao J. C.* Dynamic stability of cylindrical shells under static and periodic axial and radial loads. — AIAA J., 1963, vol. 1, N 6, p. 1391–1396.
461. *Yao J. C.* Non-linear elastic buckling and parametric excitation of a cylinder under axial loads. — Trans. ASME. Ser. E. J. Appl. Mech., 1965, vol. 32, N 1, p. 109–115.
462. *Yi-Yuan Yu.* Free vibrations of thin cylindrical shells having finite length with freely supported and clamped edges. — Trans. ASME. Ser. E. J. Appl. Mech., 1955, vol. 22, N 4, p. 547–552.
463. *Yi-Yuan Yu.* Vibrations of thin cylindrical shells analysed by means of Donnell-type equations. — J. Aero-Space Sci., 1958, vol. 25, N 11, p. 699–715.

464. *Yi-Yuan Yu*. Vibrations of elastic sandwich cylindrical shells. — Trans. ASME. Ser. E. J. Appl. Mech., 1960, vol. 82, N 4, p. 653–662.
465. *Zerna W*. Über eine nichtlineare allgemeine Theorie der Schalen. — In: Proc. IUTAM Symp. Theory Thin Elastic Shells, Delft, 1959. Amsterdam, 1960, p. 34–42.
466. *Zimcik D. G., Tennyson R. C.* Stability of circular cylindrical shells under transient axial impulsive loading. — AIAA J., 1980, vol. 18, N 6, p. 691–699.

Index